T0360791

New Era for CP Asymmetries

Axions and Rare Decays of Hadrons and Leptons

Other Related Titles from World Scientific

Fads and Fancies of Elementary Particle Physics: Selected Works of Kameshwar C Wali
edited by Kameshwar C Wali
ISBN: 978-981-123-690-7

Two-Phase Emission Detectors
by Alexander I Bolozdynya, Dmitry Yu Akimov, Alexey F Buzulutskov and Vitaly Chepel
ISBN: 978-981-123-108-7

The Physics of Experiment Instrumentation Using MATLAB Apps: With Companion Media Pack
by Dan Green
ISBN: 978-981-123-243-5
ISBN: 978-981-123-383-8 (pbk)

QCD and Heavy Quarks: In Memoriam Nikolai Uraltsev
edited by Ikaros I Bigi, Paolo Gambino and Thomas Mannel
ISBN: 978-981-4602-73-0

New Era for CP Asymmetries

Axions and Rare Decays of Hadrons and Leptons

Ikaros I Bigi
University of Notre Dame, USA

Giulia Ricciardi
Università di Napoli Federico II, Italy

Marco Pallavicini
University of Genova, Italy

World Scientific

NEW JERSEY • LONDON • SINGAPORE • BEIJING • SHANGHAI • HONG KONG • TAIPEI • CHENNAI • TOKYO

Published by

World Scientific Publishing Co. Pte. Ltd.
5 Toh Tuck Link, Singapore 596224
USA office: 27 Warren Street, Suite 401-402, Hackensack, NJ 07601
UK office: 57 Shelton Street, Covent Garden, London WC2H 9HE

Library of Congress Control Number: 2021935604

British Library Cataloguing-in-Publication Data
A catalogue record for this book is available from the British Library.

NEW ERA FOR CP ASYMMETRIES
Axions and Rare Decays of Hadrons and Leptons

ISBN 978-981-3233-07-2 (hardcover)
ISBN 978-981-3233-08-9 (ebook for institutions)
ISBN 978-981-3233-09-6 (ebook for individuals)

For any available supplementary material, please visit
https://www.worldscientific.com/worldscibooks/10.1142/10791#t=suppl

Desk Editor: Ng Kah Fee

Typeset by Stallion Press
Email: enquiries@stallionpress.com

Printed in Singapore

Dedicated to

Lev Okun
(1929 - 2015)

a true pioneer of probing fundamental dynamics in general and in particular of CP asymmetries.

Foreword

This book is dedicated to Lev Borisovič Okun – a true pioneer of fundamental dynamics and particularly of **CP** asymmetries.

Okun wrote in his book "Weak Interactions of Elementary Particles" [1] that it was crucial to continue probing **CP** violation. It was *not* 'luck': the original version in Russian was published in 1963 – i.e., clearly *before* its discovery in 1964 [2].

We give a list of very important articles of L.B. Okun that shows both his deeper and broad 'horizons' on different levels. We also mention the number of citations according to Inspire-HEP[1]: a list of citations is a very good example for the record – not only of the impact of articles.

(1) L.B. Okun, "Mirror particles and mirror matter: 50 years of speculations and search", *Phys.Usp.* **50** (2007) 380, cited 160.
(2) L.B. Okun, "C, P, T are broken. Why not CPT?", hep-ph/0210052, cited 19.
(3) V.A. Novikov, L.B. Okun, A.N. Rozanov, M.I. Vysotsky, "Extra generations and discrepancies of electroweak precision data", *Phys.Lett.* **B529** (2002) 111, cited 150.
(4) M.J. Duff, L.B. Okun, G. Veneziano, "Trialogue on the number of fundamental constants", *JHEP* **0203** (2002) 023; cited 128 [a personal comment: this article is very unusual – and therefore it is very interesting, never mind whether a reader agrees or not. We will discuss it in the Epilogue].
(5) J.D. Jackson, L.B. Okun, "Historical roots of gauge invariance", *Rev.Mod.Phys.* **73** (2001) 663, cited 92.
(6) A.D. Dolgov, A.Yu. Morozov, L.B. Okun, M.G. Shchepkin, "Do Muons Oscillate?", *Nucl.Phys.* **B502** (1997) 3, cited 48.
(7) V.A. Novikov, L.B. Okun, M.I. Vysotsky, "On the electroweak one loop corrections", *Nucl.Phys.* **B397** (1993) 35, cited 104.
(8) S.D. Drell, Lev Okun, "Andrei Dmitrievich Sakharov", *Phys.Today* **43**(8) (1990) 26.
(9) L.B. Okun, "Tests of electric charge conservation and the Pauli principle", *Sov.Phys.Usp.* **32** (1989) 543, cited 33.

[1]Citations on Jan. 15th, 2021.

(10) L.B. Okun, M.B. Voloshin, M.I. Vysotsky, "Neutrino Electrodynamics and possible effects for Solar Neutrinos", *Sov.Phys.JETP* **64** (1986) 446, cited 519.
(11) L.B. Okun, M.B. Voloshin, M.I. Vysotsky, "Electromagnetic properties of Neutrino and possible semi-annual variation cycle of the Solar Neutrino Flux", *Sov.J.Nucl.Phys.* **44** (1986) 440, cited 393.
(12) L.B. Okun, "On the Electric Dipole Moment of Neutrino", *Sov.J.Nucl.Phys.* **44** (1986) 546, cited 132.
(13) L.B. Okun, "Limits Of Electrodynamics: Paraphotons?", *Sov.Phys.JETP* **56** (1982) 502, cited 275.
(14) I.Yu. Kobzarev, B.V. Martemyanov, L.B. Okun, M.G. Shchepkin, "The Phenomenology of Neutrino Oscillations", *Sov.J.Nucl.Phys.* **32** (1980) 823, cited 116.
(15) L.B. Okun, M.B. Voloshin, "On the Electric Charge Conservation", *JETP Lett.* **28** (1978) 145, cited 55.
(16) L.B. Okun, Ya.B. Zeldovich, "Paradoxes of Unstable Electron", *Phys.Lett.* **B78** (1978) 597, cited 68.
(17) V.A. Novikov, L.B. Okun, M. A. Shifman, A.I. Vainshtein, M.B. Voloshin, V.I. Zakharov, "Charmonium and Gluons: basic Experimental acts and Theoretical Introduction", *Phys.Rept.* **41** (1978) 1, cited 833.
(18) M.B. Voloshin, L.B. Okun, "Hadron Molecules and Charmonium Atom", *JETP Lett.* **23** (1976) 333, cited 334.
(19) V.A. Novikov, L.B. Okun, M.A. Shifman, A.I. Vainshtein, M.B. Voloshin, V.I. Zakharov, "Sum Rules for Charmonium and Charmed Mesons Decay Rates in Quantum Chromodynamics", *Phys.Rev.Lett.* **38** (1977) 626, Erratum: *Phys.Rev.Lett.* **38** (1977) 79, cited 197.
(20) L.B. Okun, M.A. Shifman, A.I. Vainshtein, V.I. Zakharov, "On the $K^+ \to \pi^+ e^+ e^-$ Decay", *Sov.J.Nucl.Phys.* **24** (1976) 427, cited 105.
(21) L.B. Okun, B.M. Pontecorvo, Valentin I. Zakharov, "On the possible violation of CP-Invariance in the Decays of Charmed Particles", *Lett.Nuovo Cim.* **13** (1975) 218, cited 241.
(22) I.Yu. Kobzarev, L.B. Okun, M.B. Voloshin, "Bubbles in Metastable Vacuum", *Sov.J.Nucl.Phys.* **20** (1975) 644, cited 370.
(23) Ya.B. Zeldovich, I.Yu. Kobzarev, L.B. Okun, "Cosmological Consequences of the Spontaneous Breakdown of Discrete Symmetry", *Sov.Phys.JETP* **40** (1974) 1, cited 709.
(24) I.Yu. Kobsarev, L.B. Okun, Ya.B. Zeldovich, "Spontaneous CP violation and cosmology", *Phys.Lett.* **50B** (1974) 34, cited 52.
(25) I.Yu. Kobzarev, L.B. Okun, I.Ya. Pomeranchuk, "On the possibility of experimental observation of mirror particles", *Sov.J.Nucl.Phys.* **3** (1966) 837, cited 279.
(26) I.Yu. Kobzarev, L.B. Okun, "Gravitational Interaction of Fermions", *Sov. Phys.JETP* **16** (1963) 1343, cited 79.

Two comments:
(a) The idea of "mirror particles" was discussed in 1966 (see item 25) and again in 2007 (see item 1) – i.e., 47 years later. We may take long time to understand the impact of good ideas.
(b) Okun had more courage to discuss fundamental dynamics than most of us, see item 2.

To my parents, Pasquale and Gioia,
whose total love and encouragement never failed me,
and will always be in my heart.
They were honest, very generous of heart and faced
the challenges of life with courage and optimism.
G.R.

A Marina, ai miei genitori e a mio nonno,
un ragazzo del '900 che ha camminato tutto il secolo
fino a piangere di commozione vedendo il CERN.
M.P.

[To Marina, to my parents, and to my grandfather,
a boy of '900 that has walked along the century
till he was moved to tears seeing CERN.]

Contents

Chapter 1

Prologue

> *You don't need to know anything – except to keep your ears open and if possible, your brain awake.*
>
> Tom Stoppard

This volume is not a textbook where the reader may get a first glimpse at the subject of **CP** symmetry and its rich and complex dynamics. Although all concepts that are necessary to understand or follow the arguments are usually recalled and summarised, a previous knowledge of the fundamentals of quantum mechanics, quantum field theory, and particle and nuclear physics is required. This volume is not as well a traditional review about **CP** asymmetries (including axion dynamics) and rare decays. Instead, we try to update the progress that our community has achieved and focus particularly on new 'directions' and possible future developments, often using the past as a guide.

We have entered a new era at the beginning of the 21st century. It may be 'pictured' with the photo shown (**Fig. 1.1**), taken from one of the authors at a conference in Normandy: it shows some clouds, but also the final goal.

Fig. 1.1 Mont Saint-Michel is an island in Normandy (France) [picture taken by IIB].

Our community has discussed nuclear physics (NP) and high energy physics (HEP) since the early 1930's, finally yielding the development of the Standard Model.

The Standard Model (SM) of particle physics is based on the gauge group $SU(3)_C \times SU(2)_L \times U(1)_Y$, and it is a real quantum field theory, not a model; it has been very successful on different levels (e.g., see these textbooks [3]). It describes three fundamental interactions (electromagnetic, weak and strong interactions) acting on the fundamental constituents of matter, the quarks and the leptons. The interactions are mediated by spin-1 gauge fields, eight massless gluons for the strong interaction, one massless photon for the electromagnetic one and three massive bosons for the weak interaction. Both weak and strong interactions are non-abelian, yielding important self-interactions of the gauge fields and crucial non-perturbative dynamics.

The elementary particles occur in three generations, which are distinguished by their "flavor" quantum number . Flavor physics studies the transitions among particles in different generations. In the SM, the only source of flavor violation is described by the Cabibbo-Kobayashi-Maskawa (CKM) matrix. In particular, the CKM matrix contains a phase that generates a violation of **C**harge conjugation and **P**arity (**CP**) symmetry in the weak interactions.

We summarize below the main characteristics of the SM:

(1) Quantum Chromodynamics (QCD) is the unique Quantum Field Theory (QFT) describing strong interaction that holds quarks together *inside* the hadrons; i.e., they are "confined". It is based on the *un*broken gauge group $SU(3)_C$; C is the quantum number (QN) called "color" . Quarks carry "color 3". The strength of their interactions is given by the strong coupling α_S which has the peculiar property to be very large at low energy, namely below ~ 1 GeV; thus free quarks cannot be seen directly. But the couplings go *down* with collisions at momentum transferred larger than ~ 1 GeV, so we can observe 'almost free' quarks as jets at high energies. Yet $SU(3)_C$ symmetry is unbroken; if a quark is produced, one has to find another source of color. The only observed hadrons ($q\bar{q}$, qqq) are colorless (even in suggested exotic configurations such as $q\bar{q}q\bar{q}$ and $qqqq\bar{q}$).

(2) To describe electromagnetic and weak interactions with local gauge symmetries, one can find many candidates. However, only $SU(2)_L \times U(1)_Y$ passes the experimental tests. It is not theoretically described by a 'natural' gauge group, but it is the only one that 'works' in the sense of describing 'our' Universe.

(3) The landscape of the SM is 'complex'. The mediators of the QCD $SU(3)_C$ gauge group are the gluons that carry QN "color 8". They are massless non-abelian fields. The mediators of weak interactions, the $W^\pm$ and Z^0 bosons, are massive, namely $M_W \simeq 80$ GeV and $M_Z \simeq 91$ GeV; again, we have non-abelian dynamics forces, associated this time to the gauge group $SU(2)_L \times U(1)_Y$. In

principle this gauge symmetry gives rise to massless gauge bosons. For these bosons to become massive one invokes the Higgs mechanism. This implies that the electroweak gauge group $SU(2)_L \times U(1)_Y$ is spontaneously broken to $U(1)_{\rm em}$, i.e. $SU(2)_L \times U(1)_Y$ is still a symmetry of the Lagrangian and of the forces but not of the ground states or 'vacua'. According to the Goldstone theorem, three massless states called Goldstone bosons are generated corresponding to the numbers of broken generators. Three of the Goldstone bosons are 'eaten' by the weak gauge bosons giving masses to the $W^\pm$ and Z^0 in the SM. The photon does not acquire a mass since $U(1)_{\rm em}$ is unbroken. At the same time, by the spontaneous symmetry breaking quarks and leptons acquire masses through their couplings to the Higgs field. Their masses are proportional to the vacuum expectation values (VEV).
In a minimal version of the Higgs fields two complex $I = 1/2$ states are predicted, amounting to four degrees of freedom. Three of these act as spin-0 states for massive $W^\pm$ and Z^0, while the fourth one can be found as a neutral Higgs state H^0. It was observed in 2012 at the large hadron collider (LHC) by the ATLAS and CMS collaborations [4] with a mass of $M(H^0) \simeq 125$ GeV to be compared with the SM prediction: $M(H^0)|_{\rm SM} \sim 120 - 130$ GeV. In the following, we will discuss the impact of Higgs dynamics in several places.

(4) Not surprisingly many HEP theorists had suggested that the unification of forces could be pursued with at least another step: $SU(3)_C \times SU(2)_L \times U(1)_Y$ with $8 + 3 + 1 = 12$ generators leads to $SU(5)$ with $5^2 - 1 = 24$ generators. It means one would have 12 gauge bosons with much higher masses and a much more complex and rich dynamics. Now we know that $SU(5)$ does not work as a 'minimal' version of a "Grand Unified Theory" (GUT) [5]. Of course, we should not give up searching for a GUT that combines electroweak and strong forces into a single gauge group.

The SM has been tested with great success – yet it is incomplete for crucial reasons:

- The SM can*not* produce neutrino oscillations.
- In the standard cosmological model, the "known" matter represents 4.5 % of our Universe while Dark Matter $\sim$ 26.5 % – and even more surprisingly the 'rest' ($\sim$ 69 %) is assumed to be Dark Energy.[1]
- In the SM the "Higgs" field is 100% scalar. In all scenarios investigated by the ATLAS [6] and CMS [7] experiments, data are well compatible with the SM 0^+ hypothesis. Thus the observed spin-0 boson in the run-1 and run-2 is at least mostly a scalar. The impact of a pseudo-scalar amplitude can be seen in the future in asymmetries due to interference of scalar vs. pseudo-scalar ones.
- In principle, SM allows **CP** violating signals as large as 100%. However, the CKM parameters precisely measured in quark dynamics fail to explain the

[1] We would rather call it "vacuum energy"; yet we follow the 'fashion' in this book.

huge asymmetry in matter vs. anti-matter observed in the Universe, and by many orders of magnitude.

- The current paradigm extends the SM to include neutrino masses and describes neutrino oscillation with a mixing matrix that is called the Pontecorvo–Maki–Nakagawa–Sakata (PMNS) matrix. Neutrino mixing adds another complex phase to the Lagrangian offering at least one new mechanism for **CP** violation in the extended SM. The true value of this additional violation is still under experimental investigations, but it might well be quite large. The PMNS extension does not depend on the nature of the neutrino mass term, i.e. if Dirac or Majorana. In case of Majorana neutrinos, two additional **CP** violating phases may be present.

To summarize: the known matter

$$\text{SM} + \text{"beyond"} = SU(3)_C \times SU(2)_L \times U(1)_Y \oplus \text{"CKM"} \oplus \text{"PMNS"} \tag{1.1}$$

is characterized by the carriers of its strong and electroweak forces described by *gauge* dynamics and the *mass matrices* of its quarks and leptons. We use the symbol "$\oplus$" to emphasize that the measured peculiar pattern of fermion mass parameters is not illuminated by their gauge structure so far.

There are several regions of physics, and they do not only differ by their scales, but also by their "patterns", as discussed in the previous century. Usually the 'players' in NP are the nucleons and the pions and kaons; in HEP they are the quarks and gluons. The leptons are 'players' in both sectors, although in different ways. In this book, we focus on HEP, atomic and nuclear physics, astrophysics and cosmology.[2] It is important to build a 'bridge' between these worlds–we need alliances between different 'cultures', such as for instance between HEP and what could be analogously called Middle Energy Physics (MEP). Actually the word "MEP" does not make clear where the difference is, since it does not specify that we have to consider hadrons vs. quarks and gluons as degrees of freedom. The name of Hadrodynamics (HD) seems more satisfactory. Tools to connect HD and HEP are chiral symmetries and dispersion relations, and more active 'players' like resonances. When we work in fundamental dynamics, connections between atomic and nuclear physics, HEP and astrophysics are crucial, and often not obvious. Let us underline that:

- The best fit analyses often do not give us the best understanding of the underlying dynamics. Yet, data are the referees – in the end. At the same time, theorists should not be slaves of the data at *that* time; one should not stop thinking.[3]

[2]We give only a few comments about solid state physics, although one of the many strengths of physics is to connect different fields.

[3]There have been many instances in the past when theorists played an important role in the progress of the field; one expects this will happen again in the future.

- Important information arrives in the comparison of high energy collisions vs. transitions to low energies with high precision.[4]
- It is an important, but often under-evaluated, point that 'history' can help us to understand better fundamental dynamics. It is like this in all fields; we can go back to the Middle Ages, see **Fig. 1.2**.

Fig. 1.2 "Woman teaching geometry", Illustration at the beginning of a medieval translation of Euclid's *Elements* (c. 1310 AD), attributed to Meliacin Maste [British Library, digital collection].

- Experimentalists deal with data using state-of-the-art detectors and analyses. On the other hand, one cannot expect that theorists talk about the 'truth' all the time; speaking in 'good faith' is all one can expect from theorists.[5]

In this book, we mostly focus on **CP** asymmetries in the transitions of hadrons and leptons, but we also discuss the rare decays, and talk about axions and super-symmetry and possible connections with Dark Matter, Extra Dimensions, Baryo-genesis and Multiverse.

A few "operating instructions":

a) When we refer to 'the data', we use the numbers from PDG2020 [8], unless we give explicit references.

b) We know the numbers of two observables: $\hbar \equiv h/2\pi \simeq 6.582 \cdot 10^{-16}$ eV·s and $c \simeq 3 \cdot 10^8$ m/s. However, as usual among theorists, we use natural units $\mathbf{c} = \mathbf{1} = \hbar$.

[4]There is always a price for a prize, namely we need again thinking; after all that is our 'job'.
[5]This expectation came from Prof. Italo Manelli long time ago.

c) This book does *not* have "Problems" offered at the end of its Chapters.[6]

d) We do not always give an exhaustive list of references. When contents are well established, a small list of references is enough. On the other hand, when items are not yet established, it makes sense to give longer lists. Of course, some might disappear in the future.

e) We assume that readers 'know' QM and QFT in general, as she/he has learnt from the usual well-known books about that.

f) We will use formulas like $\sigma \cdot G \equiv \sigma_{\mu\nu} G^{\mu\nu}$ and $W_\mu W^\mu$ without much explanations.

The structure of the book is the following.

Chapter 2 should help the reader to remember what already learnt about quantum mechanics and quantum field theory. In **Chapter 3** we talk about the 'strategy' for probing fundamental dynamics with hadrons, while in **Chapter 4** we discuss the 'tactics'. In **Chapters 5, 6, 7, 8** we discuss rare decays of flavor hadrons, top quarks and charged leptons, oscillations and **CP** asymmetries.

The situation has changed, with more data and more refined analyses, after LHC has started and with the discovery of neutrino oscillations: for instance, we have a new hunting region of **CP** asymmetries in top quark transitions and we may soon have another **CP** violating mechanism within experimental sensitivity. In **Chapter 9** we talk about the dynamics of neutrinos, their oscillations and **CP** violation in the decays of τ leptons; in **Chapters 10** and **11** we discuss **CP** violation in strong forces and a 'natural' solution, the axion, to this problem.

We describe electric dipole moments in **Chapters 12** and **13** while in **Chapter 14** we discuss super-symmetry. In **Chapter 15** we examine dark matter, extra dimensions and the huge asymmetry in matter vs. anti-matter giving also a short review of a possible multiverse. We summarize plans for future searches of ND in **Chapter 16**; **Chapter 17** is the 'epilogue'.

[6] A committed reader may find them in Ref. [12]; one of the co-authors might be biased here.

Chapter 2

Quantum mechanics and quantum field theories

> Subtle is the Lord – but malicious He/She is not (we hope)!
>
> *A. Einstein*

The era of non-relativistic quantum mechanics (QM) started in the 1920's, in particular with the Schrödinger equation [9]:

$$i\frac{\partial}{\partial t}\psi(t,\vec{x}) = H_x\psi(t,\vec{x}) \ , \ \ \psi(t,\vec{x}) \equiv \langle\vec{x}|\psi;t\rangle \ , \tag{2.1}$$

based on the Hamiltonian[1]:

$$H_x = -\frac{1}{2m}\nabla^2 + V(\vec{x}) \ . \tag{2.2}$$

It was also the era of obvious love of symmetries. However, it had changed by the 1950's in subtle ways, heading towards the era of quantum field theories (QFT).

2.1 Introduction

It is often said that the goal of fundamental dynamics is the study of the elementary particles[2] (or waves) in our world, namely the quarks and leptons with spin-1/2, the gauge bosons with spin-1, the Higgs boson with spin-0 and, possibly, the graviton with spin-2. Yet, since the 50's, the focus has slowly changed: today the elementary particles are mainly seen as 'actors' playing a plot written by *symmetries*, whether they are discrete, global or local ones.[3] In the world of QFTs, symmetry transformations are based on three conditions:

(1) There is a ground state or 'vacuum',[4] which remains invariant.

[1] Remember from the Prologue that we use natural units with $\hbar = 1$; thus $\hbar^2/2m = 1/2m$.

[2] The name 'elementary particle' goes back to the Greek thinkers.

[3] The concept of Symmetry ('Συμμετρία' in Greek) is a keystone. Of course, we might be wrong, since there might be Multiverse with a basic symmetry broken generating a true huge numbers of basic ground states. Thus we are just 'lucky' in our very special Universe where symmetries work.

[4] The word 'vacuum' is not a good choice, since in QM it not empty at all!

(2) The action S, defined by a Lagrangian density $\mathcal{L}(t, \vec{x})$, remains invariant:

$$S = \int d^4x \, \mathcal{L}(t, \vec{x}) \quad \Longrightarrow \quad \delta S = 0 \tag{2.3}$$

(3) The quantization conditions remain invariant.[5]

The existence of symmetries means that physical observables are unchanged in 'acceptable' transitions. The above conditions can be relaxed in different ways, which all occur in our Universe:

- The action S changes explicitly, i.e. the symmetry is only approximate.
- The symmetry is broken *spontaneously*, which happens when condition 1) is not valid.[6]
- The symmetry is anomalous, which means that condition 3) is not met.

One can frame discussions in terms of Lagrangians: they have the nice feature of being Lorentz invariant. Often one switches to Hamiltonians, though they are *not* Lorentz scalars. The discussions of oscillation phenomena are more naturally conducted in the Hamiltonian formalism, since in the rest frame the mass of a particle corresponds to its energy and the unitarity of the theory is more transparent.[7] Furthermore, in the interaction picture the interaction terms are identical apart from their sign: $H_{\text{int}} = -\mathcal{L}_{\text{int}}$. We describe a Hamiltonian as a sum of local operators H_i

$$H_{\text{int}} = \sum_j a_j H_j + \text{h.c.} \tag{2.4}$$

with $\mathbf{CP}\, H_j \, (\mathbf{CP})^\dagger = H_j$ in most cases.[8] We underline that several features of the dynamics come from the structure of the a_j coefficients: for instance, **CP** asymmetry is connected to the complex nature of these coefficients.

In our view, symmetries give a deep understanding of fundamental dynamics, in particular after Emmy Noether's theorem, published in 1918 [10]. It connects huge classes of conservation laws to symmetries of space and time and internal observables; they include both exact ones such as electric charges in QED and "color" in QCD and broken ones such as the weak charges for the electroweak $SU(2)_L \times U(1)_Y$ group.[9]

In the following sections we discuss discrete symmetries like **P**, **C**, **CP** and Time reversal **T**, always assuming **CPT** invariance, see **Sect. 2.7**.[10]

[5] Weak local commutativity obeys the "right" statistics, namely, Bose-Einstein statistics with $[a, b]$ for identical integer spin states, and Fermi-Dirac statistics with $\{a, b\}$ for half-integer ones. We will *not* discuss para-statistics.

[6] It might be called as well a *spontaneous realization* of the symmetry, since the *dynamics* still obeys condition 2).

[7] We disagree with the words of Oscar Wilde: "Consistency is the last refuge of the unimaginative".

[8] **P** and **C** refer to Parity and Charge Conjugation transformations, respectively.

[9] If somebody would like to follow the arguments and its applications she/he can read the book "Emmy Noether's Wonderful Theorem" [11].

[10] **CPT** violation in flavor dynamics is discussed in Ref. [12] in some details.

2.2 Transformation of Parity (P) and Charge Conjugation (C)

Discrete symmetries of parity **P** and charge conjugation **C** are described by unitary operators which are also hermitian and correspond therefore directly to observables:

$$\mathbf{P}^\dagger = \mathbf{P} = \mathbf{P}^{-1}\ ,\ \mathbf{P}^2 = 1\ , \tag{2.5}$$

$$\mathbf{C}^\dagger = \mathbf{C} = \mathbf{C}^{-1}\ ,\ \mathbf{C}^2 = 1\ . \tag{2.6}$$

A reader might think that this is just 'math', but one goes to 'physics' by looking at the connection of the parity operator **P** with the position operator $\vec{X}$ in non-relativistic quantum mechanics

$$\mathbf{P}^\dagger\ \vec{X}\ \mathbf{P} = -\vec{X} \quad \text{or} \quad \{\vec{X}, \mathbf{P}\} = 0 \tag{2.7}$$

or how **P** transforms the state in the Hilbert space[11]

$$\mathbf{P}|\vec{x}\rangle = |-\vec{x}\rangle \text{ with eigenvalues } \pm 1\ . \tag{2.8}$$

Likewise for the momentum operator $\vec{P}$:

$$\mathbf{P}^\dagger\ \vec{P}\ \mathbf{P} = -\vec{P} \quad \text{or} \quad \{\vec{P}, \mathbf{P}\} = 0 \tag{2.9}$$

$$\mathbf{P}|\vec{p}\rangle = |-\vec{p}\rangle \text{ with eigenvalues } \pm 1\ . \tag{2.10}$$

Since **P** anti-commutes with both $\vec{X}$ and $\vec{P}$, it leaves the quantization condition invariant:

$$[X_a, P_b] = i\ \delta_{ab}\ . \tag{2.11}$$

Likewise for **C**:

$$\mathbf{C}^\dagger\ \vec{X}\ \mathbf{C} = -\vec{X} \quad \text{or} \quad \{\vec{X}, \mathbf{C}\} = 0\ , \tag{2.12}$$

$$\mathbf{C}^\dagger\ \vec{P}\ \mathbf{C} = -\vec{P} \quad \text{or} \quad \{\vec{P}, \mathbf{C}\} = 0\ , \tag{2.13}$$

leaving the quantization condition Eq. (2.11) invariant. Since the Hamiltonian H depends on $\vec{X}$ and $\vec{P}$, the following commutation relations hold:

$$[\mathbf{P}, H] = 0 = [\mathbf{C}, H]\ . \tag{2.14}$$

2.3 Time Reversal (T)

Time reversal – more than meets the eye

The meaning of time reversal **T** is 'complex' on different levels. *Microscopic* **T** invariance in classical physics means that the rates for processes $a \to b$ and $b \to a$ are the same if b – the final state in the first process – is arranged identically in all aspects as the initial state for the second one. Such an arrangement requires

[11] In general one uses $\mathbf{P}|\vec{x}\rangle = e^{i\delta}|-\vec{x}\rangle$ with δ being an *arbitrary* phase.

fine-tuning the initial conditions that is typically very difficult to achieve. Consider, e.g., the weak decay of a polarised muon [12–14]:

$$\mu^-(\Uparrow) \;\to\; e^-_L + \bar{\nu}_{e,R} + \nu_{\mu,L} \tag{2.15}$$

where the muon is polarized 'up', namely in a direction perpendicular to the decay plane of the final products. Next we consider a time reversed collisions of three final state particles forming a muon; they lead to muons polarized in the line-of-flight directions of the initial beams and *not* muons polarized 'down':

$$e^-_L + \bar{\nu}_{e,R} + \nu_{\mu,L} \;\to\; \mu^-(\Leftarrow) \qquad \neq \qquad \mu^-(\Downarrow)\,. \tag{2.16}$$

The key point to underline here is that we cannot usually realize the time reversed reaction, because creating the correct initial state goes well beyond the practical possibility. The underlying reason is a quantum mechanics effect, as one can see in two ways:

(i) The quantum state with spin 'up' in the x direction is *not* orthogonal to the quantum state with spin 'up' in the z direction.

(ii) The truly time reversed version of the muon decay has an initial state $e^-_L + \bar{\nu}_{e,R} + \nu_{\mu,L}$ of incoming spherical waves that furthermore have to be coherent; i.e., we have to reverse the momenta and the spins of the three leptons in all possible directions while maintaining the required phase relationships among their amplitudes! This is obviously impossible.

Instead one can resort to oscillations of neutral particle and anti-particle into each other due to QM. They can decay into different final states, and one can measure the oscillation rates in these directions. If the rates are different, we say that we have measured **T** violation. **T** violation has been found in the $K^0 - \bar{K}^0$ system by the CPLEAR collaboration [15][12] and more recently in the $B^0 - \bar{B}^0$ one by the BaBar collaboration [16].

Let us consider the time reversal operator. As done for parity in **Sect. 2.2**, we have

$$\mathbf{T}^{-1}\,\vec{X}\,\mathbf{T} = +\vec{X} \;\text{ or }\; [\vec{X},\mathbf{T}] = 0 \tag{2.17}$$

$$\mathbf{T}^{-1}\,\vec{P}\,\mathbf{T} = -\vec{P} \;\text{ or }\; \{\vec{P},\mathbf{T}\} = 0 \tag{2.18}$$

$$\mathbf{T}^{-1}\,[X_a, P_b]\,\mathbf{T} = -[X_a, P_b]\,. \tag{2.19}$$

At first a reader might be confused by the last line, since we have $[X_a, P_b] = i\,\delta_{ab}$. Then she/he will realize that **T** are "anti-unitary" operators leading to

$$\mathbf{T}^{-1}\,i\,\mathbf{T} = -\,i\,. \tag{2.20}$$

For future applications, we list two useful properties of time reversal operators (or other "anti-unitary" ones). They can be described by general unitary operators **U** and a special one, namely the complex conjugation operator K: $\mathbf{T} = \mathbf{U}K$ (see Ref. [17]); it leads to:

$$\mathbf{T}^{\dagger}\mathbf{T} = K^{\dagger}\mathbf{U}^{\dagger}\mathbf{U}K = K^{\dagger}K \tag{2.21}$$

[12] Subtle points in neutral kaon oscillations are discussed in **Sect. 6.5.7**.

$$\langle a|\mathbf{T}^\dagger\mathbf{T}|b\rangle = \langle a|K^\dagger K|b\rangle = \langle a|b\rangle^* = \langle b|a\rangle \tag{2.22}$$

They transform angular momenta $\vec{J}$ with $[J_a, J_b] = i\,\epsilon_{abc}\,J_c$ as

$$\mathbf{T}^{-1}\,\vec{J}\,\mathbf{T} = -\vec{J} \quad \text{or} \quad \{\vec{J},\mathbf{T}\} = 0\,. \tag{2.23}$$

Anti-unitary operators $\mathbf{T}$ describing time reversal have very important consequence in different regions of dynamics . The anti-linearity comes into play when an amplitude is described *through second* (or even higher) order in the effective interactions; i.e., when the final-state interactions (FSI) are included. We denote it symbolically with*out* **CP** violation:

$$\mathbf{T}^{-1}\left(\mathcal{L}_{\text{eff}}\Delta t + \frac{i}{2}(\mathcal{L}_{\text{eff}}\Delta t)^2 + ...\right)\mathbf{T} = \mathcal{L}_{\text{eff}}\Delta t - \frac{i}{2}(\mathcal{L}_{\text{eff}}\Delta t)^2 + ...$$
$$\neq \mathcal{L}_{\text{eff}}\Delta t + \frac{i}{2}(\mathcal{L}_{\text{eff}}\Delta t)^2 + ... \tag{2.24}$$

It represents transition amplitudes described by time dependent perturbation theory at order $(i)^n$. However, when we talk about weak forces, there is no real reason to go beyond 2nd order. We will discuss the impact of FSI below in **Sect. 2.10**.

2.4 Kramers' degeneracy

In non-relativistic quantum mechanics the fact that the operator $\mathbf{T}$ is anti-unitary has a very important consequence one might not have anticipated: it tells us that all possible physical states can be divided into two distinct classes. Performing time reversal twice, a state can be changed at most by phases:

$$\mathbf{T}^2 = \mathbf{U}K\mathbf{U}K = \mathbf{U}\mathbf{U}^*KK = \mathbf{U}\mathbf{U}^* = \mathbf{U}(\mathbf{U}^T)^{-1} = \eta\,, \tag{2.25}$$

where η is an $n\times n$ diagonal matrix with elements $(e^{i\eta_1}, ..., e^{i\eta_n})$. The last equality implies that $\mathbf{U} = \eta\mathbf{U}^T$ or $\mathbf{U}^T = \mathbf{U}\eta$. By combining these relations, one obtains $\mathbf{U} = \eta\mathbf{U}\eta$, satisfying $1 = e^{i(\eta_i+\eta_j)}$ $(i,j = 1, ..., n)$ for all i and j. One can set $i = j$; thus this equation must hold:

$$e^{i\eta_j} = \pm 1\,. \tag{2.26}$$

There is no such restriction for *unitary* operators, like $\mathbf{P}^2 = e^{i\alpha}$, $\mathbf{C}^2 = e^{i\beta}$, where α and β are arbitrary phases.

For $\mathbf{T}^2 = +1$ one gets a tautology, i.e., nothing new. However with $\mathbf{T}^2 = -1$, the situation changes. When the system is invariant, the operator $\mathbf{T}$ commutes with the Hamiltonian, and $|E^{(T)}\rangle = \mathbf{T}|E\rangle$ is another eigenstate with the same energy value. In that case one gets:

$$\langle E|E^{(T)}\rangle = \langle E|\mathbf{T}|E\rangle = \langle E^{(T)}|\mathbf{T}^\dagger\mathbf{T}|E\rangle = \langle E|\mathbf{T}^\dagger\mathbf{T}^\dagger\mathbf{T}|E\rangle = -\langle E|\mathbf{T}|E\rangle \tag{2.27}$$

and therefore

$$\langle E|\mathbf{T}|E\rangle = 0\,; \tag{2.28}$$

the two energy-degenerate states $|E\rangle$ and $\mathbf{T}|E\rangle$ are actually *orthogonal* to each others. This is referred to as Kramers′ degeneracy [18]. It implies that $|E\rangle$ carries an *internal* degrees of freedom with $\mathbf{T}^2 = -1$ and that $|E\rangle$ and $\mathbf{T}|E\rangle$ do *not* describe the same physical state. This statement was derived in 1930 *without* reference to spin, half-integer or otherwise. Of course, it is completely consistent with results explicitly derived using spin representations of bosons and fermions.[13] Considering a state composed by n spin-1/2 states, each with z component of the polarization s_i, one gets

$$\mathbf{T}^2|x_1, s_1; ..., : x_n, s_n\rangle = (-1)^n|x_1, s_1, ...; x_n, s_n\rangle \, , \tag{2.29}$$

which gives us two lessons about odd numbers of identical fermions:

(1) Kramers′ degeneracy is conceptual. While a world with only the eigenvalue $\mathbf{T}^2 = +1$ is conceivable, there are instead two classes of states which are realized in our Universe. All physical states fall into two classes: integer-spin states and odd-integer states, where the latter carry the eigenvalues $\mathbf{T}^2 = -1$.
(2) The second lesson is 'practical'. Kramers′ degeneracy can describe the degeneracy between odd-integer spin "up" and "down" in solid state physics. When one puts a system of electrons in an external electrostatic field, it breaks rotational invariance; neither spin nor orbital angular momentum are conserved in this situation. Yet time reversal invariance is still preserved. No matter how complicated this field is, (at least) a *two-fold degeneracy* is necessarily maintained for an *odd* numbers of electrons due to their spin-orbit couplings, but not for an *even* one; *odd* and *even* numbers of electron systems therefore exhibit very different behavior in such external fields. This degeneracy is lifted once the electrons interact through their magnetic moments with an external magnetic field, since a magnetic field is not invariant under $\mathbf{T}$. The Kramers′ degeneracy allows to probe the electronic states of paramagnetic crystals [19]. Odd number systems of electrons exhibit at least a two-fold degeneracy that remains immune to the electric fields *inside* the crystals; it is, however, lifted by magnetic fields. Measuring the energy splitting allows to deduce the local magnetic fields inside a crystal.

2.5 'Mixing' and 'oscillations'

First, an analogy from modern dancing. **Fig. 2.1** from the "Diavolo Dance Theater" describes mixing between two different states leading to 'static' situations – i.e., 'balance'. But the true goal of "Diavolo Dance Theater" is to describe 'Architecture in Motion', *not statics*; see **Fig. 2.2**.

'Mixing' (or more precisely 'state mixing') means that different states can contribute coherently in QM (or wave dynamics) and QFT; more special cases are 'oscillations', where 'moving' is crucial. Good examples are the weak decays of

[13] In the world without fermions 'we' could not exist.

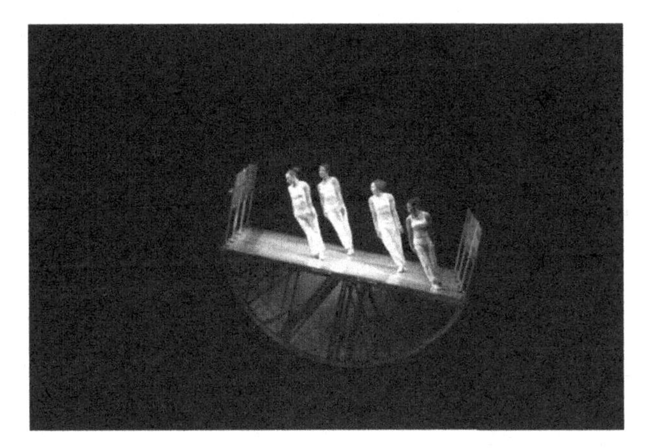

Fig. 2.1 'Balance'

Fig. 2.2 Both 'mixing' and 'oscillations' [Photos from performances of the dance company Diavolo-Architecture in Motion.]

kaons: their QCD eigenstates are super-positions of weak decays eigenstates. QCD alone yields global flavor conservation, yet weak dynamics allows $\Delta S \neq 0$ transitions for kaons.[14] In the world of three quarks, weak charged currents can be described by their Cabibbo couplings $J_\mu^{(+)} = \cos\theta_C \; \bar{d}_L \gamma_\mu u_L + \sin\theta_C \; \bar{s}_L \gamma_\mu u_L$; the couplings, expressing mixing among the families, do *not* change with time. Another example is given by the photon and the Z^0, which have the same quantum numbers and are indeed the final product of the electroweak mixing.

Oscillations are much more complex than mixing: states can produce oscillations, only if they combine states with both zero electric charge[15] and non-zero

[14]Likewise for $\Delta C \neq 0 \neq \Delta B$ for charm and beauty ones.

[15]The decay of charged states and anti-states are described by a *single* exponential functions of time t with $e^{i\Gamma t} = e^{i\bar{\Gamma} t}$ due to **CPT** invariance.

'flavor'. General 'flavor' means: strange, charm and beauty quark carry non-zero flavor; likewise for leptons and even baryons. Oscillations in the transitions of K^0, D^0, B^0 and B_s^0 mesons and neutrinos have been established. Subtle quantum mechanical effects occur when a pair of neutral flavored mesons is produced in a coherent state.[16]

How can one measure oscillations? The best way is to probe the evolution of the transitions *in time*, as we will discuss in details. The same applies to neutrinos, whose transition probabilities among mass eigenstates as function of time will be discussed in **Sect. 4.6**. It is also extensively investigated for neutrons as a tool to study baryon number violation. Neutron oscillation theory and experiments are discussed in **Sect. 6.8**.

2.6 Electric dipole moments

Let us focus on *static* observables (not related to transitions) in different systems such as an elementary particle, an atom, or a molecule. The energy shift of the system $\Delta\mathcal{E}$ due to an *external* electric field can be expanded in a power series in $\vec{E}$ [12]:

$$\Delta\mathcal{E} = d_a\, E_a + d_{ab}\, E_a E_b + d_a d_b d_c\, E_a E_b E_c + \ldots\,. \tag{2.30}$$

The *linear* term d_a is called the "electric dipole moment" (EDM), and the quadratic term d_{ab} is the "induced dipole moment".

The EDM is a measure of charge polarization within a system: it is the expectation value of the operator $\vec{d} = \sum_a e_a\, \vec{x}_a$. One gets

$$\langle j|\vec{d}|j\rangle = \langle j|\mathbf{P}^\dagger\mathbf{P}\,\vec{d}\,\mathbf{P}^\dagger\mathbf{P}|j\rangle = -\langle j|\vec{d}|j\rangle\;. \tag{2.31}$$

If $\mathbf{P}$ is conserved, this matrix element must be zero. After the discovery of parity violation in 1956 it was understood that a non-zero value of an EDM implies not only $\mathbf{P}$ violation but also $\mathbf{T}$ violation. Since $\vec{d}$ is a polar vector and $\vec{J}$ an axial one, they transform as:

$$\mathbf{T}\,\vec{d}\,\mathbf{T}^{-1} = \vec{d}\,, \quad \mathbf{T}\,\vec{J}\,\mathbf{T}^{-1} = -\vec{J}\,. \tag{2.32}$$

$\vec{d}$ is therefore *even* under $\mathbf{T}$ and $\vec{J}$ is *odd*, which implies that a term in the Hamiltonian of the form

$$\Delta\mathcal{E} = a\,\vec{J}\cdot\vec{d} \tag{2.33}$$

violates $\mathbf{T}$ invariance with a non-zero value of a. One can make the same statement with different words: a non-degenerate system in QM is characterized in the rest frame only by $\vec{J}$ as three component vector. Therefore we have

$$\vec{d} \propto \vec{J} \tag{2.34}$$

[16] Oscillations provide direct manifestations of QM fundamentally *non*-local features which were first pointed out by Einstein, Podolsky and Rosen already in 1935 [20]. Now it is fashionable to use the word 'entanglement': it points out the non-local dynamics in QM; still we prefer to use the word "EPR" following 'history'. We will discuss that in details in **Chapter 6**.

and the EDM has to vanish if **T** is conserved. The crucial statement is that the system is 'non-degenerate', not in general 'elementary'. One example is the EDM of the neutron, which is described as a bound state of three quarks: $N = [ddu]$.

One has to admire the courage of the Nobel Prize winner Ramsey, who had worked 43 years searching for neutron EDM without finding it. He ended his talk at the 1993 AIP meeting by saying [21]:

> I suppose I should be discouraged and believe that no particle dipole moment will ever be discovered and the search should be abandoned. On the contrary, I am now quite optimistic.... For the most of the past 43 years, the searches have been lonely ones. Now there are promising experiments with atoms, electrons, and protons as well. I sincerely hope that someone will hit the jackpot soon... at age 76 I cannot wait another 43 years unless there is a way to achieve real time reversal for biological clocks.

No EDM for the neutron has been observed 27 years after this statement by Ramsey, who passed away in 2011, nor for any nucleon, atom or lepton. Yet, the experimental upper limits on EDMs have been reduced significantly. Our community should have given up the search for non-zero values? Neither on the experimental nor theoretical side! Searches for EDMs constitutes an excellent example where the connection between low energy and high energy communities is crucial to understand fundamental forces. We will discuss EDMs in **Chapters 10 – 13** in details.

2.6.1 *Short comments about the impact of Ramsey's work*

The 1989 Nobel Prize in Physics was awarded to N.F. Ramsey "for the invention of the separated oscillatory fields method and its use in the hydrogen maser and other atomic clocks". Ramsey was awarded half of the Nobel Prize.[17] A simple description of Ramsey's method is given in 'Physics Today', January 2013 [22] in the first page: 'Norman Ramsey's evolving method'. His Nobel Prize is not enough to show his deep horizons; he developed novel tools and was a true pioneer about probing fundamental dynamics.

Before the 50s, our community assumed that **P**, **C** and **T** invariances, tested in QED and the strong forces, were also valid for weak dynamics. Ramsey pointed out in a very short 1950 paper that violation of parity invariance had not been probed yet with weak dynamics, and EDMs were a good place to do it [23]. He said that although there were already upper limits of $3 \cdot 10^{-18}$ e cm, he was sure he could do better with a new experiment. This was successfully completed somewhat quickly.

In 1956 the 'landscape' had changed: T.D. Lee and C.N. Yang had suggested to probe parity violation in kaon decays [24], and in 1957 parity violation was established in β decay of ^{60}Co by C.S. Wu and in the same year also in kaon decays [25]. Ramsey and his co-authors also realized that non-zero values of EDMs

[17]The remaining half was divided between H.G. Dehmelt and W. Paul "for the development of the ion trap technique".

of the neutron (or other systems) require both **P** and **T** violation. In 1957 they published a new upper limit of $5 \cdot 10^{-20}$ e cm [26] for the EDM of the neutron. This was just the beginning of the journey into the world of New Dynamics [27]; the 2020 limit for the neutron EDM is [8]

$$|\vec{d}_N| < 0.18 \cdot 10^{-25}\, e\,\text{cm}. \tag{2.35}$$

2.6.2 *Water molecules*

EDM $\vec{d}$ must be proportional to the total angular momentum $\vec{J}$ of a *non*-degenerate system. Then $|\vec{d}| \neq 0$ implies **T** (and **P**) violation. Yet for a complex system such as a molecule, there are other vectors characterizing the system.

Two hydrogen atoms bonded to an oxygen atom form a water molecule. These hydrogen atoms can move like coupled harmonic oscillators. In a symmetric state the two hydrogen atoms oscillate in the same directions, while in an anti-symmetric state they move toward each other or away from each other. The symmetric [anti-symmetric] state is in a parity $|+\rangle$ [$|-\rangle$] state. One can place a water molecule inside a constant external electric field $\vec{E} = (0, 0, E)$ and describe this novel situation with an additional Hamiltonian operator:

$$H = \begin{pmatrix} \mathcal{E}_+ & \Delta \\ \Delta & \mathcal{E}_- \end{pmatrix} \tag{2.36}$$

whose eigenvalues are

$$\mathcal{E}_{1,2} = \frac{1}{2}(\mathcal{E}_- + \mathcal{E}_+) \pm \left[\frac{1}{4}(\mathcal{E}_- - \mathcal{E}_+)^2 + \Delta^2\right]^{\frac{1}{2}} . \tag{2.37}$$

The electric field $\vec{E}$ mixes opposite parity states: $\langle +|E|-\rangle = \Delta$. If the electric field $\vec{E}$ is sufficiently weak, we can set $\Delta \ll |\mathcal{E}_- - \mathcal{E}_+|$ leading to

$$\mathcal{E}_1 \simeq \mathcal{E}_- - \frac{\Delta^2}{\mathcal{E}_- - \mathcal{E}_+} , \; \mathcal{E}_2 \simeq \mathcal{E}_+ + \frac{\Delta^2}{\mathcal{E}_- - \mathcal{E}_+} \tag{2.38}$$

$$\Delta\mathcal{E} = \mathcal{E}_2 - \mathcal{E}_1 \simeq \mathcal{E}_- - \mathcal{E}_+ + \frac{2\Delta^2}{\mathcal{E}_- - \mathcal{E}_+} ; \tag{2.39}$$

the energy shift of a water molecule is *quadratic* in $\vec{E}$ rather than linear. It corresponds to an *induced* dipole moment in the power expansion of Eq. (2.30); it is not an EDM and it does *not* imply **T** (and **P**) violation.

Ground states of an atom or molecule are non-degenerate. The above arguments thus apply: if an EDM is detected, its constituents – electrons or nucleons – must possess an EDM.

2.6.3 *Dumb-bells*

With a water molecule we have given an example of 'induced' electric dipole moments. Now we talk about a somewhat more general situation. A dumb-bell is a system where two charges q and $-q$ are connected by a rigid, but massless rod; it gets an electric dipole by definition. Yet it cannot produce **P** (or **T**) violation, since electrodynamics conserves both. This apparent paradox is solved by the system being *degenerate.*

Let us center the dumb-bell at the origin and align it with the z axis. The positive charge is located at $z = +a$ and denoted by $| \Uparrow\rangle$, while the negative charge is at $z = -a$ and denoted by $| \Downarrow\rangle$. Their energies are degenerate as any linear combination of them; in particular, we refer to parity eigenstates $|\pm\rangle = \frac{1}{\sqrt{2}}[| \Uparrow\rangle \pm | \Downarrow\rangle]$ with energy $\mathcal{E}_+ = \mathcal{E}_-$. When one applies an external electric field $\vec{E} = (0, 0, E)$ to this dumb-bell, the degeneration is lifted. The new eigenstates are linear combinations of the parity eigenstates. One gets (see Eq. (2.37)):

$$\Delta\mathcal{E} = \mathcal{E}_2 - \mathcal{E}_1 = 2\Delta \, . \tag{2.40}$$

An external electric field (even a tiny one) E_z gives a non-zero energy shift linear in the electric field. We have

$$d_z = \langle +|z|-\rangle \neq 0 \, , \tag{2.41}$$

which is fully consistent with parity symmetry: $|+\rangle$ is parity even, while z and $|-\rangle$ are odd; therefore d_z is even. Also it does *not* imply **T** violation!

When the energy difference $\Delta\mathcal{E}$ is linear in the electric field, the system can exhibit a 'structured' electric dipole moment. Schiff [28] called this a 'permanent' electric dipole moment. Other examples are atomic states such as $2s_{1/2}$, $2p_{1/2}$ of the hydrogen atom, which can possess this electric dipole moment.

In a very short summary: a system can have a 'structural' (or 'permanent') electric dipole moment that does not conflict with **P** and **T** invariance in presence of degenerate states. However, any system can have an induced dipole moment, when the energy shift is quadratic in energy (and the dipole moment is proportional to the electric field).

2.6.4 *Schiff's theorem*

As the ground state of an atom is non-degenerate, it cannot possess a structural electric dipole moment. The observation of an EDM therefore establishes **T** violation in the dynamics of its constituents. Yet EDM effects due to its constituents can be screened in an atom placed into an *external* electric field. The charges inside an atom get shifted by the external electric field; this distortion creates a polarization and thus an induced electric field in such a way that the EDM is shielded.

Schiff's theorem [28] states that this shielding is complete and an atomic EDM vanishes under the following conditions:

- Atoms consist of *non*-relativistic particles, which interact only electro *statically.*
- The EDM distributions of each atomic constituent is identical to its charge distribution.

This theorem applies to light atoms successfully. Yet in high Z atoms, valence electrons feel strong Coulomb fields when they come close to the nucleus. Their motion is highly relativistic, and one of the conditions of the Schiff's theorem breaks down. As EDMs violate parity, they generate strong mixing between opposite parity states at short distances. One might expect that the complete shielding of the electron EDM expressed by the Schiff's theorem gets translated into partial screening, when relativistic corrections are included; it turns out that the reverse is true. For heavy atoms *enhancement* factors as large as two order of magnitude are possible. For alkali metal Cesium the EDM is about a factor of 100 larger than the electron EDM.

2.7 CPT invariance

The **CPT** theorem in axiomatic quantum field theory [17,29] states that a combined transformation of **C**, **P** and **T** (taken in any order) represents a symmetry of a Lagrangian that respects unitarity.[18] It is based on three crucial assumptions:

- The dynamics is Lorentz invariant;
- The vacuum is Lorentz invariant;
- Weak local commutativity obeys the 'right' statistics, namely [., .] for local boson fields, while {., .} for local fermion ones.

Locality means that there is no action at distance, while unitarity means conservation of probability. **CPT** invariance is linked to Lorentz invariance, as can be seen using the Lagrangian density $\mathcal{L}$ and action S formalism (see above Eq. (2.3)):

$$\mathbf{CPT}\,\mathcal{L}(t,\vec{x})\,(\mathbf{CPT})^{-1} = \mathcal{L}(-t,-\vec{x}) \quad , \quad S = \int d^4x\,\mathcal{L}(t,\vec{x}) \implies S\,. \tag{2.42}$$

One can check that all possible Lorentz invariant fermion bilinears in $\mathcal{L}$ are invariant under **CPT**. Thus it is not easy to find consistent theories that violate **CPT**.[19]

Following Ref. [29, 30], **CPT** symmetry tells us that the masses, as well as the total widths or lifetimes of a particle and its anti-particle are exactly the same, The 'classical' tests of **CPT** symmetry concern these equalities, and when one looks at the Review of Particle Physics, *Tests of Conservation Laws, 3 Tests of CPT*, one finds impressive accuracy in the experimental data [8].

Some comments are in order. We think that masses provide a very *ad hoc* calibration stick for a difference in the particle-antiparticle masses, since the overwhelming bulk of the masses is *not* generated by the weak forces. While the decay widths are of weak origin, bounds as $\sim \mathcal{O}(\text{few}\cdot 10^{-4})$ – like on the normalized difference of lifetimes in pions – can hardly be called decisive *empirical* verification of this fundamental symmetry. The strengths of **CPT** invariance are rather based on our understanding of the underlying dynamics, namely on the impact of QFT.

[18] One could note that this theorem does not hold for *infinite*-component fields.

[19] We are considering conventional QFT, not extensions like string theory, non-commutative QFT or quantum gravity.

2.8 Violation of Parity and Charge Conjugation

Until the 1950s, strong and electric interactions had been thoroughly investigated without finding any sign of **P** and **C** asymmetries. It was generally assumed that they were basic symmetries of our Nature. In 1956 T.D. Lee and C.N. Yang pointed out that **P** and **C** symmetries had not been analyzed in *weak* dynamics and produced a list of new experiments [24]. Already in 1957 experimentalists found out that charged weak dynamics of hadrons violates **P** and **C** symmetries basically at 100% [25]. In principle, there is no problem with maximal parity violation for quarks and anti-quarks; data are consistent with assuming the existence of only charged weak transitions for left-handed currents of quarks, and excluding right-handed ones. The same can be said for **C** conjugation .

One can combine **P** and **C** asymmetries to get **CP** invariance as a first step and get a much better symmetry for hadron and lepton decays. As long as **CP** symmetry is conserved, 'left'- vs. 'right'-handed states represents a convention. Let us compare, for example, the weak decays of charged pions into a pair of leptons

$$\pi^- \to e^-_L \, \bar{\nu} \quad \text{vs.} \quad \pi^+ \to e^+_R \, \nu \,, \tag{2.43}$$

which are related by a **CP** transformation. One can say that a pion of a given charge will produce leptons of only one helicity ; i.e., what one means by 'left' depends on the definition of negative charge and vice versa

$$`L' = f(`-') \,. \tag{2.44}$$

This is like saying: 'The thumb is left on the right hand' – a correct as well as useless statement since circular. The situation of neutrinos and anti-neutrinos is different. Following the SM, neutrinos are *massless*; furthermore, the SM assumes the existence of ν_L and $\bar{\nu}_R$, but not that of ν_R nor $\bar{\nu}_L$ [31]. Thus for example the reaction $\nu_L \; n \to e^-_L \; p$ or $\bar{\nu}_R \; p \to e^+_R \; n$ can happen, but *not* $\nu_R \; n \to e^-_R \; p$ or $\bar{\nu}_L \; p \to e^+_L \; n$.

CP symmetry is violated in weak interactions. Even Landau, who was a late convert to parity violation, made his peace with this situation [32]. However, this feature of the SM is implemented *par ordre du mufti.*[20] It is dictated by the data with*out* any deeper reason.

With*out* assuming **CP** symmetry, **CPT** invariance tells us that the masses and the widths are the same. Even if weak forces violate parity at 100% (or close to it), **CP** symmetry is still used for classification in spectroscopy considering interactions among pions, kaons and η as pseudo-scalars, since **CP** is a good symmetry for strong and electromagnetic interactions.

[20] A 'French' decision is imposed on somebody with no explanation and no right of appeal.

2.9 CP violation

Less than a decade after **P** violation was found in weak dynamics, particle physics community entered a *novel* era.[21] **CP** violation was found in 1964 [2]. Cronin and Fitch with their collaborators Christenson and Turley had set-up an experiment to investigate an 'anomalous K_S regeneration'; however, their 1963 data seemed to make no sense. They had analyzed their data for a year in order to try to understand their findings. At a 1989 conference celebrating the XXVth anniversary of **CP** violation A. Pais stated [33] in his retrospective lecture[22]:

- I met with Jim (Cronin) and Val (Fitch) at the Brookhaven Cafeteria in early June (1964); they told me they have found $K_L \to \pi^+\pi^-$. I had said it would violate **CP** invariance! They knew that, but there it was. I asked many questions, and they said they had thought long and hard about these and other alternatives and had ruled them out one after one. After they left, I had another coffee – I was shaken by the news. With the very small 2π rate, **CP** invariance as a near miss, made the news even harder to digest.

Even with the present, refined SM we have not yet solved several puzzles in the world of hadrons with **CP** violation – like the reasons for the size of **CP** violation. We expect to deal with this crucial challenge in fundamental dynamics step by step and by the collaborations between experimentalists and theorists. One crucial challenge deals with the understanding of the final state interactions, a major problem which often obscure the fundamental dynamics, see **Sect. 2.10**.

2.9.1 *Short comments about Cronin's wide horizon*

As just mentioned, a giant leap in understanding was made in 1964, due to the discovery of **CP** violation in the kaon system [2]. Cronin and his co-authors were fully aware of the importance of this discovery and continued to search for the origin of **CP** violation [35]. In 2011 Cronin gave a personal recollection about the history of the discovery of **CP** violation [36], which includes descriptions of experimental tools like spark chambers invented in 1959, sketches from his notebook on how he adapted the di-pion apparatus for the observation of the anomalous regeneration, notes with plots derived from the **CP** violation run, etc. He also gave credit to a famous 1955 paper by Gell-Mann and Pais [37]. A few years later, Cronin summarized the part of his career in experimental particle physics that started in 1955 and ended in 1985 [38]. He discussed not only his successes, but also his failures and his bad judgments. This period was the golden age of particle physics, when the experimental possibilities were abundant and one could carry experiments with a small team of colleagues and students.

[21] Lev Okun had pointed out in his book [1] published in 1963 (in Russian) that 'our' community had to continue to probe **CP** violation.

[22] For one of the best examples of how 'our' community was shocked by the possibility of **CP** violation, see a 1964 paper [34]; we will come back to it in **Sect. 6.5**.

In the seventies Cronin also contributed significantly to fundamental physics in a more traditional way, namely by the "Production of hadrons with large transverse momentum at 200, 300, and 400 GeV" at Fermilab [39, 40].

In 1985 he moved to cosmic-ray physics. Even if the cosmic-ray experiments became imposing, yet he was able to create the conceptual design for several experiments. However, the detailed design, construction, and execution of the cosmic-ray experiments involved many physicists with technical skills. The last experiment which he was involved in was the Pierre Auger Observatory experiment that started in 2004. In 2009 he gave a talk at Blois[23] and discussed the observations of cosmic rays with energies above 10^{18} eV,[24] the measurements of the cosmic ray spectra and their arrival directions. He also reported limits for the photon and neutrino components of this cosmic radiation [41].

Before his **CP** violation years, Cronin gave a talk at the 1953 Cosmic Ray Conference, named the "Birth of Sub-Atomic Physics" [42]. The cosmic ray conference organized in 1953 in the French Pyrénées by Patrick Blackett and Louis Leprince-Ringuet[25] was a seminal one. It marked the beginning of sub-atomic physics and its shift from cosmic ray research to research at the new high energy accelerators. The knowledge of the heavy unstable particles found in the cosmic rays was essentially correct in its interpretation and defined the experiments that needed to be carried out with new accelerators. A large fraction of the physicists who had been using cosmic rays for their research moved to accelerators physics.

2.10 T-odd distributions and final state interactions (FSI)

We have already discussed **T** violation in general in **Sects. 2.3** and **2.4**. Here we consider the weak transition of an initial state i to a final state f

$$T(i \to f) \propto ...\langle f|H_W|i\rangle \tag{2.45}$$

mediated by the weak Hamiltonian H_W which can be described by a series of local terms, as in Eq. (2.4). It is crucial to have non-trivial interactions, since **CP** and **T** violations manifest themselves through complex phases of the underlying dynamics.

2.10.1 T *invariance and the Watson's theorem*

Let us consider the weak non-leptonic decays of kaons. In the rest frame of a kaon energy conservation allows FS with 2 and 3 pions, while $K \to 4\pi$ is forbidden. Furthermore, there is a symmetry called the G-parity, which tells us that a state of even numbers of pions can*not strongly* evolve into a state with odd numbers even

[23]Blois is a city in France with true culture: there are many reasons to visit it.

[24]That is beyond the 'ankle' of cosmic rays fluxes = 1 particle per km^2-year!

[25]Leprince-Ringuet always wrote in French regardless of the nationality of the recipient; the sole exception was the joint letter drafted by Blackett to the commissioners. It tells us something about the world after World War II.

in the presence of FSI[26]:

$$K \xrightarrow{H_{weak}} 2\pi \overset{H_{str}}{\not\Longleftrightarrow} 3\pi \tag{2.46}$$

The amplitudes for $K^0 \to 2\pi$ are described by a series of operators and their matrix elements

$$H_W = \sum_j a_j H_W^j + \text{h.c.} \qquad \langle (2\pi)_I; \text{"out"}|H_W^j|K^0\rangle = |A_I^j| e^{i\Phi_I} \tag{2.47}$$

where I denotes the isospin of the 2π states, namely $I = 0, 2$. The two pions emerging from the weak decay $K \to 2\pi$ are not asymptotic states yet; they undergo re-scattering due to strong FSI, which generate phases Φ_I. In Eq. (2.47) we assume, with*out* loss of generality, that H_W^j are **T** invariant: $\mathbf{T} H_W^j \mathbf{T}^{-1} = H_W^j$. Thus the phases of the coefficients a_j are the 'portal' for **T** (and **CP**) asymmetries; we will come back to it below. With **T** being an *anti*-unitary operator – using $\langle a|\mathbf{T}^\dagger \mathbf{T}|b\rangle = \langle b|a\rangle$ – one gets

$$\langle (2\pi)_I; \text{"out"}|H_W^j|K^0\rangle = \langle (2\pi)_I; \text{"out"}|\mathbf{T}^\dagger \mathbf{T} H_W^j \mathbf{T}^{-1}\mathbf{T}|K^0\rangle^* = \langle (2\pi)_I; \text{"in"}|H_W^j|K^0\rangle^*$$

For a single particle state – K^0 in this case – there is no distinction between an "in" and "out" state. Then one insert a complete set of "out" states:

$$\langle (2\pi)_I; \text{"out"}|H_W^j|K^0\rangle = \sum_n \langle (2\pi)_I; \text{"in"}|n; \text{"out"}\rangle^* \langle n; \text{"out"}|H_W^j|K^0\rangle^* \tag{2.48}$$

$\langle n; \text{"out"}|(2\pi; \text{"in"}\rangle$ is a $S(=\textit{Scattering})$ matrix element which contains the energy momentum conserving delta function. The allowed hadronic states in the sum over n are 2π and 3π states; furthermore G-parity enforces $\langle 2\pi; \text{"in"}|3\pi; \text{"out"}\rangle = 0$. Therefore only the 2π "out" states can contribute to the sum. That coincides with the S wave 2π phase shifts δ_I taken at the energy M_K; it is the 1954 "Watson theorem" [43].[27] It is called *elastic unitary*:

$$S_{\text{elastic}}((2\pi)_I \to (2\pi)_I) = \langle (2\pi)_I; \text{"out"}|(2\pi)_I; \text{"in"}\rangle \equiv e^{2i\delta_I} \tag{2.49}$$

where δ_I denote the 2π phase shifts (see **Fig. 2.3**).

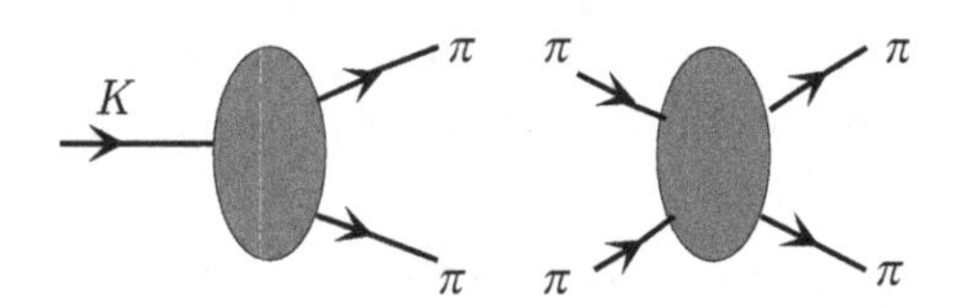

Fig. 2.3 Only 2π states contribute to the sum of intermediate states to the amplitudes.

Replacing $n \equiv (2\pi)_I$ in Eq. (2.48) and using Eq. (2.49), one gets

$$\langle (2\pi)_I; \text{"out"}|H_W^j|K^0\rangle = e^{2i\delta_I} \langle (2\pi)_I; \text{"out"}|H_W^j|K^0\rangle^* \tag{2.50}$$

$$\arg\left(\langle (2\pi)_I; \text{"out"}|H_W^j|K^0\rangle\right) = 2\delta_I - \arg\left(\langle (2\pi)_I; \text{"out"}|H_W^j|K^0\rangle\right) \tag{2.51}$$

[26]The situations are more complex for the weak decays of B and D: FSI leads to $2\pi \to 4\pi$..., but also to $\pi\pi \leftrightarrow \bar{K}K$.

[27]It is not obvious how one can apply Watson's theorem to the decays of strange *baryons*.

Therefore we obtain:

$$\langle (2\pi)_I; \text{"out"} | H_W^j | K^0 \rangle = |\langle (2\pi)_I; \text{"out"} | H_W^j | K^0 \rangle | e^{i\delta_I} \, . \tag{2.52}$$

This reasoning holds for all H_W^j after having the strong phase shift factored out, and the amplitude for a neutral kaon decaying into two pions with total isospin I can be expressed as

$$\langle (2\pi)_I; \text{"out"} | H_W | K^0 \rangle = \mathcal{A}_I e^{i\delta_I} \tag{2.53}$$

Likewise for $\bar{K}^0 \to \pi\pi$:

$$\langle (2\pi)_I; \text{"out"} | H_W | \bar{K}^0 \rangle = \bar{\mathcal{A}}_I e^{i\delta_I} \tag{2.54}$$

As long as H_W conserves **T**, the two amplitudes $\mathcal{A}_I$ and $\bar{\mathcal{A}}_I$ remain "real" together. However, when H_W violates **T** invariance, additional phases appear in $\mathcal{A}_I$ and $\bar{\mathcal{A}}_I$ that are most likely different.

2.10.2 *FSI and partial widths*

In general FSI become indispensable in making **CP** asymmetries observable in partial widths.[28] Consider a weak decay channel $H \to f$ receiving contributions from two coherent processes. The amplitudes read as follows:

$$A(H \to f) = e^{+i\varphi_1^{\text{weak}}} e^{+i\delta_1^{\text{FSI}}} |\mathcal{A}_1| + e^{+i\varphi_2^{\text{weak}}} e^{+i\delta_2^{\text{FSI}}} |\mathcal{A}_2| \tag{2.55}$$

$$A(\bar{H} \to \bar{f}) = e^{-i\varphi_1^{\text{weak}}} e^{+i\delta_1^{\text{FSI}}} |\mathcal{A}_1| + e^{-i\varphi_2^{\text{weak}}} e^{+i\delta_2^{\text{FSI}}} |\mathcal{A}_2| \tag{2.56}$$

where $\varphi_{1,2}^{\text{weak}}$ are weak phases that change sign under **CP** conjugation, while $\delta_{1,2}^{\text{FSI}}$ due to strong (or electromagnetic) FSI do not. The **CP** asymmetry can be expressed by the ratio:

$$\frac{\Gamma(H \to f) - \Gamma(\bar{H} \to \bar{f})}{\Gamma(H \to f) + \Gamma(\bar{H} \to \bar{f})} = -\frac{2\text{sin}(\Delta\varphi_W)\text{sin}(\Delta\delta^{\text{FSI}})|\mathcal{A}_2/\mathcal{A}_1|}{1 + |\mathcal{A}_2/\mathcal{A}_1|^2 + 2|\mathcal{A}_2/\mathcal{A}_1|\text{cos}(\Delta\varphi_W)\text{cos}(\Delta\delta^{\text{FSI}})} \, , \tag{2.57}$$

where $\Delta\varphi_W = \varphi_1^{\text{weak}} - \varphi_2^{\text{weak}}$ and $\Delta\delta^{\text{FSI}} = \delta_1^{\text{FSI}} - \delta_2^{\text{FSI}}$. There are two requirements to satisfy at the same time for a decay to reveal **CP** asymmetries:

- Obviously one needs a non-trivial weak phase difference $\Delta\varphi_W$: that is the easy part for a theorist using SM and/or New Dynamics (ND).
- FSI induce a non-trivial strong phase shift $\Delta\delta^{\text{FSI}}$; it happens, in particular, when the two transitions amplitudes differ in their isospin content. Yet, here one cannot give even semi-quantitative predictions; one has to depend on the data.

The asymmetry gets larger, if the two interfering amplitudes are of comparable size: $|\mathcal{A}_2/\mathcal{A}_1| \sim \mathcal{O}(1)$. FSI affect also the decays of *heavy flavor* hadrons, yet we *cannot* apply Watson's theorem there blindly. There is no reason why *elastic* unitarity should apply in two-body or even quasi-two-body FS of beauty hadrons since the intermediate states in Eq. (2.48) will involve light flavor hadrons with more phases.

[28] A notable exception are oscillations.

2.11 CPT constraints in general

Strong re-scatterings happen with non-perturbative QCD and produce a significant impact on **CP** asymmetries for strange, charm and beauty hadrons. Our control of non-perturbative QCD is limited so far. In order to deal with this challenge we can use tools following constraints coming from symmetries (broken or not). FSI has a large impact on direct **CP** asymmetries assuming **CPT** invariance, as it was discussed first in the Refs. [44–46].

Let us consider the amplitude of a generic decay and its **CP** conjugate:

$$T(H \to f) = e^{i\delta_f} \left[T_f + \sum_{f \neq a_j} T_{a_j} \, i T^{\rm resc}_{a_j f} \right] \tag{2.58}$$

$$T(\bar{H} \to \bar{f}) = e^{i\delta_f} \left[T^*_f + \sum_{f \neq a_j} T^*_{a_j} \, i T^{\rm resc}_{a_j f} \right] \tag{2.59}$$

where $T^{\rm resc}_{a_j f}$ describes (mostly strong) FSI with **CP** invariance between the final state f and the intermediate *on*-shell states a_j. The second line follows from **CPT** invariance. We have assumed that all states have the same weak coupling and therefore can re-scatter into each other: they connect through FS. Thus one gets:

$$\Delta\gamma(f) = |T(\bar{H} \to \bar{f})|^2 - |T(H \to f)|^2 = 4 \sum_{f \neq a_j} T^{\rm resc}_{a_j f} \, {\rm Im}\, T^*_f T_{a_j} \; . \tag{2.60}$$

They have to vanish upon summing over all f between *sub-classes* of partial widths:

$$\sum_f \Delta\gamma(f) = 4 \sum_f \sum_{f \neq a_j} T^{\rm resc}_{a_j f} \, {\rm Im}\, T^*_f T_{a_j} = 0 \; , \tag{2.61}$$

since $T^{\rm resc}_{a_j f}$ and ${\rm Im} T^*_f T_{a_j}$ are symmetric and anti-symmetric, respectively, in the indices f and a_j. **CPT** symmetry imposes equalities not only between total widths of particles and antiparticles, but also *regional* ones, between subclasses of partial widths.

Eqs. (2.58,2.59,2.60) apply to amplitudes in the world of hadrons, namely the bound states of quarks and anti-quarks (and gluons). It is a challenge to connect the amplitudes of hadrons with those of q and $\bar{q}$. The weak dynamics is described by couplings of quarks with gauge bosons, providing the weak phases for **CP** violation. Furthermore for direct **CP** asymmetries one needs re-scattering. That gives the imaginary part needed in FSI – however the situation is rather 'complex' here. The 'roads' are quite different depending on the FS.

We conclude this **Chapter** by giving a few comments about the history of our understanding fundamental dynamics:

- Ref. [44] gave a general introduction to **CP** asymmetries in B transitions, including penguins diagrams. In the end the authors used a refined model for strong forces leading to a semi-quantitative estimate of $A_{\bf CP}(B^0 \to K^+\pi^-) \sim -0.1$.

Wolfenstein in Ref. [46] used both tree and penguin diagrams with quarks, including absorptive part of penguin graphs, namely

$$b \to u + \bar{u} + s \tag{2.62}$$

$$b \to c + \bar{c} + s \; . \tag{2.63}$$

He used perturbative QCD. However, when one considers the re-scattering of the amplitudes as $B \to D_s^{(*)} \bar{D}^{(*)}$, it is unlikely one can give semi-quantitative predictions using perturbative QCD.

- The landscape has changed very much *even before* the "Higgs" state was established in 2012: very large **CP** violation was found in $B^0 \to \psi K_S$ decays by the BaBar and Belle collaborations in 2001, just at the very early beginning of the third century. We will discuss it in **Sect. 6.6**. One has to go after accuracy to understand the lessons the data give us.
- Finally, it is *crucial* to probe **CP** asymmetries in three- and four-body FS of beauty and charm hadrons, since two-body FS are small parts there.

Chapter 3

Connecting the worlds of hadrons, quarks and gluons

In this **chapter** we discuss the dynamics of hadron weak transitions and the relevant tools used to describe them. There are many theoretical tools we can use; some are obvious, some others are not.

It is said that the king Ptolemy once approached the resident 'sage' (Euclid) with the request to be educated in 'geometry', but in a 'royal way', since he was busy with many obligations. Whereupon Euclid replied with admirable candor: "There is *no* 'royal way' to 'geometry'." One can say the same for QFT.

In this **chapter**, we first mention quark models, which are still used in the 21st century; they have systematic uncertainties, whose value is usually determined by hand-waving arguments. Then we briefly discuss renormalized QFTs, where one can better control uncertainties in systematic ways. Lastly, we follow the tradition and analyze effective field theories in different 'dimensions'; sometimes they were *not* labelled effective field theories, although they are.

We focus both on **CP** asymmetries and rare decays–what was found in the past and what we expect for the future. As stated in the **Prologue**, the SM is incomplete. The available experimental data, however, do not suggest any clear path to follow and our community, rightly so, is seeking for new dynamics along diverse directions.

3.1 Quark *models*

The idea of "quarks" was introduced by Gell-Mann [47] and independently by Zweig [48] in 1964 as 'constituents' of hadrons.[1] The three quarks u, d and s belong to the fundamental representation of the global Lie group $SU(3)$.[2] For a long time quarks were seen mostly as a mathematical tool to describe the spectroscopy of hadrons, not as physical states [49].[3] In the early times, the goal was to describe the spectroscopy of *light* mesons and baryons with*out* claiming that 'quarks' are physical states. The landscape has started to change for mainly two reasons:

[1] The name of "quark" was coined by Gell-Mann, the name suggested by Zweig was "ace".

[2] One can learn how to apply Lie group in physics in many examples [49, 50].

[3] M. Gell-Mann in his 1994 book *The Quark and the Jaguar* said: "As I postulated the existence of quarks, I thought from the beginning for some reason that they exist in closed form. I named them as 'mathematical' ones."

(a) analyses of inclusive $eN \to eX$ first from SLAC data showed that quarks are physical states as parts of nuclei. (b) QCD, based on quarks and gluons as part of a QFT, was established as the theory of strong forces.

Quark models have greatly helped to understand the impact of strong forces. We can group quarks into three classes, depending on the ranges of the values of the masses of the quarks compared to $\bar{\Lambda} \sim 1$ GeV. Masses of 'constituent' quarks are phenomenological parameters to be fitted to hadron mass spectra. The first class refers to the 'light' quarks, namely u, d and s. Their 'constituent' masses are below $\bar{\Lambda}$: $m_u^{\rm const} \simeq m_d^{\rm const} \sim 0.33$ GeV and $m_s^{\rm const} \sim 0.5$ GeV. The second class includes the b quark, whose mass $m_b^{\rm const} \sim 5$ GeV is 'heavy' because it is much larger than $\bar{\Lambda}$. It often includes also the c quark which has mass $m_c^{\rm const} \sim 1.5$ GeV and acts as a somewhat 'heavy' one when one does not go for accuracy. In some cases the c quark may be modeled as a 'light' one. These two classes are often used in non-relativistic models, that represents the larger part of the so called 'constituent' quark models. One of the success of the 'constituent' quark models has been the reproduction of the mass spectra of heavy quarkonia, namely a non relativistic $\bar{Q}Q$ system. The third class includes the 'ultra-heavy' top quark. It decays *before* forming top hadrons [51]. Obviously there is no good reason to define 'constituent' top quarks.

In quark models we have a special baryon, namely $\Omega^- = [sss]$ with spin-3/2 that decays only weakly. In order to give the observed charge and spin to this baryon, the simplest configuration of the baryon would require all three constituent quarks of Ω^- to be aligned with their spins. Since they are in a S-wave, their spatial wave-functions are symmetric, and we have imposed that their spin and flavor wave-function factors be symmetric, too. This creates a problem with the Fermi-Dirac statistics since baryons, having an odd number of spin-1/2 components, should have totally anti-symmetric wave-functions. To solve this problem, first it was suggested the three strange quarks combine in a P-wave to produce Ω^-. Another suggestion was to look at a broken $SU(3)_{\rm flav}$. Both suggestions proved to be wrong: they were a 'band-aid' at best. We know now that the solution consists in adding an additional quantum number to the quarks: "color". The Fermi-Dirac problem is solved if we assume that baryon wave-functions are anti-symmetric under this new quantum number. Since the wave-functions are symmetric under space, spin and flavor, anti-symmetry in color gives totally anti-symmetric wave-functions. The need for such quantum number had arisen even before the emergence of QCD as the theory for the strong interactions.

From the quark pioneering era we learn the following lessons:
(a) When one enters a new era of fundamental dynamics, one uses models. Specific models are better to probe the correlations among different regions of dynamics. One needs good thinking about which models should be used.
(b) When one has large data sets, it is better to use model-independent analyses.
(c) That is *not* the end of the 'road': refined theoretical tools describe the data with good control. Although best theories do not necessarily give the best fitted values of the data, their strengths often rely on the correlations among transitions.

3.2 Local gauge interactions

One important lesson about fundamental dynamics that we have learnt for the past fifty years: basic states are given by local gauge interactions based on symmetries. In a visual description one can say that quantum fields are the 'actors' on the 'stage', while the 'writers' are the "symmetries". The SM is described by local QFT based on $SU(3)_C \times SU(2)_L \times U(1)_Y$ in the beginning, and by the unbroken local $SU(3)_C \times U(1)_V$ with QED $= U(1)_V$ in the end.[4] We have three massive gauge bosons $W^\pm$ and Z^0 and nine massless gauge boson: eight gluons and the photon.

QED is a powerful tool for describing electromagnetic dynamics with an amazing record of true precision in different 'landscapes'. It allows for instance an accurate description of $e^+e^- \to e^+e^-$ and $e^+e^- \to \mu^+\mu^-$ transitions, including photons in the FS (and actually even in the definition of a beam of incoming electrons). Another crowning achievement of QED is the calculation of the anomalous magnetic moments. The electron and muon anomalous magnetic moments $(g-2)$ belong to the most precisely measured quantities in particle physics.[5]

The introduction of non-abelian QFT was first based on local symmetries due to their 'beauty', not on an urgency from the data. A quote from C.N. Yang says: "We did not know how to make the theory fit the experiments. It was our judgment, however, that the beauty of the idea alone merited attention".

The Lagrangian of local non-abelian gauge dynamics $SU(N)$ (N = 2, 3, ...) is given by:

$$\mathcal{L}_{\rm SU(N)} = -\frac{1}{4} G^a_{\mu\nu} G^{\mu\nu,a} + \bar{q}\, i\, \gamma_\mu D^\mu q - \bar{q}_R \mathcal{M} q_L + \text{h.c.}\,; \tag{3.1}$$

$G^a_{\mu\nu}$ denote the field strength tensor of gauge bosons

$$G^a_{\mu\nu} = \partial_\mu A^a_\nu - \partial_\nu A^a_\mu + g f^{abc} A^b_\mu A^c_\nu\,, \tag{3.2}$$

where the covariant derivative $D_\mu = \partial_\mu - i\, g\, t^a A^a_\mu$ is expressed with the gauge fields A^a_μ and the generators t^a with $a = 1, 2, ..., N^2 - 1$; q represents quark fields with their spin $\frac{1}{2}$, and $\mathcal{M}$ is their mass matrix. It shows not only that three and four gluon fields interact with themselves, but also describes the connection with g and g^2 in a calculated way, using a set of well-known numbers f^{abc} called "structure constants".

At the beginning, local $SU(N)$ theories were seen as a nice idea, but useless: it had been pointed out by Pauli[6] that unbroken $SU(N)$ gauge bosons are massless, therefore unable to describe short-range weak interactions. Later on, the landscape changed in two different directions:

[4]We will explain the meaning of the subscripts C, L, Y and V later.

[5]The $g-2$ values for the electron and the muon show a small discrepancy between calculation and data; for a recent review on the subject see Ref. [8].

[6]In one of Pauli's letters to Pais in 1950, see Ref. [52].

- The idea of "Higgs" forces was suggested for 'broken' electroweak dynamics in the 1960's, initially to give masses to charged W bosons.
- Unbroken local "color" symmetry emerged providing strong dynamics to produce hadrons from quarks, with non-zero masses, and massless gluons. There was only one candidate among local QFTs to do the 'job', namely QCD = $SU(3)_C$.

3.3 Quark *masses* in quantum field theories (QFT)

In QED one adopts the pole mass for the electron, which is defined as the position of the pole in the electron Green function; actually it is the beginning of the cut, to be more precise. It is gauge invariant and can be measured. An electron never appears 'alone'; yet one can define the mass of an 'isolated' electron and compare this definition with its experimental value.

The situation with quarks is very different, even qualitatively, because of their confinement within hadrons: they are never seen as asymptotic states, there are no actual poles in the S-matrix. Their pole masses cannot be defined outside perturbation theory. Quark masses are *not* observables; one uses the word 'scheme' to describe this situation, where the notion of a quark mass relies on a theoretical construction.

When one looks at PDG, one finds the values for "current" quark, without mentioning that, especially when pertain to light quarks, "current" quarks are based on our understanding of strong forces with QCD=$SU(3)_C$. In quantum field theory, the bare masses are parameters in the Lagrangian, entering on the same footing as, say, the bare coupling constant, with the difference that the masses carry dimension. Bare parameter are never observable, so it is a necessary step to move to the renormalized quark masses. Then an appropriate renormalization prescription, as well as a scheme, has to be defined.

According to the mass values, one generally distinguishes again three classes of quarks: light quarks u, d and s; the obviously heavy quark b with the quark c, which mostly acts as heavy; the top quarks , which are hyper-heavy and decay *before* they can produce top hadrons, but still carry "color", that is unbroken.

Light quarks have "current" masses which are well below $\bar{\Lambda}$,[7] but still not zero. In PDG 2020 we find [8]: $m_u = 2.16^{+0.49}_{-0.26}$ MeV, $m_d = 4.67^{+0.48}_{-0.17}$ MeV, $m_u/m_d \sim 0.47^{+0.06}_{-0.07}$, $m_s = 93^{+11}_{-5}$ MeV and $2m_s/(m_d + m_u) = 27.3^{+0.7}_{-1.3}$. One can compare these values to the ones on PDG 2000: $m_u = (1 - 5)$ MeV, $m_d = (3 - 9)$ MeV, $m_u/m_d \sim 0.2-0.8$ and $m_s = (75-170)$ MeV and $m_s/m_d \sim 17-25$; the comparison shows the very good progress made in understanding strong forces for the light quarks.

Computational convenience can suggest to use the pole mass for a heavy quark Q. We can identify the location of the pole in the quark propagator whether or not

[7] We assume as a conservative value of the QCD scale $\bar{\Lambda}$ the upper limit of about 1 GeV.

we identify this mass with the renormalized mass in the QCD Lagrangian. While not measurable per sé because of confinement, the pole mass is infrared (IR) stable order by order in *perturbative* theory. The pole mass is however *not* IR stable in *full* QCD, which leads to a theoretical uncertainty. Since the quarks are confined in the full theory, the theoretical uncertainty in the case of a B meson is intuitively of the order

$$\delta m_Q^{\rm pole}/m_Q \sim \mathcal{O}(\bar{\Lambda}/m_Q)\,. \tag{3.3}$$

Another way to illustrate the theoretical uncertainty due to this instability is to consider the energy stored in the chromo-electric field in a sphere of a radius $R \gg 1/m_Q$ around a static source of mass m_Q:

$$\delta\mathcal{E}_{\rm color\ Coul}(r) \propto \int_{1/m_Q\leq |x|\leq R} d^3x\, \vec{E}^2_{\rm color\ Coul} \propto {\rm const} - \frac{\alpha_S(R)}{\pi}\frac{1}{R} \tag{3.4}$$

This is the energy that one adds to the bare mass of a heavy particle to determine what will be the mass including the QCD interactions. The definition of the pole mass amounts to $R \to \infty$; in evaluating the pole mass one integrates the energy density associated with the color source over all space assuming it has a Coulomb form. However, color interactions become strong at around $1/R \sim \bar{\Lambda}$. At such distances the chromo-electric field has nothing to do with the Coulomb tail and the region beyond it cannot be included in a meaningful way.

One can go beyond hand-waving argument: it exhibits an *irreducible theoretical* uncertainty called a renormalon ambiguity,[8] which we briefly discuss in **Sect. 3.8**.

Top quarks plays a special role in the SM and beyond: as said above, top quarks can*not* produce top *hadrons* due their high mass and very short lifetime [51]; furthermore, Yukawa coupling to the Higgs boson(s) are order of unity. However, QCD is *unbroken*, as discussed below; top (or anti-top) quarks carry "color 3" (or "$\bar{3}$"), thus one has to find another source of "$\bar{3}$" (or "3") in their decays.

3.3.1 "$\overline{\rm MS}$", "*kinetic*", "*PS*"

The most common definition of quark mass is the $\overline{\rm MS}$ mass $\bar{m}_Q(\mu)$, where $\overline{\rm MS}$ stays for 'Modified Minimal subtraction scheme'. It represents a quantity of computational convenience, in particular when calculating perturbative contributions in "dimensional regularization". It does not necessarily mean that we understand the underlying dynamics. In this scheme the mass parameter entering the quark propagator is defined in such a way that just divergent terms (and no finite contributions) are absorbed such that the quark propagator is finite (after wave function renormalization). Hence the name 'minimal'. 'Modified minimal' subtraction also subtracts off $\log(4\pi)$ and γ_E factors. For $\mu \geq m_Q$ it basically coincides with the running mass in the Lagrangian and a natural normalization is at $\mu \sim m_Q$.

[8]It first pointed out in [53], followed by [54]; see also [55].

It is appropriate for describing heavy-flavor *production* like $Z^0 \to \bar{b}b$ and now also $H \to \bar{b}b$.[9]

The mass $\bar{m}_Q$ diverges logarithmically for $\mu \to 0$:

$$\bar{m}_Q(\mu) = \bar{m}_Q(\bar{m}_Q) \left[1 + \frac{2\alpha_S}{\pi} \log \frac{\bar{m}_Q}{\mu}\right] \to \infty \quad \text{as} \quad \frac{\mu}{\bar{m}_Q} \to 0 \tag{3.6}$$

Thus one needs another definition of quark masses to describe weak dynamics of hadrons. It is crucial to include the impact of non-perturbative QCD. There are two good options:

- The "kinetic" mass of the heavy quark (so-called since it enters in the kinetic energy) – introduced in Refs. [54, 63] – is defined by introducing an explicit factorization scale, and subtracting the physics at scales below this scale from the quark mass definition.
  It is regular in the infrared regime [53, 63–65]:

  $$\frac{dm_Q^{\text{kin}}(\mu)}{d\mu} = -\frac{16}{9}\frac{\alpha_S}{\pi} - \frac{4}{3}\frac{\alpha_S}{\pi}\frac{\mu}{m_Q} + \mathcal{O}(\alpha_S^2) \,. \tag{3.7}$$

  It is well suited for describing weak decays of heavy hadrons. It can be shown that for b quarks $\mu \sim 1$ GeV is an appropriate scale for this purpose. Using $\mu \sim m_b$ instead leads to higher-order perturbative corrections that are artificially large for which one has *no* control [63]. For charm quarks on the other hand this distinction disappears since m_c exceeds the 1 GeV scale by a small amount only.
- The PS mass (PS = 'potential-subtracted') [66] and the kinetic mass are similar in the sense that they both subtract out the troublesome infrared part by introducing an explicit factorization scale, but are quite different already on the conceptual level. The PS mass is based on the properties of non-relativistic quark-antiquark systems. The dynamics of heavy quarkonium is determined by the Schrödinger equation, which includes the total static energy of two heavy quarks (at a distance r): $E = 2m_b + V(r)$. Because E is a physical quantity, it is well-defined and should not suffer from a renormalon ambiguity. The infrared sensitivity of the long-distance quark-antiquark potential cancel that of the pole mass. In the PS scheme, the contribution to the Fourier integral of the potential from the region of small momenta, identified as the source of the

[9] For the record: the pole and $\overline{\text{MS}}$ masses are related by a series that starts as

$$m_Q^{\text{pole}} = \bar{m}_Q(\bar{m}_Q) \left[1 + \frac{4\alpha_S(\bar{m}_Q)}{3\pi} + (1.56b - 3.73)\left(\frac{\alpha_S}{\pi}\right)^2 + ...\right] \tag{3.5}$$

The relation between the $\overline{\text{MS}}$ and pole mass for a heavy quark has been computed to the fourth order in the strong coupling α_s [56–60]. This relation is affected by infrared renormalons [53,61,62]. Such a badly behaved perturbation expansion can of course be avoided by directly extracting the $\overline{\text{MS}}$ mass from data without extracting the pole mass as an intermediate step.

leading renormalon, is subtracted from the mass through the definition:

$$m_b^{PS}(\mu_f) = m_b - \delta m(\mu_f) \,, \qquad \delta m(\mu_f) = -\frac{1}{2}\int_{|\vec{q}|<\mu_f} \frac{d^3\vec{q}}{(2\pi)^3} V_F(q) \,, \tag{3.8}$$

where V_F is the Fourier transform of V.

3.3.2 *'1S'*

There is another scheme: the '1S' threshold scheme. Its observables are re-expressed in terms of its b quark mass defined as $m_b^{1S} \simeq M_{\Upsilon(1S)}/2$ in *perturbation* theory [67]. PDG2020 reviews had focused on the '1S' scheme,[10] while basically ignoring "kinetic" and "PS" schemes.[11]

It is often claimed that all these threshold schemes give the same information on underlying dynamics, but there are legitimate concerns that this statement is *not* correct. In the '1S' scheme there is a mismatch with the usual perturbation theory. As Uraltsev pointed out [69], the m_b^{1S} differs from the usual pole mass at the second order in α_S

$$m_b^{1S} = m_b^{\rm pole}[1 - C_F^2(\alpha_S^2/8) + \mathcal{O}(\alpha_S^3, \beta_0\alpha_S^3 \log\alpha_S)] \,. \tag{3.9}$$

The last terms $\propto \alpha_S^3 \log\alpha_S$ are infrared (IR) divergent, hence the relation is not infrared finite, in the conventional terminology; i.e., the m_b^{1S} is *not* well-defined at the *non*-perturbative level.[12]

3.4 Effective quantum field theories

For a long time our community has focused mostly on renormalized theories like QED, QCD, the SM, $SU(5)$ and others. Nevertheless, there are many good reasons to describe QFTs with 'effective' theories under different circumstances . An effective field theory is a quantum field theory (non necessarily a renormalizable one) which describes the dynamics in a simpler way, although in a limited range of energy and momenta. Effective QFTs have limits, some of which are obvious. Chiral symmetry tells us how we can describe strong forces for low energy collisions of pions and, with less precision, for kaons: it is QCD *there*! We can still use perturbative expansion, but one needs 'judgement' about its limits. It is described in **Sect. 3.5**. There is another example, namely to apply Heavy Quark Expansion (HQE) to the decays of beauty hadrons, see **Sect. 3.6**: still there are non-trivial challenges. For

[10] Due to 'par ordre du Mufti' (= no right of appeal).

[11] Another threshold mass definition is the renormalon-subtracted (RS) mass, which removes from the pole mass the pure renormalon contribution [68].

[12] To get around, a working tool – the so-called Υ expansion – has been postulated. It considers a number of terms appearing to the k-th order in perturbation theory, $c_k\alpha_S^k(m_b)$, to be actually of a lower order, $c_n\alpha_S^n(m_b)$ with $n < k$. Since the power of the strong coupling is explicit, this is done by introducing an ad hoc factor $\epsilon = 1$, making use of the property that unity remains unity raised to arbitrary power. The validity of this expansion has been questioned, already in the simplest toy analogue of B decays, the muon β-decay [69].

charm hadrons, because of their intermediate mass, there is no specific effective theory and many qualitative arguments are normally adopted. It depends on the situation whether one can use Wilsonian OPE for amplitudes, mostly in case of inclusive decays, see **Sect. 3.7**.

3.4.1 *Going beyond renormalized Lagrangians*

As we discuss in **Chapter 4**, the SM is a true theory based on local symmetries and a very successful one looking to the data.[13] It describes the landscape of known matter except where it assumes that neutrinos are massless. The discovery of neutrino oscillations shows they have masses $\neq 0$, and that the three flavors are mixed. The SM is *not* wrong – it is incomplete.[14]

One can treat the SM as an effective theory valid at energies below a scale of Λ where only the SM fields $\phi_{\rm SM}$ produce *dynamical* degrees of freedom. This is best expressed through a Wilsonian OPE, which has an excellent record to move us forward:

$$\mathcal{L}_{\rm eff}(\phi_{\rm SM}) = \mathcal{L}_{\rm SM}(\phi_{\rm SM}) + \sum_{n=1}^{\infty} \frac{1}{\Lambda^n} \mathcal{O}^{(4+n)}(\phi_{\rm SM}) \; ; \tag{3.10}$$

the $\mathcal{O}^{(n)}(\phi_{\rm SM})$ are polynomial in the SM fields (with*out* new fields) of operators with dimensions $4+n$ that are consistent with gauge symmetries of the SM and with Lorentz symmetry. The scale Λ calibrates the observable impact of the anticipated new dynamics (ND) at different energies through an expansion in powers of $\sqrt{s}/\Lambda$. Thus it disappears for $\Lambda \to \infty$.[15] Our community has mostly focused on leading contributions in the sums. For flavor-changing transitions it starts with $n = 2$ local operators of six dimensions like $(\bar{\psi}_i \Gamma \psi_i)(\bar{\psi}_j \Gamma \psi_j)$. In some situations one can start with operators of five dimensions like $\bar{\psi} i \sigma_{\mu\nu}[\gamma_5]\psi B_{\mu\nu}$ or $\bar{\psi}[\gamma_5]\psi \bar{H} H$, where $B_{\mu\nu}$ is a tensor and H is a spin-zero field. In the future we should go beyond that; to be realistic one can hardly go beyond $n = 3$ in general situations. In special cases one can think about $n = 5$, namely about operators of dimension 9; one example is probing the decays of a nucleon into three leptons or more.

So far no clear manifestations for ND have surfaced in *quark* dynamics. It has lead to the speculation that ND by *itself* might be 'flavor neutral' at some 'high' scale; i.e., the dynamics driving all flavor-*non*diagonal transitions – including **CP** breaking effects – are related to the known structure of the SM Yukawa couplings. This conjecture has been labelled as 'minimal flavor violation' (MFV), since we infer from the successes of the SM that its dynamical elements do exist in our world. We do not intend to give a detailed discussions of the technicalities involved;

[13] Some may say, not in a 'beautiful' way.

[14] We are not claiming that gravity can be included into the world of local QFTs.

[15] Since the operators $\mathcal{O}^{(n)}(\phi_{\rm SM})$ carry dimensions higher than four, they represent non-renormalizable terms with coupling proportional to $1/\Lambda^n$. This does not cause problems, since we apply this ansatz only at energies well below Λ. What one means with 'well below' requires judgment: the value of Λ shows how far (or close) one can set the thresholds for the impact of ND.

for interested readers we refer to the ample literature that can be tracked through two references [70].

3.5 Chiral symmetry

One of the striking feature of strong interactions in low energy particle physics is the approximate spontaneously breaking of $SU(N) \times SU(N)$ global chiral symmetries, which manifest itself, in particular, in the small masses of the pseudoscalar mesons. There are many applications of chiral symmetry; for example, it can be used to calculate mass differences between heavy-quark doublets, the so-called 'hyperfine splitting'.

In the fifties, the vector-axial theory of Feynman, Gell-Mann and others emphasized the importance of chiral symmetry in particle physics. The Goldberger-Treiman relation [71] goes back to 1958, well before the appearance of QCD as a gauge interaction theory and actually at the beginning of the era of QFT. It was derived on the basis of the $(V-A)\times(V-A)$ nature of the currents of weak interactions.[16] The Goldberger-Treiman relation connects the *strong* coupling of pions to the nucleons with the axial vector current that determines the *weak* decay of the neutron, which is amazing.

3.6 Heavy Quark Expansion (HQE)

In the limit of an infinitely heavy quark, the mass of the quark is separate from its dynamics; it is the source of a static 'color' Coulomb field that is independent of the heavy-quark spin.[17] That is the core of Heavy Quark Symmetry (HQS). There are consequences on the heavy-light system spectrum of mesons $[Q\bar{q}]$ and baryons $= [Qq_1q_2]$.[18] The spin of the heavy quark Q *decouples*, and the spectra of the heavy-flavor hadrons are described in terms of the angular momentum by their light components. One can restrict oneself to a fixed number of heavy quarks Q and/or $\bar{Q}$, while $\bar{Q}Q$ fluctuations can be ignored.

This situation can be depicted in an intuitive way: an hadron H_Q with a heavy quark Q of mass m_Q in the center, surrounded by a 'cloud' of *light* degrees of freedom carrying quantum numbers of an anti-quark $\bar{q}$ or di-quarks qq.[19] The light component of $\mathcal{L}_{\rm HQE}$, see Eq.(3.11), has a complicated structure: the soft modes of the light fields are strongly coupled and strongly fluctuate; typical frequencies are $\mathcal{O}(\bar{\Lambda}) \sim 1$ GeV. One can visualize them as a soft medium. Sometimes one can use symmetries about 'light' components like chiral symmetry.

[16] V: vector, A: axial.

[17] One can say the same with different words: heavy quark fields are *not* integrated out, since we discuss a sector with heavy-flavor *non-zero* charge.

[18] We will not discuss exotic hadrons $Q\bar{q}_1q_2\bar{q}_3$ or $Qq_1q_2q_3\bar{q}_4$, where the situations are quite different.

[19] This cloud is often referred to – somewhat disrespectfully – as 'brown muck', a term coined by the late Nathan Isgur. Still it seems to be an unfair name for the 'soft' components of the quark and gluon fields; perhaps 'we' have not yet unveiled their beauty.

Of course, we have to talk about the real world; the mass of a heavy quark is finite, but also *not* close to zero: $m_Q \gg \bar{\Lambda} \simeq 1$ GeV. One can use $\bar{\Lambda}/m_Q$ as an expansion parameter in a heavy quark expansion (HQE) to compare predictions with present and future data.

The advantages of the limit $m_Q \to \infty$ in QCD were first emphasized by Shuryak [72]. The next step was the observation of the HQS [73]. The heavy quark theory in QCD was finally formalized in Ref. [74] where the systematic $1/m_Q$ expansion of the H_Q dynamics was cast as an *effective* Lagrangian. Several aspects of HQE have been analyzed and discussed since, and a vast literature has ensued; aspects regarding the effective field theory correlation with QCD have been discussed in details in Refs. [54, 75].

In HQE the strong dynamics is described very similarly to Eq.(3.1):

$$\begin{aligned}\mathcal{L}_{\rm HQE} &= \sum_Q \bar{Q}(i\gamma_\mu D^\mu - m_Q)\,Q - \frac{1}{4}G^a_{\mu\nu}G^{\mu\nu,a} + \sum_q \bar{q}\,i\gamma_\mu D^\mu\, q \\ &\equiv \sum_Q \bar{Q}(i\gamma_\mu D^\mu - m_Q)\,Q + \mathcal{L}_{\rm light}\end{aligned} \tag{3.11}$$

with the gluon fields A^a_μ and $SU(3)$ generators t^a and $a = 1, \ldots, 8$. The heavy quarks Q has a mass term m_Q that is larger than $\bar{\Lambda} \simeq 1$ GeV; for simplicity, the light quarks $q = u, d$ and s are assumed to be massless. HQE describes heavy quarks in general. In the world of hadrons we have only one true heavy quark – the beauty quark – and one semi-heavy one – the charm. Even from the latter we get highly non-trivial lessons (using some judgment).

Expansions are common in QFT. In QED an electron never appears 'alone'. It is always surrounded by a cloud of soft photons; several methods have been developed in QED long time ago, such as the Pauli expansion and the Foldy-Wouthuysen transformation [76], to give a more complete and realistic description of the electron.

HQE describes two classes of observables: the masses of heavy flavor hadrons and their widths. The hadron masses control the phase spaces of the heavy quarks; in a general formulation, they obey an expansion that include a linear term:

$$M_{H_Q} = m_Q[1 + \bar{\Lambda}/m_Q + \mathcal{O}(1/m_Q^2)]\,. \tag{3.12}$$

$\bar{\Lambda}$ was introduced as a HQE parameter [77], representing a contribution which it is entirely due to the light degrees of freedom; the values are different between mesons and baryons. The situation is more subtle for hadron widths, see below.

3.7 Wilsonian Operator Product Expansion (OPE)

The number of applications of HQE grows dramatically when one include *external* (non-QCD) interactions of heavy quarks, namely electroweak forces and beyond. Many challenges can be formulated best in the language of the *Wilsonian* operator product expansion [54, 78][20]; to make it short, we use the word "OPE".

[20] The Nobel Prize in Physics 1982 was awarded to Kenneth G. Wilson "for his theory for critical phenomena in connection with phase transitions".

Inclusive heavy quark decays are one of the best studied examples. One describes the decay rate into an inclusive final state f in terms of the imaginary part of a forward scattering operator evaluated to second order in weak interactions [79, 80]:

$$\hat{\mathrm{T}}(Q \to f \to Q) = \mathrm{Im} \int d^4x\, i\, [\mathcal{L}_W(x)\mathcal{L}_W^\dagger(0)]_T \tag{3.13}$$

where the subscript T denotes the time-ordered product and $\mathcal{L}_W$ is the relevant weak Lagrangian. The time-space separation x is actually fixed by the inverse energy release, not necessarily $1/m_Q$. If it is sufficiently large in the decay, one can express the non-local operator $\hat{\mathrm{T}}(Q \to f \to Q)$ as an infinite sum of local operators O_i of increasing dimensions. The width for $H_Q \to f$ is obtained by averaging $\hat{\mathrm{T}}(Q \to f \to Q)$ over the heavy flavor hadron H_Q (normalizing $\langle H_Q|X|H_Q\rangle$ with $2M_{H_Q}$):

$$\begin{aligned}\frac{\langle H_Q|\mathrm{Im}\hat{\mathrm{T}}(Q \to f \to Q)|H_Q\rangle}{2M_{H_Q}} &\propto \Gamma(H_Q \to f)\\ &= G_F^2|V_{\mathrm{CKM}}|^2 m_Q^5 \sum_{i=3}^{\infty} \bar{c}_i^{(f)}(\mu)\frac{\langle H_Q|O_i|H_Q\rangle_{(\mu)}}{2M_{H_Q}}\ ;\end{aligned}$$

V_{CKM} denotes the appropriate combination of CKM parameters . Furthermore:

- The parameter μ is the normalization point, and it indicates the evolution from m_Q down to μ. The effects of momenta below μ are lumped into the matrix elements of local operators O_i.
- Using μ shows the *limits* of our calculation power of amplitudes and/or diagrams, in particular of the differences between long and short distance dynamics. Setting the scale needs judgement and attention to correlations with other amplitudes.
- The coefficients $\bar{c}_i^{(f)}$ contain powers of the $1/m_Q$ that go up with the dimension of the operator O_i. One can introduce dimensionless coefficients $c_i^{(f)} = m_Q^{d_i-3}\bar{c}_i^{(f)}$ with the dimension d_i of the operator.

For semi-leptonic and non-leptonic decays, treated through order $1/m_Q^3$, one gets

$$\begin{aligned}\Gamma(H_Q \to f) = \frac{G_F^2}{192\pi^3}|V_{\mathrm{CKM}}|^2 m_Q^5 \Bigg[& c_3^{(f)}(\mu)\frac{\langle H_Q|\bar{Q}Q|H_Q\rangle_{(\mu)}}{2M_{H_Q}}\\ & + \frac{c_5^{(f)}(\mu)}{m_Q^2}\frac{\langle H_Q|\bar{Q}\frac{i}{2}\sigma\cdot GQ|H_Q\rangle_{(\mu)}}{2M_{H_Q}}\\ & + \sum_i \frac{c_6^{(f)}(\mu)}{m_Q^3}\frac{\langle H_Q|(\bar{Q}\Gamma_i q)(q\Gamma_i Q)|H_Q\rangle_{(\mu)}}{2M_{H_Q}} + \mathcal{O}\left(\frac{1}{m_Q^4}\right)\Bigg] .\end{aligned} \tag{3.14}$$

There is a particular interest in applying OPE here. There is *no linear* correction of $1/m_Q$: $c_4^{(f)} = 0$ as pointed out in Ref. [81]. There are three reasons for the *absence* of a linear $1/m_Q$ contribution to the total widths:

(a) The expectation value of the QCD operator $\bar{Q}Q_{(\mu)}$ of dimension "three" is corrected only by $\mathcal{O}(1/m_Q^2)$.
(b) There is no independent QCD operator of dimension "four" for forward matrix elements of heavy flavor H_Q.
(c) Since the coefficients functions $c_5^{(f)}$, $c_6^{(f)}$ etc. are purely short-distance, infrared effects cannot penetrate into them. Maybe there is a more illuminating statement about this absence: bound-state effects in the *initial* state produce $1/m_Q$ corrections; likewise for the hadronization in the *final* states. Yet local unbroken color symmetry demands that they cancel each other out in total widths.

The operator $\bar{Q}Q$ has been expanded through $1/m_Q^2$ with a covariant derivative $\vec{\pi} = -i\vec{D}$:

$$\frac{\langle H_Q|\bar{Q}Q|H_Q\rangle_{(\mu)}}{2M_{H_Q}} = \left[\frac{\langle H_Q|\bar{Q}\gamma_0 Q|H_Q\rangle_{(\mu)}}{2M_{H_Q}} + \frac{\mu_\pi^2(\mu)}{2m_Q^2}\right] + \frac{\mu_G^2(\mu)}{2m_Q^2} + \mathcal{O}(1/m_Q^3) \quad (3.15)$$

$$\mu_\pi^2(\mu) = \frac{1}{2M_{H_Q}}\langle H_Q|\bar{Q}(-i\vec{D})^2 Q|H_Q\rangle_{(\mu)} \quad (3.16)$$

$$\mu_G^2(\mu) = \frac{1}{2M_{H_Q}}\langle H_Q|\bar{Q}\frac{i}{2}\sigma_{\mu\nu}G^{\mu\nu}Q|H_Q\rangle_{(\mu)} \quad (3.17)$$

On the right side of Eq. (3.15) one has to combine two items to get an invariant one.

For pseudo-scalar P_Q one gets, using the notation from the OPE approach

$$\frac{\langle P_Q|\bar{Q}Q|P_Q\rangle_{(\mu)}}{2M_{P_Q}} = 1 - \frac{\mu_\pi^2(\mu)}{2m_Q^2} + \frac{3}{8}\frac{M_{V_Q}^2 - M_{P_Q}^2}{m_Q^2} + \mathcal{O}(1/m_Q^3)\,, \quad (3.18)$$

and using the mass of a vector meson V_Q, as stated in the Refs. [54, 82].

The 'landscapes' in the *spectra* of the inclusive semi-leptonic final states of H_Q are even more 'complex' than those of their weak widths: we have to deal with different challenges, and consider terms up to order of $1/m_Q^2$ [83] (at least). To describe inclusive spectra in $H_Q \to \ell\bar{\nu}_l X_q$ with $q = c, u$, we need the same set of operators. However, novel features arise: these series are singular at the endpoint of the lepton spectra; thus one has to interpret the results with care.

For $Q \to q\ell\nu$ with $m_q = 0$ one gets [83]:

$$\frac{1}{\Gamma_0}\frac{d\Gamma}{dy} = 2y^2\left[3 - 2y - \left(\frac{5}{3}y + \frac{1}{3}\delta(1-y) - \frac{1}{6}(2y^2 - y^3)\delta'(1-y)\right)K_Q + \left(2 + \frac{5}{3}y - \frac{11}{3}\delta(1-y)\right)G_Q\right] \quad (3.19)$$

where

$$\Gamma_0 = \frac{G_F^2 m_Q^5}{192\pi^3}|V_{qQ}|^2\,, \quad y = \frac{2E_l}{m_Q}\,, \quad K_Q = \frac{\mu_\pi^2}{m_Q^2}\,, \quad G_Q = \frac{\mu_G^2}{m_Q^2}\,. \quad (3.20)$$

We note the infinite functions $\delta(1-y)$ and $\delta'(1-y)$, which reflect the previously mentioned singular nature of the expansion at the endpoint. Due to these singular terms, the expression given above can be identified with the observable spectrum, and allow predictions, only *outside* a finite neighborhood at the endpoint region. Yet this neighborhood does *not* represent true 'terra incognita' in the Wilsonian approach: integrating our expression over this kinematic region yields a finite and trustworthy result that can be confronted with the data:

$$\Gamma(E_l) = \int_{E_l}^{E_{max}} dE_l \frac{d\Gamma}{dE_l} \; , \; E_l \leq E_{\rm QCD} \leq E_{max} \; . \tag{3.21}$$

E_{max} denotes the maximal energy allowed by kinematics, while $E_{\rm QCD}$ is the maximal energy for which one can trust the QCD expansion; it value depends on the size of K_Q and G_Q. Clearly $\Gamma(0) = \Gamma_{SL}$ has to hold. Γ_{SL} is deduced from a regular expansion in $1/m_Q$, whereas $\Gamma(0)$ contains these singularities; thus the singular terms are essential for recovering the correct width.

Similar statements apply to $Q \to q\ell\nu$ with $m_q \neq 0$. The 'landscape' is even more 'complex' [83]. There are two scales of mass, $m_q^2 \ll m_Q^2$, which affect the differential width through new parameters

$$f = \frac{\rho}{1-y} \; , \qquad \rho = \frac{m_q^2}{m_Q^2} \tag{3.22}$$

In the real world one can focus on semi-leptonic "moments" (see Ref. [84]). As most experiments can detect the leptons only above a certain threshold in energy, the charged-lepton energy moments are defined as

$$< E_l^n >= \frac{1}{\Gamma_{E_l > E_{cut}}} \int_{E_l > E_{cut}} E_l^n \frac{d\Gamma}{dE_l} dE_l \tag{3.23}$$

where E_l is the lepton energy in the $H_Q \to l\bar{\nu}_l X_q$ decays, n is the order of the moment, $\Gamma_{E_l > E_{cut}}$ is the semileptonic width above the energy threshold E_{cut} and $d\Gamma/dE_l$ is the differential semileptonic width as a function of E_l. Other moments (and cuts on other observables) can be defined in a similar way.

Summarizing, Wilsonian approach to OPE provides a powerful theoretical tools of wide applications, although we should not forget that there are different realizations of OPE. Their connections are both crucial and subtle.

3.8 ♠ Quark renormalons ♠

In 1952 it was pointed out by Freeman Dyson that the series in e^2 in QED could not be convergent, since the analytic continuation to negative e^2 produces a theory with unstable vacuum [85]. Thirring [86] and Lipatov [87] came to the same conclusion: the perturbative series are asymptotic and characterized by the *factorial divergence of the form*

$$Z = \sum_k C_k \alpha^k k^{b-1} A^{-k} k! \tag{3.24}$$

with $\alpha \equiv e^2/4\pi$, $k \gg 1$ is the number of loops, C_k's are numerical coefficients of order one, and b and A are numbers. The factorial divergence of the perturbative series can be traced back to the large number of multi-loop Feynman diagrams, which grows as the factorial of the number of loops. Thus QED and QCD perturbative series, after renormalization, have all their coefficients finite, but the expansion does not converge. To deal with these features of QFT one introduces the Borel transform:

$$B_Z(\alpha) = \sum_k C_k \alpha^k k^{b-1} A^{-k} ; \tag{3.25}$$

where the k-th term of the expansion (3.24) is divided by $k!$. That implies that the singularity of $B_Z(\alpha)$ closest to the origin of the α plane is at a distance A, and thus $B_Z(\alpha)$ is *convergent*. One recovers the original function Z by the Laplace transformation

$$Z(\alpha) = \int_0^\infty dt\, e^{-t} B_Z(\alpha t) . \tag{3.26}$$

The integral representation is well-defined – *provided* that $B_Z(\alpha)$ has no singularities on the real positive semi-axis in the complex α plane. However, if $B_Z(\alpha)$ has singularities on the real positive semi-axis – as is the case if the coefficients C_k are all positive or all negative – the integral in the Eq. (3.26) becomes *ambiguous*. This ambiguity is of the order of $e^{-A/\alpha}$; more information is needed from the underlying dynamics.

In the situations with weak couplings this further physical information is given by classical solutions with non-vanishing action, such as instantons and instanton-anti-instanton pairs. These may be treated consistently by semi-classical methods, based on a series of repeated cancellations of ambiguities in the physical observables [88].

The real problem is due to another type of divergence of the Borel series, the so-called renormalons [89]. One can identify another source of the factorial divergence – unique diagrams of a special type present in Yang-Mills theories which produce a factorial growth not because there are many of them, but because a single graph with n loops is factorially large. In other terms, there exists a class of isolated graphs, in which each diagram grows factorially as we increase the number of loops. These graphs induce renormalons. As suggested first by 't Hooft, the renormalons arise from the procedure of renormalization and bring in ambiguities in the perturbative formulation of the theory when the coupling is strong. The features responsible for the renormalon factorial divergence is the logarithmic running of the effective coupling constant. It is easy to see that, if the ambiguity is of the order of $e^{-A/\alpha}$, in the weak coupled QED these corrective terms are extremely small and not very important in practice. On the contrary in QCD $\alpha = \alpha_s(Q^2) \sim 1/(b \log Q^2/\Lambda^2)$ and the ambiguous terms are of order $(1/Q^2)^n$, so they are power suppressed.

Let us underline the argument above is just for naïve illustration. Renormalons can be seen in bubble diagrams as the one depicted in **Fig. 3.1**. It is not obvious that the external momentum Q^2 sets the scale of all virtual momenta in loops [55].

The QCD coupling runs as

$$\alpha_S(Q^2) = \frac{\alpha_S(\mu^2)}{1 - \frac{\beta_0 \alpha_S(\mu^2)}{4\pi} \log(\mu^2/Q^2)} \tag{3.27}$$

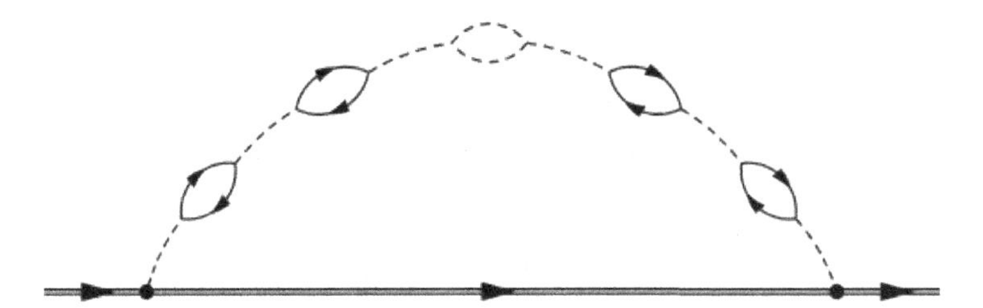

Fig. 3.1 Perturbative diagrams leading to the IR renormalon uncertainty in $m_Q^{\rm pole}$ of order $\bar{\Lambda}$. The number of bubble insertions in the gluon propagator is arbitrary. The horizontal line at the bottom is the heavy quark Green's function.

with $\beta_0 = \frac{11}{3}N_C - \frac{2}{3}N_{\rm flav} = 11 - \frac{2}{3}N_{\rm flav}$, and it uses the energy scale μ to calibrate $\alpha_S(Q^2)$.

Let us also observe that with $Q^2 \ll \mu^2\, \alpha_S(Q^2)$, the QCD running gets larger and larger; thus QCD gives us true strong forces at low scales. One might say it goes to infinite, but that is too naive to understand the impact of Eq. (3.27). One has to stop at $\mu \sim 1$ GeV, a non-perturbative scale, in a context based on perturbative QCD. How can we do that?

The general answer is that one can use renormalons to capture some non-perturbative effects. The first example of the 'renormalon guidance', that later was extended to many other analyses, was the so-called heavy quark pole mass. It was routinely used in analyzing data, since it is IR stable and unambiguous to any finite order in perturbation theory. These aspects gave the impression that the pole mass was well-defined in general, until the 1990's. It was pointed out first in 1994 that the pole mass is *not* well-defined at the non-perturbative level [53] (as mentioned in **Sect. 3.7**). Furthermore a rather powerful renormalon-based tool was suggested for evaluating the corresponding non-perturbative contribution. Indeed, the pole mass is sensitive to large distance dynamics, although this fact is not obvious in perturbative calculations. IR contributions lead to an *intrinsic uncertainty* in the pole mass of order Λ – i.e., a Λ/m_Q power correction. It comes from the factorial growth of the high order terms in the α_S expansion corresponding to a singularity residing at the $2\pi/\beta_0$ in the Borel plane. Thus one cannot see it as a correction.

A display of the effects of IR renormalons occurs in a simple way by estimating the perturbative diagram in **Fig. 3.1**. Order by order one has [90]:

$$\frac{\delta m_Q^{(n+1)}}{m_Q} \sim \frac{4}{3}\frac{\alpha_S(\mu^2)}{\pi}\, n!\, \left(\frac{b\,\alpha_S(\mu^2)}{2\pi}\right)^n \tag{3.28}$$

with $b = \frac{11}{3}N_C - \frac{2}{3}N_f = 11 - \frac{2}{3}N_f$. The coefficients grow factorially ("$n!$") and contribute with the *same* sign. Thus the sum of these contributions is not defined, even with the Borel summation, and one is left with the only option to truncate the series, originating an uncertainty. Moreover, those perturbative contributions affect the value of the b quark mass numerically [90]:

$$\begin{aligned} m_b^{\rm pole} &= m_b(1\,{\rm GeV}) + \delta m_{\rm pert}(\leq 1\,{\rm GeV}) \\ &\simeq 4.55\,{\rm GeV} + 0.25\,{\rm GeV} + 0.22\,{\rm GeV} + 0.38\,{\rm GeV} + 1\,{\rm GeV} + 3.3\,{\rm GeV}...\,, \end{aligned}$$

where $\delta m_{\rm pert}(\leq 1\,{\rm GeV})$ is the perturbative series taking account the loop momenta down to zero. It is evident that the corrections start to blow up already at rather low orders.

In conclusion we summarize some points emphasized by Shifman in 2013 [55], which are still actual and not obvious.

(a) The renormalon counting remains the only known method for evaluating non-perturbative corrections in processes with*out* OPE. Of course, the appearance of new method or even just novel ideas would be a progress.

(b) OPE with its explicit separation scale μ conceptually solves the problem of factorial divergence of the perturbative series, at least at $N \to \infty$. Factorial divergence of the $(\Lambda/Q)^k$ series – with $k \gg 1$ being the number of loops – emerges in OPE and needs further exploration and an appropriate theoretical description.

(c) The analysis of "hidden" structure of perturbative theory and the consequent intertwine between perturbation and non-perturbative physics are among the topics of the "resurgence" research program (see [91, 92]).

3.9 Heavy Quark Effective Theory (HQET)

HQET is another example of an effective field theory for heavy quarks. It is based also on HQS, since it describes the hadron dynamics in the heavy-quark limit. In HQET all the light degrees of freedom have to have momenta of the order QCD, i.e., the momentum p of the heavy quark inside a heavy meson moving with velocity $v = p_{Meson}/M_{Meson}$ is decomposed as $p = m_{quark}v + k$, and all components of the residual momentum k are assumed to be of the order of QCD low energy scale.

Often it is claimed that HQE and HQET are saying the same thing in different ways. However, they are *not*, in principle.[21] The spin of the heavy quark Q decouple, and the spectra of the heavy-flavor hadrons are described in terms of the angular momentum by their light components. Therefore, to leading order of accuracy, one obtains no hyperfine splitting and simple scaling laws:

$$M_B \simeq M_{B^*} \quad , \quad M_D \simeq M_{D^*}\,; \tag{3.29}$$

$$M_{B^*} - M_B \sim \frac{m_c}{m_b}(M_{D^*} - M_D)\,; \tag{3.30}$$

$$M_B - M_D \sim m_b - m_c\,. \tag{3.31}$$

In a HQET framework it is possible to find $\sim$ zero in Eq. (3.31); in HQE it is not. The systematic framework provided by HQE is based on the assumption that the energy release of the decaying quark is large, and works best in approximations where

[21] The last author of Ref. [3] consistently avoided using the term HQE.

all the other quarks are lighter. In contrast, the HQET in its ordinary formulation is based on the idea that both the b and the c quarks are heavy.[22]

HQET papers want to show the impact of non-perturbative physics:

$$\text{“observable”} = \text{perturbative forces} + \text{non} - \text{perturbative forces}. \tag{3.32}$$

In the standard HQET perturbative corrections are simply added to what is considered non-perturbative physics. Yet there is a deeper, separation scale-dependent, description, as pointed out in Ref. [54]:

$$\text{“observable”} = \text{short} - \text{distance dynamics} + \text{long} - \text{distance dynamics}, \tag{3.33}$$

which is, formally, a different procedure. A Lagrangian can be applied to perturbative calculation of Feynman graphs with heavy quarks, where the characteristic virtual momenta flowing through all lines in the graphs are less than μ, the scale separating the hard and soft domains.[23] Contribution of all virtual momenta above μ is explicitly included in the coefficients of the effective Lagrangian. In the HQET version, the parameters of the Lagrangian (perturbatively defined) correspond to tending $\mu \to 0$, after appropriate subtraction of perturbation theory, at a certain order. In the HQE version one has to stop there; the parameters are μ dependent.

3.10 Duality

It is well known that physicists (or in general scientists) have a limited dictionary of words. One example: the complex landscape of 'duality'. From the beginning of quantum mechanics one talks about 'duality of particle-wave'. Here we discuss the concept of quark-hadron duality, which goes back to the early days of the quark model; we use the term “duality” for short. A quark level description should provide a good description for transition rates that involve hadrons – if one sums over a 'sufficient' numbers of channels. In the 1970's it was a rather vague formulation: how many channels are 'sufficient'? How good an approximation one expects? No precise definition of duality had been given for a long time, and the concept has been used in many different incarnations. A certain lack of intellectual rigor can be of great heuristic value in the early going – but not forever. A true definition of “duality” requires theoretical control over perturbative as well as non-perturbative dynamics. Limitations of duality have to be seen as effects *over and beyond* uncertainties due to truncation in perturbative and non-perturbative expansions.

To be more explicit: duality violations are due to correlations *not* accounted for in the truncation in the expansions. The best requirement is to have a Wilsonian OPE treatment of the process under study, since otherwise we have no unambiguous or systematic inclusion of non-perturbative corrections. While we have no complete theory for duality and its limitations, we have moved *well beyond* the 'folkloric'

[22]Therefore it works best in the infinity mass limits, where both quarks are heavy but their ratio is otherwise arbitrary, or in the Small Velocity (SV) limit, where the c quark is emitted with only a small velocity in the b rest frame.

[23]We take a conservative value of the typical hadronic scale $\mu \sim 1$ GeV.

stage. The challenge is how to compare QCD-based amplitudes with the hadronic measured (or measurable) probabilities. Quarks and gluons producc transitions, while hadrons describe asymptotic states.

By 1994 it has clearly been realized that fundamental questions were still unanswered, as summarized by Shifman [93] in this list:

- What energy is considered to be high enough for the quark-hadron duality to set in, and what accuracy is to be expected?
- What weight functions are appropriate for averaging experimental transitions?
- If the theoretical prediction includes only perturbative theory, should one limit oneself to some particular order in the α_S series?
- Do we have to include known non-perturbative effects (like condensates) in the theoretical predictions?
- Given a definition of quark-hadron duality, can one estimate deviations from duality and how?

Systematic explorations of these and related issues started around that time, prompted by increased precision of data, mostly associated with the b-quark physics. It is often heard that "duality" represents an additional *ad hoc* assumption. The problem with that statement is that it is not even wrong – it just misses the point.

The main point is that this duality is *not* an addition – it is deep rooted in QCD, but in a 'complex' way. One could say that under the assumption of duality the true hadronic observables coincide with what one ultimately obtains in the quark-gluon language, provided all possible sources of corrections to the parton picture stemming from QCD itself are properly accounted for. The practical validity of duality thus depends on the theoretical tools available for treating QCD dynamics.

Duality actually represents a very natural concept – however, in highly non-trivial ways it describes transitions from the initial to final states based on local operators. Duality is based on the picture that processes with hadrons evolve in three steps:

(1) The initial states can be two hadrons that collide or a heavy flavor hadron that decays weakly. These hadrons are built from quarks and gluons that couple due to QFT for strong forces.
(2) In a first stage, short distance dynamics proceeds in the 'femto-universe' that is characterized by large scales like momentum transfer $\sqrt{Q^2}$ for the jets or the high mass m_Q in the weak decays of an heavy flavor hadron H_Q with an heavy quark Q. It implies a time scale of a femtosecond, namely 10^{-15} sec, which is shorter than the lifetimes of beauty and charm hadrons.
(3) Subsequently, quarks and gluons transmogrify themselves into hadrons; since this transformation is driven by soft dynamics; this final step is characterized by much larger distance scales. We cannot control this step; we have to depend on relations with other data.

The primary criterion for addressing duality violation is the existence of the Wilsonian OPE for the particular observable. The OPE defines a systematic framework to compute the correlators at high Q^2 in QCD. In principle, if one could calculate the correlators in the Euclidean domain exactly, one could analytically continue the result to the Minkowski domain, and then take the imaginary part. The spectral densities obtained in this way would present the exact theoretical prediction for the measurable hadronic cross section and there would be no need for duality [93]. In practice, our calculation are approximate, and the series truncated. In the Minkowski region, there will be uncertainties in the OPE-based predictions that go beyond those due to neglected orders in α_s or in $1/m_Q$. Such effects are referred to as violations of quark-hadron duality.

Wilsonian OPEs can be constructed in Euclidean domains and then extrapolated to Minkowskian ones. One can identify mathematical portals through which duality can be applied and where violations can enter. There are two classes of models that help us to apply duality: instanton-based and resonances/thresholds-based ones [93].

When one probes jets of hadrons at high collisions, one can use energies (or other kinematic variables) that refer to duality.

One example of inclusive observable is the total cross section of e^+e^- annihilation into hadrons, where the parton ansatz yields an energy-independent ratio with the total cross section of e^+e^- annihilation into $\mu^+\mu^-$.

The experimental cross section, however, exhibits manifest resonance structure up to relatively high energies. In this domain duality between the QCD-inferred cross section and the observed one at a given fixed energy looks problematic. The equality between the two may be restored if averaging over an energy interval or smearing is applied. If the equivalence of the averaged resonance and scaling structure functions holds over restricted regions, a *local duality* is said to exist. We have to use duality averaged over an energy interval, when we discuss weak transitions of H_Q.[24]

Following the three steps (1), (2) and (3) outlined above, we focus on heavy flavor hadrons H_Q (with a single heavy quark Q) that decays into light flavors (step (1)). While the typical time scale for the second step is provided by $1/m_Q$, for the third step it is by $1/\Lambda_{QCD}$ in the rest frame of a final state quark and thus by m_Q/Λ^2_{QCD} in the original rest frame. We expect the second step controls gross characteristics, like total rates or the directions of energetic jets, which should be well established by that 'time', namely in the 'femto universe'. The third step determines the composition of the final states of hadrons. They are produced by quarks and anti-quarks that are far removed from each other – but still connected by the unbroken "color". The impact of final state interaction (FSI) shows, as well as long distance dynamics, which is non-perturbative in general. Other details on this admittedly complex subject can be found by the committed reader in Ref. [94].

[24] It does not mean that one talks about true local duality; often one refers to semi-local ones close to thresholds etc.

Certain features of local duality have been clarified over the last years:

- To address duality violation it is very important to apply (Wilsonian) OPE for particular observables.
- Duality cannot be exact at finite masses. It represents an approximation, the accuracy of which increases with the energy scales in a way that depends on the process in question.
- Effects of violation of local duality can only have the form of an oscillating function of energy (m_Q, E_r, ...) or have to be exponentially suppressed. Duality violations cannot be blamed for a systematic excess or deficit in the decay rates. Furthermore, *no* local duality violation can convert m_Q into M_{H_Q} in the total width.
- The oscillating component of local duality violation may be only power suppressed. In real QCD it nevertheless is to become exponentially suppressed as well at large enough energy, fading out as $e^{-(E/M)^{\gamma}}$ with a positive γ. The onset of that regime, however, can be larger than the typical hadronic scale ~ 1 GeV – for example, it might grow with increasing N_C. The details of the asymptotic states, in particular the power of energy by which duality violation is suppressed, depend on the underlying strong dynamics and on the concrete process. This power is rather high in *total semi-leptonic* widths of heavy quarks.
- OPE equally applies to semi-leptonic and non-leptonic total decay rates. Likewise, both widths are subject to the violation of local duality. The difference here is quantitative rather than qualitative. At finite heavy quark masses, corrections are generally larger in the non-leptonic decays. In particular, local duality violation in non-leptonic decays can be boosted by accidental presences of narrow hadronic resonances. Similar effects are extremely suppressed for semileptonic decays.[25]
- It is not necessary to have a proliferation of decay channels to reach the onset of duality, either approximate or asymptotic. A successful example is provided by the so-called small velocity (SV) kinematics in the semi-leptonic decays [79].
- A divergence in the power expansion of the 'practical' OPE underlying violations of local duality is related to the presence of singularities in the quark or gluon propagators at finite (or even infinitely large) space time intervals. This is in contrast to finite-order OPE terms which account for the singularities of interactions for the perturbative corrections or for the expansion of propagators near zero space-time intervals. Some class of non-perturbative effects (presumably strongly suppressed) comes from small-size instantons, which are neglected in the simplest version of the 'practical' OPE. They are *not* specific to local duality and are similar in both Euclidean and Minkowskian amplitudes.

[25]This is another example of the landscapes of theorists vs. experimenters: the first prefer to focus on semi-leptonic decays, where they have better control, while experimenters prefer non-leptonic ones, where they have more data and more control on the backgrounds.

Such instantons contribute to the Wilson coefficients in the OPE computed in the short-distance expansion.

- The presence of such finite-distance singularities is a general rather than exceptional feature of theories with nontrivial self-interaction. In particular, they are *not* directly related to quark-gluon confinement. The latter, however, may essentially influence the nature of the singularities.

3.10.1 ♠ *Tools for probing duality* ♠

Here we briefly discuss two specific approaches to the description of duality violations, the one based on instantons and the other on the so-called 't Hooft model.

Instantons provide a not obvious dynamic realization of non-perturbative physics generating full OPE series which, in principle, can be evaluated to sufficiently high order. Such models assume that quarks and gluons propagate *not* in the perturbative vacuum, but in the background of instanton configurations of a typical size. We can learn from the instanton ansatz about duality and its violations.
(i) It was claimed that while non-leptonic widths suffer from duality violations, semi-leptonic ones do not. However, it had been explicitly shown that the instanton ansatz leads to duality violations in total semi-leptonic widths. The instanton calculus has demonstrated a quantitative rather than qualitative distinction between the two processes.
(ii) The instanton ansatz illustrated that finite-distance singularities in the Green functions lead to divergences of the OPE and to violations of local duality: those are common rather than exceptional features of strong interacting systems. It shows that duality violations are not intrinsically tied to confinement – contrary to what a historical perspective might suggest. The instanton ansatz exhibits some general features: oscillations and fast decrease with energy, strong suppression upon averaging, larger effects in non-leptonic widths compared to the semi-leptonic ones.
(iii) It was shown that conventional instantons cannot induce any appreciable duality violating effects in total semi-leptonic widths of B hadrons, regardless of uncertainties in the model parameters. Even boosting up the possible instanton density leaves the effect below the 10^{-3} level. While having a minor effect on Euclidean short-distance observables at energy scales around m_b, they would be manifest at a lower scale $\sim m_c$ even in the Euclidean domain.

The 't Hooft model [95] (a $(1+1)$-dimensional analogue of QCD with $N_C \to \infty$) is an attractive theorist's laboratory in exploring various complicated aspects of non-perturbative dynamics in QCD. Being solvable, it allows in principle to derive precise numerical values of the model's counterparts of actual hadronic characteristics. It is possible to confront them with the results of particular approximations employed in real QCD to test their viability. This model is particularly appealing for studying duality and its limitations. It automatically respects the basic underlying features of QCD related to gauge invariance. including its rigorous *sum rules*, that we discuss in **Sect. 4.1.6**; *ad hoc* models typically fail in that respect.

When an analytic solution of total QCD was found, discussion about duality and its violations would be obsolete. The exact asymptotic behavior of inclusive cross sections would be known, and the very concept of duality violation would become irrelevant (except for talking about the history of fundamental dynamics). Can this 'miracle' happen – maybe? The reader can guess what our bet is.

3.11 Summary of this Chapter

A reader should remember four points about connecting the worlds of hadrons and quarks and gluons:

- Drawing Feynman diagrams is the first (and second) step to understand the underlying dynamics. Subsequently, one needs more thinking and judgement, in particular on non-perturbative forces.
- Heavy quark theory describes the masses of heavy flavor hadrons in the general form
$$M_{H_Q} = m_Q[1 + \bar{\Lambda}/m_Q + \mathcal{O}(1/m_Q^2)] \tag{3.34}$$
- Wilsonian OPE leads to the total widths:
$$\Gamma(H_Q) \propto m_Q^5 \left(1 + \mathcal{O}(1/m_Q^2)\right) . \tag{3.35}$$
The practical implementation of the OPE expresses the widths as an expansion in the $1/m_Q$ (or the inverse of the energy release) with the coefficients shaped by short-distance physics accounted for perturbatively. We can use the general notation
$$\mathcal{A}(m_Q) = \sum_{k=0}^{\infty} c_k \frac{(\mu_k)^k}{m_Q^k} , \qquad c_k = \sum_{l=0}^{\infty} a_l^{(k)} \frac{\alpha_S^l}{\pi} , \tag{3.36}$$
where $\mathcal{A}$ is a generic (dimensionless) quantity and $(\mu_k)^k$ are parameters related to the non-perturbative expectation values.
Apart normalization, for semi-leptonic total widths one gets
$$\Gamma(m_Q) = c_0 + \sum_{k=2}^{\infty} c_k \frac{(\mu_k)^k}{m_Q^k} ; \tag{3.37}$$
i.e., no impact of c_1.
- To investigate asymptotic expansions one introduces Borel transforms. From the 1990's it has been pointed out that the pole mass is not well-described on the non-perturbative level and that there are singularities in the Borel plane.

Chapter 4

The Standard Model and beyond

In **Chapter 3** we have discussed 'strategies' for understanding the impact of the SM and its possible existence of new dynamics, while in this **Chapter** we talk about 'tactics'.

The Lagrangian of the SM[1] is based on *local* $SU(3)_C \times SU(2)_L \times U(1)_Y$ symmetries: C refers to 'color' as an unbroken local non-abelian symmetry , L to left-handed currents in the broken non-abelian one and Y to the weak hypercharge $Y = 2(Q - T_3)$ for the abelian one. We have 'Up'-type quarks (u, c, t) with electric charge " + 2/3"; they are described in two ways: left-handed quarks with weak isospin T_3 = " + 1/2" and Y = " + 1/3" or right-handed ones with weak isospin T_3 = "0" and Y = " + 4/3". Similarly for 'Down'-type quarks (d, s, b), with an electric charge " − 1/3": left-handed quarks with T_3 = " − 1/2" and Y = " + 1/3" or right-handed ones with T_3 = "0" and Y = " − 2/3". The situation with leptons is totally different. Left-handed neutrinos (ν_e, ν_μ, ν_τ) have T_3 = " + 1/2" and Y = " − 1", while in the SM one has no right-handed neutrinos. Left-handed charged leptons (e, μ, τ) have T_3 = " − 1/2" and Y = " − 1", while right-handed ones have T_3 = "0" and Y = " − 2".

The Lagrangian $\mathcal{L}_{\rm SM}(\phi_i(x), \partial_\mu \phi_i(x))$ represents fields of spin-1/2 (quarks and leptons), spin-1 gauge bosons and the scalar Higgs boson(s). As already mentioned in our **Prologue**, the SM has been very successful, but it is incomplete:

- It fails to describe the huge difference in matter vs. anti-matter observed in the Universe.
- In the standard cosmological model the "known" matter represents about 5% of our Universe: the Dark Matter about 26%. Its existence and its features have been established by gravitational considerations, but *so far* not beyond. Here we will hardly discuss Dark Energy, which is assumed to cover the remaining $\sim 69\%$.

[1]It is just another example that scientists are short of words to explain what they are doing. The words 'Standard Model' are used in many fields such as HEP, cosmology, nuclear physics etc., although their exact meaning often change. Furthermore the 'present' SM of HEP is not a model – it is based on a true theory for fundamental dynamics.

- Three neutrinos ν_1, ν_2 and ν_3 have been established, and they are *not* massless. Data from neutrino oscillations tell us there are three different neutrino masses with $\Delta_{ij} = m_i^2 - m_j^2 \neq 0$ $(i \neq j)$. We take $m_1 < m_2$ by definition and thus $\Delta_{21}^2 = m_2^2 - m_1^2 > 0$. Then we look at two options: $m_1 < m_2 \ll m_3$ with $\Delta_{31}^2 > 0$ – 'Normal Hierarchy' – or $m_3 \ll m_1 < m_2$ with $\Delta_{31}^2 < 0$ – 'Inverted Hierarchy'.[2]

4.1 QCD × QED

In addition to QED describing electrodynamics with quarks, there is only one known candidate to describe the strong forces between quarks and gluons with an unbroken local gauge interaction: $SU(3)_C$ = QCD. We have therefore established two unbroken gauge forces: QCD with eight massless gauge boson – gluons[3] – and QED with one massless photon; the former is described by a non-abelian theory, the latter by an abelian one.

4.1.1 *Color*

In this **section** we list general milestones,[4] that we discuss in details in different places.

- At the beginning, quarks have been seen as just a mathematical 'trick' to describe the properties of light hadrons with one family: u and d quarks with electric charge $q_u = 2/3$ and $q_d = -1/3$. Later it was realized that quarks are physical states with color quantum number (QN) "3" plus 8 gluons with color QN "8", while the initial and final states are hadrons with color QN "zero". Quarks and gluons are real but they do not show in the open because of the "confinement".[5]
- The ratio of $\sigma(e^+e^- \to \text{hadrons})/\sigma(e^+e^- \to \mu^+\mu^-)$ can be shaped over the ratio $\sigma(e^+e^- \to \bar{q}q)/\sigma(e^+e^- \to \mu^+\mu^-)$ involving quarks. The three $\bar{u}u$, $\bar{d}d$ and $\bar{s}s$ channels give a ratio of $N_C \cdot 2/3$ (N_C is the number of *quark* colors) around $\sqrt{s} \sim 1.2$ GeV at e^+e^- collisions. One can add the $\bar{c}c$ channel leading to $N_C \cdot 10/9$ at $\sqrt{s} \sim 5$ GeV and the $\bar{b}b$ channel at $\sqrt{s} \sim 11$ GeV with $N_C \cdot 11/9$ for a threshold of a new flavor. In **Chapter 3** we have discussed the connections between the worlds of hadrons and quarks.[6]

[2] 'Normal ordering' and 'Inverted ordering' would probably be a better choice. We stick to the common slang, however.

[3] The bosons of glue-on → gluons.

[4] Old statement: "What the 'gods' can do, the 'bulls' cannot do" (in latin "Quod licet Iovi, non licet bovi"). Here we talk about the 'gods = Symmetries/Pauli/C.N. Yang/...'.

[5] The word "confinement" has a deeper reason than another term sometimes used, infrared 'slavery'.

[6] Speaking of connections: Jack Steinberger started his amazing career in 1949 as a theorist to predict $\Gamma(\pi^0 \to \text{"}\bar{p}p\text{"} \to 2\gamma)$, which is based on an anomaly [96]; as an experimenter, in 1950, he showed 'evidence for the production of neutral mesons by photons'!

- At a later stage it was realized that the couplings of unbroken non-abelian theories in general go *down* with short-distance dynamics. Not only that: their values are connected, because of unbroken local color symmetry. In particular, self-couplings of three and four non-abelian bosons and their coupling with a quark were predicted with accuracy [97–99].
- Top quarks are 'somewhat free', since they decay *before* they can produce top *hadrons* [51]. A subtle point is that a top quark carry "color"; it is unbroken local symmetry; it has to be transferred to other (anti-)quarks. We will detail the situation in **Chapter 8**.

The gluon coupling depends on the scale:

$$\alpha_S(Q^2) = \frac{g_S^2(\mu)}{4\pi} \simeq \frac{\alpha_S(\mu^2)}{1 + \frac{\beta_0 \alpha_S(\mu^2)}{4\pi}\log(Q^2/\mu^2)} \tag{4.1}$$

$$\beta_0 = \frac{11}{3}N_C - \frac{2}{3}N_{\text{flav}} = 11 - \frac{2}{3}N_{\text{flav}} \tag{4.2}$$

i.e., $\alpha_S(Q^2) \to 0$ with $Q^2 \to \infty$ 'slowly', namely on the logarithmic scale. It gives "asymptotic freedom" at high energies. To be more precise: interacting of the eight gauge bosons of QCD $= SU(3)_C$ produces asymptotic freedom, unless we have too many flavor quarks. So far we have six flavor quarks: u, d, s, c, b, t; thus $\beta_0 = 11 - 4 = 7$ at most. To lose asymptotic freedom one needs 17 flavors leading to $\beta_0 = -1/3$. We are 'safe' about short-distance dynamics.

4.1.2 *Non-perturbative and perturbative strong forces*

In Eq.(3.1) we described non-abelian QFT with $SU(N)$ in general. Now we discuss the strong forces with $SU(3)_C$ and eight gluons of color QN "8". Thus $G^a_{\mu\nu}$ denotes the gluon field strength tensor and its covariant derivative

$$G^a_{\mu\nu} = \partial_\mu A^a_\nu - \partial_\nu A^a_\mu + g_S f^{abc} A^b_\mu A^c_\nu \qquad D_\mu = \partial_\mu - i g_S t^a A^a_\mu \tag{4.3}$$

with the gluon fields A^a_μ and $SU(3)$ generators t^a, $a = 1, ..., 8$. The couplings of quark [anti-quark] fields with color QN "3" ["$\bar{3}$"] and their flavor reincarnation are flavor-diagonal. Once we define a set of Hermitian fields $t^a A^a_\mu$, g_S must be "real". In QFT one has to specify the normalization point μ where all operators are defined; likewise the coupling g_S and the masses of the quarks are functions of the scale μ.

4.1.3 *Chiral perturbative theory*

Chiral symmetry, already mentioned in **Sect. 3.5**, was embraced in the previous century, namely in 1958; yet it is not old 'stuff'. Perturbative QCD with quarks and gluons is very successful in describing the production of jets of hadrons. On the other hand, chiral symmetry represents QCD at low energy, in particular non-perturbative dynamics. Amplitudes involve pions (and also kaons) at low energies, where the strong coupling becomes large. Chiral symmetry is at the basis of an effective theory (see **Sect. 3.4**) applied to QCD. Chiral 'perturbation theory' (ChPT) has been successful in describing the dynamics of meson at low energies and still is. Furthermore, it is a good tool to probe ND in different directions and at different thresholds. Yet it does not provide a foolproof algorithm; one needs judgment: the meaning of the word 'perturbation theory' can vary according to different situations.

Consider QCD with one family doublet of u and d quarks: $m_u < m_d \ll \bar{\Lambda} \sim 1\,\text{GeV}$ (see **Sect. 3.3**). By looking at the Lagrangian, one might think that QCD possesses a global $U(2)_L \times U(2)_R$ symmetry.[7] Indeed, the vectorial component $U(2)_{L+R} = U(2)_V$ is conserved, even after quantum corrections. The axial part $SU(2)_{L-R}$ is known as spontaneously broken leading to the emergence of a triplet of Goldstone bosons – the pions. Those actually acquire a mass due to $m_u \neq 0 \neq m_d$. What about the remaining $U(1)_{L-R}$? A subtle, but crucial puzzle arises regarding this axial current. For now we ignore it; however, we will come back to that in **Chapter 10**. We can go one step more, adding the strange quark s with $0 < m_u < m_d \ll m_s \ll \bar{\Lambda}$. The meson spectra is augmented of four kaons $(K^-, \bar{K}^0, K^0, K^+)$, for a total of eight pseudoscalar mesons, including η^0.[8] Then one can discuss global $SU(3)_L \times SU(3)_R$ symmetry.

If we aim at evaluating cross-sections in QCD at low energies, the degrees of freedom are no longer the quarks and gluons, but the light mesons and maybe also light baryons themselves: π, K, η, p, n, Λ. An expansion is made not in the terms of coupling constants, but in the terms of their *momenta*, that are small compared to the scale $\bar{\Lambda} \simeq 1$ GeV, provided we stay at low energies. To be more explicit: the broken $SU(3)_{flavor}$ by the 'enhanced' symmetry[9]

$$SU(3)_{flavor} \implies SU(3)_L \times SU(3)_R \tag{4.4}$$

in the world of quarks is defined by the mass matrix M for the light quarks:

$$M = \begin{pmatrix} m_u & 0 & 0 \\ 0 & m_d & 0 \\ 0 & 0 & m_s \end{pmatrix} . \tag{4.5}$$

Indeed, the massless Lagrangian L_{QCD} is invariant under

$$q_H = \begin{pmatrix} u \\ d \\ s \end{pmatrix}_H \implies U_H \begin{pmatrix} u \\ d \\ s \end{pmatrix}_H \quad , \quad H = L, R \tag{4.6}$$

separately for both helicities, left (L) and right (R). The matrices U_H are unitary 3×3 matrices. There is a *global* $SU(3)_L \times SU(3)_R$ *chiral symmetry* of QCD. According to Noether theorem this symmetry gives rise to 16 conserved currents and thus 16 conserved charges. The vector charges $Q^i_V \equiv Q^i_L + Q^i_R$ also annihilate

[7] The subscript R and L refer, as usual, to right-handed and left-handed states, respectively.

[8] To be more precise about spectroscopy: $|\eta_8\rangle = |\bar{u}u + \bar{d}d - 2\bar{s}s\rangle/\sqrt{6}$. We will observe in **Chapter 10** that the η meson does not fit this situation perfectly.

[9] It has been suggested to describe the strong forces with local $SU(3)_{flavor}$ for light flavors; this idea can hardly reach off the floor – not unusual for theorists.

the 'vacuum', but the axial charges $Q_A^i \equiv -Q_L^i + Q_R^i$ do not:

$$Q_V^i \left|0\right\rangle = 0 \quad \text{and} \quad Q_A^i \left|0\right\rangle \neq 0 \,. \tag{4.7}$$

One can describe the same situation with different words: the dynamics is symmetric in $SU(3)_L \times SU(3)_R$, while the 'vacuum' is not. This is known as "spontaneous chiral symmetry breaking" (χSB). According to Goldstone theorem the 8 spontaneously broken axial symmetries require the existence of 8 Goldstone bosons which are pseudo-scalars and at first massless. Pions are excellent candidates for these roles, since their masses are small compared with 1 GeV; one can say that somehow also for kaons and eta, completing the octet.

Let us now introduce the effective Lagrangian L_{ChPT}. The Goldstone bosons have zero interactions at zero energy. The expansion of ChPT has the form

$$L_{ChPT} = L_2 + L_4 + L_6 + \ldots \tag{4.8}$$

where L_n contain terms with n-derivatives. The derivatives in the amplitudes turn into low momenta of the pseudo-scalar mesons. Thus expansions are done in powers of momenta. Quark masses are present, and nonzero energies allow nonzero values for masses and interactions; e.g., the pion mass is an intricate mixture of the current quark mass, explicit χSB, and the quark vacuum expectation value of spontaneous χSB [100]. At leading order:

$$M_\pi^2 = -(m_u + m_d)\frac{\langle 0|\bar{u}u|0\rangle}{F^2} \tag{4.9}$$

where F is a non-perturbative parameter of ChPT with dimension of mass. The expansion of Eq.(4.8) is done in powers of momenta and of small quark masses, that is m_u, m_d for $SU(2)$ ChPT and m_u, m_d and m_s for $SU(3)$ ChPT. While $m_u, m_d \ll \Lambda_H$, Λ_H being a typical meson mass, m_s is larger, and the expansion is known to converge much slower in the case of $SU(3)$ ChPT than of $SU(2)$ ChPT.

4.1.4 *Dispersion relations*

The idea to use "Dispersion Relations" as a powerful nonperturbative tool came from theorists. They were applied already in classical electrodynamics before the era of QFTs. Even before QM they were encountered in many branches of physics (and even of engineering) in different contexts.

Here, we focus on the general validity of central statements in QFT, going back to a paper from 1954 titled 'Use of Causality Conditions in Quantum Theory' [101]. In the S (= scattering) matrix theory one assumes:

- "unitarity" (= conservation of probability),
- "Lorentz invariance",
- "crossing symmetries"[10]
- "analyticity" in various energy variables as external momenta.

[10]Crossing symmetry relates the amplitudes where the "crossed" particles are replaced by their anti-particles, swapping them from the initial to the final state.

These requirements lead to dispersion relations, which connect the dispersive (real) part in an amplitude with its absorptive (imaginary) part, which is often better accessible. Dispersion relations were first derived by Kramers and Kronig in 1926 [102,103]; their name follows from this first applications to the quantum theory of dispersion. We give one example from classical electrodynamics, the Kramers-Kronig relation:

$$\mathrm{Im}\,\epsilon(\omega)/\epsilon_0 = -\frac{2\omega}{\pi}\,P\int_0^\infty d\omega'\frac{\mathrm{Re}\,\epsilon(\omega')/\epsilon_0 - 1}{(\omega')^2-\omega^2} \tag{4.10}$$

where P denotes the principal part of the integral. The dielectric constant ϵ is frequency dependent: $\vec{D}(\vec{x},\omega) = \epsilon(\omega)\vec{E}(\vec{x},\omega)$.

The S matrix can be parametrized as:

$$S = \mathbf{1} + i\mathcal{T} \tag{4.11}$$

where $\mathcal{T}$ is the transition operator. From the unitarity of the S operator one gets:

$$\mathbf{1} = S\,S^\dagger = \mathbf{1} + i\,(\mathcal{T}-\mathcal{T}^\dagger) + |\mathcal{T}|^2 \quad \text{or} \quad 2\,\mathrm{Im}\,\mathcal{T} = |\mathcal{T}|^2 = \mathcal{T}\sum_n |n\rangle\langle n|\mathcal{T}^\dagger\,; \tag{4.12}$$

i.e., it shows the famous *optical theorem*! By virtue of this theorem, the imaginary parts of the amplitude is related to total cross sections, which may be determined experimentally. In general, we have

$$\mathrm{Im}\,\mathcal{M}(A\to A) = 2E_{CM}|\vec{p}|\sum_X \sigma(A\to X)\,. \tag{4.13}$$

Scattering amplitudes and vertex functions contain real and imaginary parts. The imaginary one is due to the propagation of on-shell intermediate states. One finds a general form for dispersion relations

$$\mathrm{Re}\,f(s) \;=\; \frac{1}{\pi}P\int_0^\infty \frac{ds'}{s'-s}\,\mathrm{Im}\,f(s') \tag{4.14}$$

with the identity

$$\frac{1}{s-s_0-i\epsilon} = P\frac{1}{s-s_0} + i\pi\delta(s-s_0)\,. \tag{4.15}$$

It is derived starting from $f(s+i\epsilon)$ expressed in integral form according to the Cauchy theorem. Using the same theorem one can write an amplitude as an integral over its imaginary part:

$$f(s) = \frac{1}{\pi}\int_0^\infty \frac{ds'}{s'-s-i\epsilon}\,\mathrm{Im}f(s')\,. \tag{4.16}$$

This integral involves all values of s': to know $f(s)$ at small s, one needs to know Im $f(s')$ also at large s'. The Cauchy theorem implies an integration in the two dimensional complex plane z along three paths: real and imaginary linear paths, both running along the direction where the function $f(z)$ has a branch cut, and a circular path joining them. On the last one, $f(z)$ goes to zero as $z\to\infty$; in principle one wants that $f(z)$ is going to zero quickly enough to give zero contributions 'as a good behavior at infinity'. Yet one has to be realistic: it can give systematic uncertainties. In order to cancel possible finite terms, one applies a dispersion relation as a 'subtracted' relation at a fixed point,

$$f(s) = f(0) + \frac{s}{\pi} \int_0^\infty \frac{ds'}{s'(s' - s - i\epsilon)} \mathrm{Im} f(s') \ . \tag{4.17}$$

One could continue with 'doubly subtracted' ones, and so on. These equations mean that the function can be reconstructed at any point s from the knowledge of its absorptive part along the branch cut. Sometimes one can use both Eqs. (4.16,4.17), but mostly only Eq. (4.17) is used. There is a rich literature with a long history from different 'cultures'; see a short list in [104, 105]. Again, it needs judgment where to apply and to what extent; one example is to apply to pion-nucleon cross sections.

Likewise one can relate the values of two-point functions $\Pi(q^2)$ in a transition of $ab \to cd$ in a QFT at different complex values of q^2 to each other through an integral representation [106]; q denotes a four-momentum. In particular one can evaluate $\Pi(q^2)$ for large *Euclidean* values $-q^2 = q_0^2 + |\vec{q}|^2$ with the help of a (Wilsonian) OPE. Then one can relate the coefficients I_n^{OPE} of local operators O_n to observables (like in the transitions of $e^+e^- \to$ hadrons or the inclusive transitions of $B \to l\bar{\nu} X_{c,u}$) in the physical – i.e., Minkowskian – domain through an integral over the discontinuity along the real axis:

$$I_n^{\mathrm{OPE}} \simeq \frac{1}{\pi} \int_0^\infty ds \frac{s}{(s + q^2)^{n+1}} \cdot \sigma(s) \ . \tag{4.18}$$

Such a procedure implies that there are only *physical* singularities – poles and cuts – on the real axis of q^2: one can calculate two-point functions for large Euclidean values of q^2, and will not pick up extra nonphysical contributions from poles etc. Dispersion relations are used to calculate transition rates in the HQE and to derive new classes of sum rules, that we mention in **Sect. 4.1.6**. The item of 'dispersion relations' is one significant example of different 'cultures', namely in the worlds of hadrons vs. quarks and gluons.

4.1.5 $1/N_C$ *expansion*

For several reasons the number of colors N_C has to be three (and not two or four in particular). Yet in the limit of $N_C \to \infty$ QCD non-perturbative dynamics becomes tractable [107]: only planar diagrams contribute to hadronic scattering to leading order, and the asymptotic states are $\bar{q}q$ = mesons and qqq = baryons; in this respect, "confinement" is proved. Furthermore the Zweig/OZI rules hold [108]. Starting from $N_C = 3$, multiquark hadrons configurations may be anticipated to exist. States as $\bar{q}q\bar{q}q$ and $qqqq\bar{q}$ 'exist' and they can be analyzed and constrained in the large N_C limit.

One can treat short-distance dynamics with $1/N_C$ expansions in this manner: first one keeps $N_C = 3$ fixed and perturbatively evolves down to the appropriate scale to derive effective Lagrangian, conservatively around $\mu \sim 1$ GeV. Then one evaluates hadronic matrix elements, which are shaped by long-distance dynamics. One expands those in powers of $1/N_C$. This expansion of $N_C \to \infty$ has often given us good ideas and more, namely it has shown the 'roads'. For example, it helped us to deal with two-body non-leptonic decays of charm mesons [65, 109].

Model uncertainties are defined *inside* those models. Yet uncertainties of true theories can be decreased systematically; of course that is not easy to achieve. The situation for $1/N_C$ expansion is between: look at weak amplitudes in the heavy meson decay $H_Q \to f$:

$$\langle f|\mathcal{L}_{\text{eff}}|H_Q\rangle \propto b_0 + \frac{b_1}{N_C} + \mathcal{O}(1/N_C^2) \tag{4.19}$$

this is not a true expansion, since it can hardly go beyond b_1. The $1/N_C$ expansions do not enable us to decrease the uncertainties systematically.

Usually people talk about $1/N_C$ expansions, but there is another 'landscape' where a fixed parameter is sent to infinity, which is even more subtle and connects with duality, which we have discussed in **Sect. 3.10**. The leading term in the semi-leptonic width of heavy flavor hadrons depends on the fifth power of the heavy quark mass. An analysis of this width shows one can deal with n as a free parameter and set $n = 5 \to \infty$ [63]. The resulting expansion elucidates why the small velocity treatment is relevant for inclusive semi-leptonic $b \to c$ transitions. The large-n treatment also explains why the scales of order m_b/n are appropriate. One can use short-distance mass normalized at a scale around $m_b/n \sim 1$ GeV [63].[11]

4.1.6 *QCD sum rules*

First our community had used perturbative QCD to describe the production of 'jets' of hadrons, including features that relate "color" or "flavor" symmetries. Later the impact of non-perturbative QCD was included as a crucial part. "Sum rules" are ubiquitous tools in many branches of physics that involve sums or integrals over observables such as rates and moments and dispersion relations. Prototype QCD sum rules started already at the time of current algebra, in 1960, before QCD, but the main flow originated with the celebrated case of the SVZ QCD sum rules, named as Shifman, Vainshtein and Zakharov [110], which allow low-energy hadronic quantities to be expressed through basic QCD parameters. The physical process is described by building an OPE (discussed in **Sect.3.7**), and non-perturbative dynamics is parametrized through condensates $\langle 0|\bar{q}q|0\rangle$, $\langle 0|GG|0\rangle$ etc.; they are zero in perturbative QCD. Such condensates are treated as free parameters, the values of which are fitted from certain observables.

There is a vast literature: examples of QCD sum rules have been discussed in general (see e.g. Ref. [54]) or in special cases, like Bjorken sum rules [64, 111], "small velocity" [75] and "spin sum rules" [112]. At least they give us novel lessons of QCD about how to combine perturbative and non-perturbative forces.

[11] Maybe one of the co-authors of this book is somewhat biased here.

4.2 Lattice QCD (LQCD)

As already said in **Chapter 3**, QCD is the only candidate among local field theories to describe the strong forces. Monte Carlo simulations of QCD on the lattice – or LQCD for short – provide a different framework to deal with the complementary features of "asymptotic freedom" in the UV and "confinement" in the IR.[12] The four-dimensional space-time continuum is replaced by a discrete lattice with spacing a between lattice sites. This is usually viewed *not* as representing physical reality, but providing mathematical tools to deal with long-distance dynamics through an expansion in the *inverse* coupling. Expansions in the inverse of the coupling constant demonstrate quark confinement in the strong coupling limit, which occurs at large distances. Distances $\sim a$ and smaller cannot be treated in this way. That means that the finite spacing introduces an UV cut-off $\sim \pi/a$ for the lattice version of QCD. Short-distance dynamics is treated by perturbative QCD. Considerable care has to be applied in matching the two theories at distance scales $\sim a$. One uses the technique of effective field theory to incorporate short-distance dynamics cut-off by the finite lattice spacing; the discretization effects are described through an expansion in powers of a:

$$\mathcal{L}_{\text{eff}} = \mathcal{L}_{\text{QCD}} + a\mathcal{L}_1 + a^2\mathcal{L}_2 + ... \tag{4.20}$$

with the $\mathcal{L}_{1,2,...}$ containing operators of dimension higher than four. Those are non-renormalizable; this poses no problem since they are constructed to describe long-distance dynamics. All local operators allowed by the underlying lattice symmetries will appear. At leading order, the Lagrangian is just the continuum QCD Lagrangian [114].

Putting fermions on the lattice create problems between the 'Scylla' of 'fermion doubling' and the 'Charybdis' of vitiating chiral symmetry. Theorems tell us that in four dimensions chiral symmetry is either violated for $a \neq 0$ or maintained at the price of getting too many fermions [115].

At the beginning of the LQCD era one talked about 'quenching approximation'; loops of fermions are neglected. Until the 1990s, most of lattice simulations were still limited by this approximation, inducing uncontrolled systematic errors. Now the LQCD community has gone well beyond that era. It has produced simulations with u and d, s and even c loops.[13]

Some interesting applications of lattice gauge theory arise in heavy quark physics. Measured in lattice units, the beauty and charm quark masses can easily be large. For heavy quarks one actually needs $a\,m_Q \ll 1$. Then lattice spacing effects are more challenging for heavy quarks than for light quarks. However, it is important to work on the connections of LQCD with OPE and HQE (see Eq. (3.14)). In the SM we can calculate the coefficients $c_i^{(f)}(\mu)$ based on perturbative QCD, while one can use LQCD for the matrix elements of local operators.

[12]There is a very good introduction with the title "Lattice QCD for Novices" from Lepage [113].
[13]Old history told us how Alexander the Great 'solved' the puzzle of the 'Gordian knot' – by just cutting it off; however, we know that the experts of LQCD are much more subtle than that.

Often LQCD is perceived as *panacea*. However, the situation is 'complex' in different directions: (a) the LQCD community (while 'orthodox' in 'substance') exhibits a different 'sociology'.[14] (b) FSI with many-body final states are not seen as a strength of LQCD, in particular for the weak decays of *beauty* hadrons. Even the situation with *strange* hadrons is not trivial with two-body non-leptonic FS.

In weak transitions of kaons we know from data that the $\Delta I = 1/2$ amplitudes are much larger than for $\Delta I = 3/2$ ones. The SM points to the correct direction at least semi-quantitatively, since the $\Delta I = 1/2$ amplitude is enhanced by perturbative QCD and the impact of penguin amplitudes. There seems to be no real problem for LQCD analyses of $\Delta I = 3/2$. The situation of $\Delta I = 1/2$ is more complex for LQCD. Its determination has been a long-standing challenge for the lattice community; only in the last few years first true LQCD progress has been attained [116]. One can compare lattice results with the analyses from local dynamics [117–119]; is there a sign of the impact of ND in direct **CP** asymmetry in $\Delta S = 1$ transitions? We will come back to this item in **Sect. 7.2**.

We know from the data that two-body FS in weak decays produces only small parts in non-leptonic $\Delta C \neq 0$ transitions and only tiny ones for $\Delta B \neq 0$. On the other hand known technology of LQCD allows one-to-one ($a \to b$) or one-to-two ($a \to b\ c$) hadrons in different regions, but not Dalitz plots etc.; in general, re-scattering cannot be described with local operators employed for transitions.

4.2.1 *Chiral gauge theories on the lattice*

Lattice tool has been successful and still continue to be in describing vector gauge theories like QCD.[15] LQCD is manifestly gauge invariant, and this implies that gauge degrees of freedom decouple in the regulated theory, i.e. at any momentum scale. On the other hand, the lattice formulation of chiral gauge theories (ChGTs) is much more complicated when one considers fermions: due to chiral anomaly there is no exact chiral γ_5 invariance on the lattice [120].

To keep a symmetry chiral (meaning each fermion contributes to the chiral anomaly), chiral symmetry has to be broken on the lattice. If the chiral symmetry is to be gauged, it means breaking gauge invariance, and one has to worry about the gauge degrees of freedom. It seems the lattice community has been closer to satisfactory constructions of lattice ChGTs; however, the job is not done yet [121].

Many different approaches have been pursued. An interesting scenario has been proposed in the 1990s.

Lattice theory for *massive* interacting fermions in $2n+1$ dimensions can be used to simulate the behavior of *massless* chiral fermions in $2n$ dimensions, if the fermion mass has a step function shape in the extra dimension. The massless states arise

[14]It has not developed a true inquisition and deals with 'heretics' in a rather gentle way.

[15]As it is well known, the characteristic feature of the chiral gauge theories is that the left- and right-handed components of the fermion fields do not couple to the gauge fields in the same way. All other gauge theories (such as QCD) are referred to as vector-like, since the gauge fields only couple to vector currents.

as zero modes bound to the mass defect, and all doublers can be given large gauge invariant masses. Chiral anomalies in the $2n$ dimensional subspace correspond to charge flow into the extra dimension [122]. The 'travel' on this 'road' can continue; constructing a non-perturbative lattice regulator for chiral gauge theories is a true challenge. Gauge-invariant solutions were proposed based on domain wall fermions, which are defined by extending space-time to five dimensions; a chiral mode on a domain wall is induced by a mass defect along the extra dimension [123]. Progress has been achieved in the connections of lattice and chiral gauge theories, but authors are honest to say that a lot of work is needed to reach their goal.

4.3 U- vs. V-spin (broken) symmetries and their connections

U- and V-spin symmetries are connected by QCD, but also have impact on flavor dynamics. At the beginning of the 1950s we had discovered three baryons that cannot decay by strong or electric forces – p,[16] n and Λ – and six mesons that decay only with weak forces – $\pi^\pm$, K and $\bar{K}$ – plus three neutral mesons – π^0, η and η' – that decay with QED. A global $SU(3)_{\rm flav}$ symmetry was introduced to describe this situation with light flavor quarks – *up*, *down* and *strange* ones – to produce baryons with $|q_1q_2q_3\rangle$ and mesons with $|\bar{q}_1q_2\rangle$. One also uses operators $I-$, $U-$ and $V-$spins, related to three sub-groups, $SU(2)_I$, $SU(2)_U$ and $SU(2)_V$, respectively. For $SU(2)_I$ one gets

$$[I_i, I_j] = i\epsilon_{ijk}I_K \tag{4.21}$$

that leads to

$$[I_3, I_\pm] = \pm I_\pm \qquad [I_+, I_-] = 2I_3 \tag{4.22}$$

These relations mean that the operators $I_\pm$ move up or down isospin. The algebra is the same for $SU(2)_U$ and $SU(2)_V$; in particular

$$[U_+, U_-] = 2U_3 \qquad [V_+, V_-] = 2V_3 \tag{4.23}$$

Thus U_+ and U_- move up and down for U-spin and likewise V_+ and V_- for V-spin. In the 60s, it was pointed out in Ref. [47] how to organize the non-leptonic decays of $\pi^\pm$, K and $\bar{K}$ with the "eightfold way".

For mesons one gets *strong*[17] hypercharge $Y_{\rm str} = S$ with electric charge $Q = I_3 + \frac{S}{2}$. Thus one can 'paint' it in the S and I_3 plot[18]:

- $S = +1$; $I = \frac{1}{2}$: K^0 K^+
- $S = 0$; $I = 1$ and 0 : π^- $\pi^0[\eta_8]$ π^+

[16]So far data tell us that protons are stable.
[17]*Strong* hypercharge $Y_{\rm str}$ is defined as the baryon number + strange number S. One can test the masses for the three baryons $|\Lambda^0\rangle \equiv |sud\rangle$, $|n\rangle \equiv |dud\rangle$ and $|p\rangle \equiv |udu\rangle$, that is $M(\Lambda^0) \simeq 1115.7$ MeV, $M(n) \simeq 939.6$ MeV and $M(p) \simeq 938.3$ MeV, and compare with the constituent masses $m_s^{\rm const} \sim 500$ MeV, $m_d^{\rm const} \sim 314$ MeV and $m_u^{\rm const} \sim 312$ MeV. It is not bad for a model of 'constituent' quarks, but it does not work for light pseudoscalar mesons.
[18]The mass of η_8 is larger than for π^0.

- $S = -1; I = \frac{1}{2}$: K^- $\bar{K}^0$

It shows the pattern of mesons with the same charge, namely along the direction "↘" in the S-I_3 plane. Isospin I has a long successful history: we know $SU(2)_I$ is close to a symmetry.

One can also 'paint' the landscape with U-spin symmetry in the Q and U_3 plot:

- $Q = +1; U = \frac{1}{2}$: K^+ π^+.
- $Q = 0; U = 1$ and 0 : K^0 $\pi^0[\eta_8]$ $\bar{K}^0$.
- $Q = -1; U = \frac{1}{2}$: π^- K^-.

It shows the pattern of mesons with the same charge, namely along the direction "⟶". However the situations are very different: members of the same U-spin 'symmetry' have quite different masses, which is not small on the hadron scale $\mathcal{O}$(1 GeV). Similar for V-spin symmetry in the S and V_3 plot:

- $\Delta S = +1; V = \frac{1}{2}$: K^0 π^+.
- $\Delta S = 0; V = 1$ and 0 : K^- $\pi^0[\eta_8]$ K^+.
- $\Delta S = -1; V = \frac{1}{2}$: π^- $\bar{K}^0$.

It shown the pattern of mesons with the same charge, along "↘". It makes good sense to use U-spin symmetry about *spectroscopy* of beauty and charm hadrons. Let us focus on weak decays of B^0 vs. B^0_s mesons, in particular on **CP** asymmetries. In 2006 direct **CP** asymmetry has been established in $B^0 \to K^+\pi^-$, with $A_{\mathbf{CP}}(B^0 \to K^+\pi^-) \sim -0.1$ as expected, but only a limit was given for $B^0_s \to \pi^+K^-$. In 2005 Lipkin had suggested a 'robust' test of the SM, based on U-spin symmetry, by checking the equality [124]

$$\Delta = \frac{A_{\mathbf{CP}}(B^0 \to K^+\pi^-)}{A_{\mathbf{CP}}(B^0_s \to \pi^+K^-)} + \frac{\Gamma(B^0_s \to \pi^+K^-)}{\Gamma(B^0 \to K^+\pi^-)} = 0 \tag{4.24}$$

that connects these two asymmetries. **CP** asymmetry in $B^0_s \to \pi^-K^+$ has been established in 2014. The values reported by PDG2020 [8] are:

$$A_{\mathbf{CP}}(B^0 \to K^+\pi^-) = -0.083 \pm 0.004 \quad , \quad A_{\mathbf{CP}}(B^0_s \to \pi^+K^-) = 0.221 \pm 0.015 \; . \tag{4.25}$$

The total LHCb data from run-1 lead to [125]:

$$\Delta = -0.11 \pm 0.04 \pm 0.03 \; . \tag{4.26}$$

With the present experimental precision one cannot conclude that there is evidence that a deviation from zero is observed. Let us add a few comments:

- In the world of quarks V-spin symmetry is defined by $s \leftrightarrow d$. Obviously it is broken: $m_d \simeq 4.7$ MeV vs $m_s \sim 93$ MeV, as discussed in **Sect. 3.3**. Likewise for V-spin symmetry: $m_u \simeq 2.2$ MeV vs $m_s \sim 93$ MeV.
- Does U-spin symmetry explains the ratio of $\tau_{B^0} \simeq 1.520 \cdot 10^{-12}$ s vs. $\tau_{B^0_s} \simeq 1.51 \cdot 10^{-12}$ s – i.e., within 1% uncertainty? One should realize that this simple statement is not trustworthy: look at the ratio of $\tau_{B^+}/\tau_{B^0} \simeq 1.08$; it suggests isospin symmetry has uncertainties of $\sim 8\,\%$.

- SM $b \to d$ penguin diagrams have less impact than $b \to s$ ones. Indeed, $BR(B^0_s \to \pi^+ K^-) \simeq 0.56 \cdot 10^{-5}$, i.e. less than $BR(B^0 \to K^+ \pi^-) \simeq 2 \cdot 10^{-5}$, which is not surprising qualitatively.
- Does $\Delta \simeq 0$ show the strength of U-spin symmetry – see Eq. (4.24) – or $\Delta \sim -0.1$ shows the impact of re-scattering [126]?
- FSI have important impact on weak amplitudes in general and in particular on **CP** asymmetries. Therefore when looking for precision we cannot ignore the correlations with V symmetry. To say it differently: we can*not* focus only on two-body FS and even more with only charged ones in weak transitions. As shown in Eqs. (2.58,2.59,2.60), intermediate two states strongly re-scatter into FS with two-, three-, four-body etc. states either in the world of hadrons or quarks. The substantial point is that weak and strong forces have very different time scales. Re-scattering makes the differences between U and V symmetries very 'fuzzy'; it is risky to just focus on U-spin symmetry and its uncertainties when one analyses weak decays and probe the impact of ND.

As said above: I, U and V symmetries are fine for spectroscopy even for beauty and charm hadrons; however, the situations are different for weak transitions with $\Delta B \neq 0 \neq \Delta C$. Non-leptonic weak decays of strange hadrons have two- and three-body FS, but $\Delta B \neq 0 \neq \Delta C$ transitions have mostly many-body FS.[19] For inclusive decays, produced by sums of exclusive decays with hadrons,[20] one can use quarks. Thus the violations of U- and V-spin symmetries are small, and tiny for I-spin one. We can deal with inclusive rates and asymmetries of beauty and maybe charm hadrons using effective operators in the world of quarks.

The connections of *inclusive* with *exclusive* hadronic rates are not obvious, in particular in a quantitative way. The violations of I, U and V symmetries in *exclusive* decays in the world of hadrons are expected to scale by the differences in pion and kaon masses, which are *not* small compared to the hadron scale $\bar{\Lambda}$. This is even more crucial while dealing with direct **CP** violation and the impact of FSI on amplitudes . One reason is that suppressed decays in the world of hadrons consist of a large numbers of states in the FS, where FSI have great impact with opposite signs. Furthermore the worlds of hadrons are controlled by FSI due to *non*-perturbative QCD; they show the strongest impact on exclusive ones.

4.4 The Glashow-Salam-Weinberg (GSW) theory

In the previous century the SM was introduced[21] to describe electroweak dynamics with the $SU(2)_L \times U(1)_Y$ gauge group and four gauge bosons [127]: photon, $W^\pm$ and Z^0. There is an obvious challenge: the non-abelian gauge bosons $W^\pm, Z^0$

[19] The FS are mostly produced by 'intermediate resonances'.
[20] We have first mentioned "duality" in **Sect. 3.10**.
[21] The 1979 Nobel prize of physics was given to Glashow, Salam and Weinberg based on their papers from 1968.

are not massless, as it was known even before $W^{\pm}$ and Z^0 were found.[22] It has been suggested since the 1960s that we have to add a $SU(2)$ doublet of scalar Higgs fields (at least): this produces non-zero masses for weak bosons with 70 GeV $< M_{W^{\pm}} < M_{Z^0} <$ 100 GeV plus a neutral Higgs field with non-zero mass, 'the Higgs boson'. Indeed for a local $SU(2)_L \times U(1)_Y$ theory we have first four massless gauge bosons with spin $= (1, \pm 1)$, namely eight states. Then we add complex doublet fields (H^-, H^0), $(\bar{H}^0, H^+)$. After mixing we have still a massless photon, but we get the massive $W^{\pm}$ and Z^0 plus 'the' Higgs field. **CPT** invariance tells us that $M(W^+) = M(W^-)$.

In other terms: three of the four Higgs fields re-appear as the additional degrees of freedom for the massive fields $W_0^{\pm}$ and Z_0^0, while the fourth one appears as 'the' Higgs field.[23] That is the somewhat 'easy' part to describe "electroweak" interactions, since we have followed the gauge 'road'.

4.4.1 *Yukawa couplings of Higgs fields with quarks*

The landscape of the Higgs dynamics with fermions is more 'complex', since it is not based on gauge symmetry.[24] The SM produces the masses of the quarks as a consequence of Yukawa couplings of Higgs fields with quarks. The quarks come in families of left-handed doubles $Q_L = (U, D)_L$ and right-handed singlets U_R and D_R.

Yukawa interactions of Higgs fields with *Up*-type $= (u, c, t, ...)$ quarks and *Down*-type $= (d, s, b, ...)$ ones are described by matrices G_U and G_D. The SM admits only three families of quarks. In the SM the neutral scalar Higgs field H^0 acquires a non-zero vacuum expectation value (VEV): $\langle \phi^0 \rangle = v \neq 0$ that leads to quark mass matrices: $\mathcal{M}_U = vG_U$ and $\mathcal{M}_D = vG_D$.

Since the Yukawa couplings are quite arbitrary, so are the mass matrices, and in general they can contain complex elements. In 2012 'the' Higgs has been established by ATLAS and CMS experiments at LHC [4], with a measured mass $M_{H^0} \simeq 125$ GeV, close to the SM expected value. In the SM the H^0 amplitude is 100% scalar. Present data show that H^0 is at least mostly scalar, but do not exclude a non-zero pseudo-scalar amplitude yet.

The $SU(2)_L \times U(1)_Y$ theory with Higgs dynamics is not gorgeous (in the views of theorists), but it 'works' with gauge bosons $W^{\pm}$, Z^0 and quarks, as said in the previous century. We have learnt that **CP** violation has emerged from this sector of the SM Lagrangian. This question is analyzed next.

[22] It depends whether one calls them "heavy" or "light". We know that their masses are 'just' below of 100 GeV; then they were called "heavy" 'then'; 'now' one could call them as "light" on the scale of $\mathcal{O}(1000)$ GeV that is probed in LHC collisions at CERN.

[23] Adding more Higgs doubles is easy; it leads to phenomenological consequences like the existence of charged Higgs fields.

[24] An exception is given by super-symmetry that we discuss in **Chapter 14**.

4.5 KM ansatz

> The KM phase is like the 'Scarlet Pimpernel': "Sometimes here, sometimes there, sometimes everywhere!"
> Sir Percival Blakeney

Kobayashi and Maskawa had pointed out in their famous paper published in 1973 [128] that there are three classes of QFT to describe **CP** asymmetries, characterized by: *non-minimal* Higgs dynamics, *right-handed* charged weak currents[25] – or three (or more) families of quarks. The beauty quark was discovered five years later at Fermilab; the top quark were established directly by the CDF/D0 collaborations in 1995. Of the three options Kobayashi and Maskawa listed, their names has been attached only to the last one as 'the' KM description.

CP violation can enter SM dynamics only through the quark mass matrices $\mathcal{M}_U$ and $\mathcal{M}_D$. The physical interpretation is more transparent when we write the Lagrangian in terms of the mass eigenstates of quarks. We diagonalize the mass matrices with two unitary matrices each – $T_{U,L}$, $T_{U,R}$ and $T_{D,L}$, $T_{D,R}$ – acting in the family space[26]:

$$T_{U,L}\mathcal{M}_U T^\dagger_{U,R} = \mathcal{M}_U^{\rm diag} \; , \quad T_{D,L}\mathcal{M}_D T^\dagger_{D,R} = \mathcal{M}_D^{\rm diag} \; . \tag{4.27}$$

Lagrangians can be described in terms of mass or flavor eigenstates of quarks. The eigenstates are related by:

$$U^m_{L[R]} = T_{U,L[R]} U_{L[R]} \quad , \quad D^m_{L[R]} = T_{D,L[R]} D_{L[R]} \; . \tag{4.28}$$

where the superscript m refers to mass eigenstates. T_L and T_R can be found by diagonalizing the Hermitian matrices $\mathcal{M}^\dagger\mathcal{M}$ and $\mathcal{M}\mathcal{M}^\dagger$:

$$\mathcal{M}^2_{diag} = T_R \mathcal{M}^\dagger \mathcal{M} T^\dagger_R = T_L \mathcal{M}\mathcal{M}^\dagger T^\dagger_L \; . \tag{4.29}$$

The expressions for *neutral* currents – electroweak as well as strong ones – remain manifestly the same whether they are expressed in terms of *flavor* or *mass* eigenstates, and *no* flavor-changing neutral currents arise on the tree-level. This is referred to as GIM mechanism [131] and follows from the unitarity of T_L and T_R:

$$T_{U,L}T^\dagger_{U,L} = \mathbf{1} = T_{D,L}T^\dagger_{D,L} \quad , \quad T_{U,R}T^\dagger_{U,R} = \mathbf{1} = T_{D,R}T^\dagger_{D,R} \; . \tag{4.30}$$

For *charged* left-handed currents of quarks we find, in terms of mass eigenstates:

$$\bar{U}_L \gamma_\mu D_L = \bar{U}^{\rm m}_L \gamma_\mu \mathbf{V} D^{\rm m}_L \quad , \quad \mathbf{V} = T_{U,L} T^\dagger_{D,L} \; . \tag{4.31}$$

The unitary matrix $\mathbf{V}$ reflects weak universality. There is no reason for $\mathbf{V} = \mathbf{1}$. Even if some speculative dynamics were to enforce an alignment between the U and D quark fields at some high scales causing their mass matrices to get diagonalized, this alignment would likely get upset by renormalization down to the electroweak scale.

[25] There was a short remark by Mohapatra in a 1972 paper invoking the need for right-handed currents to induce **CP** violation [129].

[26] There is a theorem called "Singular Value Decomposition" [130] that says: Any complex $m \times n$ ($m \geq n$) matrix $\mathcal{M}$ can be diagonalized as $U\mathcal{M}V^\dagger = D$, where D is a diagonal $n \times n$ matrix with positive elements, V is a $n \times n$ unitary matrix and U is a $n \times m$ matrix that is unitary for $n = m$.

4.5.1 *Describing weak phases through* unitary triangles

Quark mass matrices and thus also **V** contain phases: it does not mean that all of them generate observables. Apart from a possible term $G \cdot \tilde{G}$ (discussed in **Chapter 10**), **CP** violation can enter SM dynamics only through the mass matrices $\mathcal{M}_U$ and $\mathcal{M}_D$. Kobayashi and Maskawa analyzed the question under which conditions *not all* of these phases can be rotated away.

- A general $n \times n$ complex matrix contain $2n^2$ real parameters. Unitary implies $\sum_j V_{ij} V^*_{jk} = \delta_{ik}$ and yields n constraints for $i = k$ and $n^2 - n$ for $i \neq k$. It constraints n^2 independent real parameters.
- The phases of the quark fields can be rotated freely:

$$U_i^m \to e^{i\phi_i^U} U_i^m \qquad D_j^m \to e^{i\phi_j^D} D_j^m \tag{4.32}$$

  implying that **V** is multiplied by two diagrams matrices whose elements are pure phases. Since overall phases are irrelevant, $2n - 1$ *relative* phases can be removed from **V** this way. Accordingly, **V** contains $(n-1)^2$ independent observables.
- An *orthogonal* $n \times n$ matrix is constructed to give independent rotation angles:

$$N_{\text{angles}} = \frac{1}{2} n(n-1) \, . \tag{4.33}$$

  The number of independent phases in **V** is:

$$N_{\text{phases}} = (n-1)^2 - \frac{1}{2} n(n-1) = \frac{1}{2}(n-1)(n-2) \, . \tag{4.34}$$

- For two families of quarks we have $N_{\text{angles}} = 1$ – just the Cabibbo angle – while $N_{\text{phase}} = 0$; i.e., no **CP** violation through **V**. For three families we have

$$N_{\text{angles}} = 3 \qquad N_{\text{phase}} = 1 \; ; \tag{4.35}$$

  i.e., in addition to the three mixing angles we have one single irreducible phase that represents **CP** violation. Beyond $n = 3$ there is a rapid proliferation of physical parameters; for $n = 4$ we would have six angles and three phases.
- If all *Up*-type quarks were mass-degenerate, we could not distinguish them. Any linear combination of them would still be a mass eigenstate; thus one could make the transformation $\bar{U}_L^m \to \bar{U}_L^m \mathbf{V}^\dagger$ to remove the effect of **V**. Therefore no **CP** asymmetry could be found. Furthermore, if two *Up*-type quarks were degenerate, one could make $\bar{U}_L^m \to \bar{U}_L^m A^\dagger$, where A is a block-diagonal matrix which mixes the two degenerate quarks; the parameters in A can be adjusted to remove the complex phase in **V**; therefore no **CP** asymmetry. Likewise for *Down*-type quarks.

Already with three families there is an infinity of ways to express the elements of **V** in terms of three rotation angles and one phase.[27] One representation has been

[27] See the reference to the 'Scarlet Pimpernel' in the motto of **Sect. 4.5**.

'sanctioned' by the Particle Data Group (PDG):

$$\mathrm{V}_{\mathrm{CKM}} = \begin{pmatrix} c_{12}c_{13} & s_{12}c_{13} & s_{13}e^{-i\delta} \\ -s_{12}c_{23} - c_{12}s_{23}s_{13}e^{i\delta} & c_{12}c_{23} - s_{12}s_{23}s_{13}e^{i\delta} & s_{23}c_{13} \\ s_{12}s_{23} - c_{12}s_{23}s_{13}e^{i\delta} & -c_{12}s_{23} - s_{12}c_{23}s_{13}e^{i\delta} & c_{23}c_{13} \end{pmatrix} \tag{4.36}$$

where $c_{ij}[s_{ij}] = \cos[\sin]\theta_{ij}$ for the Euler angles θ_{ij} with i and j being families. Different parametrizations lead to the same physics, if they are consistent, and choosing one over another is a matter of convenience.

Of course, phases by themselves are not observables unless they are connected with others.

By changing the phase convention for the quark fields, one has to change the phase of a given CKM matrix element and even rotate it away; it will re-appear in other matrix elements. For example: $|s\rangle \to e^{i\delta_s}|s\rangle$ leads to the CKM matrix elements $V_{qs} \to e^{i\delta_s}V_{qs}$ with $q = u, c, t$.

For the case of three families, a geometric representation can facilitate an intuitive understanding. The unitarity of the CKM matrix leads to two relations:

(1) The overall couplings of each quark to all the possible quarks in *charged* currents (*Up*-type to *Down*-type quarks or *Down*-type to *Up*-type quarks) are of universal strength

$$\sum_{i=1}^{3} |V_{ij}|^2 = 1 \; ; \; j = 1, 2, 3. \tag{4.37}$$

These relations are usually referred to as *weak universality*.

(2) Weak phases are essential for **CP** violation

$$\sum_{i=1}^{3} V_{ji}V_{ki}^* = 0 = \sum_{i=1}^{3} V_{ij}V_{ik}^* \quad ; \quad j, k = 1, 2, 3, j \neq k \tag{4.38}$$

in the complex plane, one can describe these relations as triangles. The six triangles have different shapes, but all show the *same area*:

$$\text{area (every triangle)} = \frac{1}{2}J = \frac{1}{2}|\mathrm{Im}V_{km}^* V_{lm} V_{kn} V_{ln}^*| = \frac{1}{2}|\mathrm{Im}V_{mk}^* V_{ml} V_{nk} V_{nl}^*|$$

irrespective of the indices k, l, m, n!

J is called the Jarlskog invariant in the quark sector: it is independent of changes of arbitrary phases and parametrization in the CKM matrix. This relation is the geometric translation of the algebraic result that with three families there exists only a single irreducible phase.

Three brief comments follow naturally: (a) In the first decade of this century the Belle and BaBar collaborations have established large **CP** violation in $B^0 \to J/\psi K_S$; thus one names the corresponding triangle the 'golden' one, as we will discuss below. (b) Now we are in the era of precision physics, when it is crucial to probe other triangles, and their correlations. (c) Triangles can be constructed in several ways, according to data. For example, one can determine a triangle knowing three sides, two sides plus an angle, one side plus two angles and two sides plus its height – if data are consistent.

CP violation cannot be implemented unless holds:

$$m_u \neq m_c \neq m_t \qquad m_d \neq m_s \neq m_b \tag{4.39}$$

$$\theta_{12}, \theta_{13}, \theta_{23} \neq 0, \frac{\pi}{2} \qquad \delta \neq 0, \frac{\pi}{2} \tag{4.40}$$

The first conditions means that no pair of *Up*-type can be degenerate, and the same for *Down*-type pairs [132]. When one of these pairs becomes degenerate, we somewhat return back to the two generation case, the CKM matrix becomes real, and does not express **CP** violation. The second condition can be explicitly checked by expressing J in the above parametrization:

$$J = |\mathrm{Im}(V_{i\alpha} V_{j\beta} V^{\star}_{i\beta} V^{\star}_{j\alpha})| = s_{12}\, s_{13}\, s_{23} c_{12} c_{13}^2 c_{23} \sin\delta \tag{4.41}$$

These conditions can be summarized in a compact way:

$$iC \equiv [\mathcal{M}_U \mathcal{M}_{\mathcal{U}}{}^{\dagger}, \mathcal{M}_D \mathcal{M}_{\mathcal{D}}{}^{\dagger}] \tag{4.42}$$

leading to

$$\det C = -2J(m_t^2 - m_c^2)(m_c^2 - m_u^2)(m_u^2 - m_t^2)(m_b^2 - m_s^2)(m_s^2 - m_d^2)(m_d^2 - m_b^2) \tag{4.43}$$

A non-vanishing $\det C$ is a necessary condition for this **CP** violation [132]. The parameter J is a tiny dimensionless number around 3×10^{-5} measured with three families of quarks and present data. Yet the interpretation of $\det C$ is not quite so straightforward, since this parameter carry a high mass dimension: $\det C \propto M^{12}$! Depending on the relevant scale M for a certain process, $\det C$ can vary significantly. Arguing that baryogenesis at the electroweak scale is characterized by $M \sim 100$ GeV makes CKM dynamics utterly irrelevant for this process. However, in many cases there are several mass scales M, and the relevant ones may be much smaller – like M_K, M_B, ΔM_K and ΔM_B – not ~ 100 GeV; furthermore observable **CP** asymmetries in K and B decays are *ratios* of CKM parameters. One could use instead:

$$\tilde{C} \equiv J \cdot \log\left(\frac{m_t}{m_c}\right) \log\left(\frac{m_c}{m_u}\right) \log\left(\frac{m_u}{m_t}\right) \log\left(\frac{m_b}{m_s}\right) \log\left(\frac{m_s}{m_d}\right) \log\left(\frac{m_d}{m_b}\right) \tag{4.44}$$

This parameter also vanishes, if any pair of *Up*-type or *Down*-type are mass degenerate with different landscape. $\tilde{C}$ is dimensionless; in some cases, it may reflect the possible impact of **CP** violation better than $\det C$.

Let us remark that sometimes it is claimed that **CP** violation analyses are based on general parametrization of the CKM matrix. However when one looks inside the papers from experimental groups or theorists, one realizes that their results mostly depend on the (little refined) Wolfenstein parameterization [133], as we will discuss with some details in **Sect. 4.5.3.**

4.5.2 *'Landscapes' of* P *and* C *violation vs.* CP *one*

P violation can be defined in the Lagrangians – at least in the SM – with charged bosons W^μ: $g_L\ (\bar{q}_L\gamma_\mu q'_L)\ W^\mu_L$ vs. $g_R\ (\bar{q}_R\gamma_\mu q'_R)\ W^\mu_R$. Then maximal violation is obvious: $g_L \neq 0$, $g_R = 0$ (or $M(W_R) \to \infty$). When one gets beyond maximal one, one has options, namely $g_L \neq 0 \neq g_R$ or $M(W_L) \neq M(W_R)$ or combinations of both: $g_L^2/M^2(W_L)$ vs. $g_R^2/M^2(W_R)$ at low energies.[28] Likewise for charge conjugation **C** violation, when one uses **CP** invariance.

The situation is quite different for **CP** asymmetries, namely it is much more subtle. We give a somewhat simple example with SM neutrinos[29]: one might think that maximal **CP** violation might mean that ν_L and $\bar{\nu}_R$ exist, while ν_R and $\bar{\nu}_L$ do not. Alas – **CPT** invariance already told us we can have ν_L and $\bar{\nu}_R$ with*out* ν_R and $\bar{\nu}_L$.[30]

What about maximal **CP** asymmetry in our Universe? The measured asymmetry between known "matter" vs. known "anti-matter" gives:

$$\Delta_{\mathbf{C/CP}} \equiv \frac{\Gamma_{\text{“matter”}} - \Gamma_{\text{“anti-matter”}}}{\Gamma_{\text{“matter”}} + \Gamma_{\text{“anti-matter”}}} \simeq \frac{\Gamma_{\text{“matter”}}}{\Gamma_{\text{“matter”}}} = 1\ ; \qquad (4.45)$$

i.e., it is basically 100%. It is a true challenge to understand the underlying dynamics. We will come back to that, but just now we point out that the *present* asymmetry – see Eq. (4.45) – was tiny *a long time ago* in our Universe: $\Gamma_{\text{“matter”}} \simeq \Gamma_{\text{“anti-matter”}}$. We realized in the 1980s that **CP** violation in flavor forces has nothing to do with the present huge asymmetry in "matter" vs. "anti-matter".

CPT invariance implies the same mass and width for particle P and anti-particle $\bar{P}$. This theorem in QFT is rather simple, although its consequences are far-reaching.[31] Both **CP** and **T** violations are expressed through couplings with *weak* phases.

Let us turn to the world of hadrons, and precisely to the weak decays of strange, charm and beauty hadrons, where the initial state is a single hadron, and the final state in general has two or more hadrons with different flavors. We have[32]:

$$\Gamma(P \to f; t) \propto e^{-\Gamma t} G_f(t) \quad , \quad \Gamma(\bar{P} \to \bar{f}; t) \propto e^{-\Gamma t} \bar{G}_{\bar{f}}(t)\ . \qquad (4.46)$$

Obviously **CP** invariance is violated if $G_f(t) \neq \bar{G}_{\bar{f}}(t)$. We distinguish two categories of FS f exhibiting different phenomenology:

- A "flavor-specific" mode emerges when $P \to f$ is allowed, but not $P \to \bar{f}$; likewise $\bar{P} \to \bar{f}$, but not $\bar{P} \to f$. Then (at least) two different amplitudes for

[28] It is not the only option: one can think about parity violation in the "vacuum" (ground state).
[29] The measured neutrinos are different from SM ones, as we discuss below.
[30] To be more precise: ν_L and $\bar{\nu}_R$ couple to weak gauge bosons, ν_R or $\bar{\nu}_L$ do not. One can talk about 'sterile' neutrinos; those could connect with Dark Matter.
[31] Some general statements on amplitudes are listed in **Sect. 2.11**.
[32] **CPT** symmetry gives $\Gamma = \bar{\Gamma}$.

$P \to f$ with $\Delta_{\text{flavor}} = 1$ can interfere; likewise for $\bar{P} \to \bar{f}$. It can produce $G_f \neq G_{\bar{f}}$ time *independently*; that is called *direct* **CP** asymmetry. It can originate from baryons and charged mesons; one can find it also in neutral mesons decays.

- "Flavor-*non*specific" modes can be **CP** eigenstates, but not only that. They come from neutral mesons that can give both *direct* and *indirect* **CP** violation due to $\Delta_{\text{flavor}} = 1$ *and* 2. It can produce $G_f(t) \neq \bar{G}_f(t)$ time *dependently*:

$$\frac{d}{dt}G_f(t) \neq \frac{d}{dt}\bar{G}_f(t)\,. \tag{4.47}$$

It can be demonstrated that **CP** violation is established by finding the decay rate of a neutral meson P^0 into a **CP** *eigenstate* $f_\pm$ to be different from any *single pure* exponential, that is:

$$\frac{d}{dt}\left[e^{\Gamma t}\,\text{rate}(P^0 \to f_\pm; t)\right] \neq 0\,, \forall\,\text{real}\,\Gamma \implies \textbf{CP}\text{ violation}\,. \tag{4.48}$$

4.5.3 *Smart Wolfenstein parametrization*

Wolfenstein devised a good parametrization based on the expansion in the Cabibbo angle $\lambda = \sin\theta_C \simeq 0.223 \ll 1$, with parameters A, ρ and η of the order of unity [133]

$$\text{V}_{\text{CKM}} \simeq \begin{pmatrix} 1-\frac{1}{2}\lambda^2 & \lambda & A\lambda^3(\rho - i\eta) \\ -\lambda & 1-\frac{1}{2}\lambda^2 & A\lambda^2 \\ A\lambda^3[1-(1-\rho-i\eta) & -A\lambda^2 & 1 \end{pmatrix} + \mathcal{O}(\lambda^4)\,. \tag{4.49}$$

This matrix is the first step in the expansion; one has to go further, up to terms of $\mathcal{O}(\lambda^6)$:

$$\text{V}_{\text{CKM}} \simeq \begin{pmatrix} 1-\frac{1}{2}\lambda^2-\frac{1}{8}\lambda^4 & \lambda & A\lambda^3(\rho-i\eta) \\ -\lambda+\frac{1}{2}A^2\lambda^5[1-2(\rho+i\eta)] & 1-\frac{1}{2}\lambda^2-\frac{1}{8}\lambda^4(1+4A^2) & A\lambda^2 \\ A\lambda^3[1-(1-\frac{1}{2}\lambda^2)(\rho+i\eta)] & -A\lambda^2+\frac{1}{2}A\lambda^4[1-2(\rho+i\eta)] & 1-\frac{1}{2}A^2\lambda^4 \end{pmatrix} \tag{4.50}$$

As said above, with three families one has six unitarity triangles with the same area $\frac{1}{2}J \neq 0$. They can be organized into three classes[33]:

$$\text{Old triangle 1.1}: \; V_{ud}V^*_{us}\,[\mathcal{O}(\lambda)] + V_{cd}V^*_{cs}\,[\mathcal{O}(\lambda)] + V_{td}V^*_{ts}\,[\mathcal{O}(\lambda^5)] = 0 \tag{4.51}$$
$$\text{Old triangle 1.2}: \; V^*_{ud}V_{cd}\,[\mathcal{O}(\lambda)] + V^*_{us}V_{cs}\,[\mathcal{O}(\lambda)] + V^*_{ub}V_{cb}\,[\mathcal{O}(\lambda^5)] = 0 \tag{4.52}$$
$$\text{Old triangle 2.1}: \; V_{us}V^*_{ub}\,[\mathcal{O}(\lambda^4)] + V_{cs}V^*_{cb}\,[\mathcal{O}(\lambda^2)] + V_{ts}V^*_{tb}\,[\mathcal{O}(\lambda^2)] = 0 \tag{4.53}$$
$$\text{Old triangle 2.2}: \; V^*_{cd}V_{td}\,[\mathcal{O}(\lambda^4)] + V^*_{cs}V_{ts}\,[\mathcal{O}(\lambda^2)] + V^*_{cb}V_{tb}\,[\mathcal{O}(\lambda^2)] = 0 \tag{4.54}$$
$$\text{Old triangle 3.1}: \; V_{ud}V^*_{ub}\,[\mathcal{O}(\lambda^3)] + V_{cd}V^*_{cb}\,[\mathcal{O}(\lambda^3)] + V_{td}V^*_{tb}\,[\mathcal{O}(\lambda^3)] = 0 \tag{4.55}$$
$$\text{Old triangle 3.2}: \; V^*_{ud}V_{td}\,[\mathcal{O}(\lambda^3)] + V^*_{us}V_{ts}\,[\mathcal{O}(\lambda^3)] + V^*_{ub}V_{tb}\,[\mathcal{O}(\lambda^3)] = 0 \tag{4.56}$$

We have underlined the dependence of λ, which determines the size of each side. The pattern in the six triangles shows the impact of **CP** violation.

[33]The naming of the triangles will be clarified in the next section.

As we will discuss in **Chapter 6** in details, the 'long-lived' neutral kaon K_L mostly decays into non-leptonic FS made by 3 pions, which is a **CP** *odd* eigenstate. However, in 1964 [2] we have learnt that K_L can decay into **CP** *even* eigenstate:

$$\mathrm{BR}(K_L \to \pi^+\pi^-) = (1.967 \pm 0.010) \cdot 10^{-3} \neq 0 \ . \tag{4.57}$$

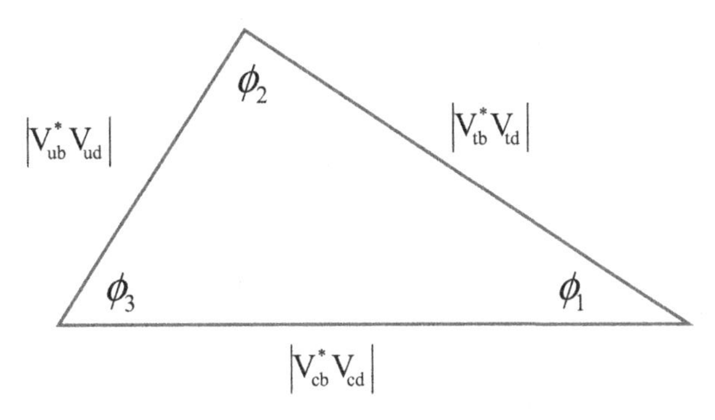

Fig. 4.1 The 'golden' CKM unitarity triangle for B transitions.

In the SM one can describe that on the basis of the 'Old' triangle 1.1, as pointed by Kobayashi and Maskawa in 1973 [128] in a qualitative way. At the time, it was crucial to probe **CP** violation in very different situations. One of us (together with Sanda) has predicted in 1981 **CP** violation in the transitions of $B^0 \to J/\psi K_S$ in the region of (10–20)%, based on the 'Old' triangle 3.1 [134]. Then we realized we need a *ultra*-heavy top quark to describe large $B^0 - \bar{B}^0$ oscillations – and our community found it. Thus **CP** violation could arrive close to 100%. The weak decays of $B_{d,u}$ mesons have successfully been described, see Eq. (4.55). It is obvious to call it the 'golden' triangle, see **Fig. 4.1**: the 'Old' triangle 3.1 shows that the sizes of the three sides are quite similar. The angles ϕ_1, ϕ_2, ϕ_3 are opposite the sides with $\bar{u}u$, $\bar{c}c$, $\bar{t}t$. Other people name its angles β, α, γ, respectively. It is crucial to probe the correlations in the triangles. **Fig. 4.2** shows that truly large **CP** violation in $B^0 \to \psi K_S$ is connected with very small one in $K_L \to \pi\pi$ transitions and the ratio of $B^0 - \bar{B}^0$ and $B^0_s - \bar{B}^0_s$ oscillations due to $\Delta M_{B^0}/\Delta M_{B^0_s}$; i.e., those observables (mostly) come from three triangles. We will discuss that in details in **Chapter 6**.

The Wolfenstein parameterization has been successful, yet it has weak points: measured decays give $A \simeq 0.81$, which is consistent with $\mathcal{O}(1)$, but both $\eta \simeq 0.35$ and $\rho \simeq 0.12$ are not close to unity. With three families of quarks one gets six triangles to describe the weak transitions of strange, charm and beauty hadrons and maybe also top quarks; four of those six ones one can be probed directly.

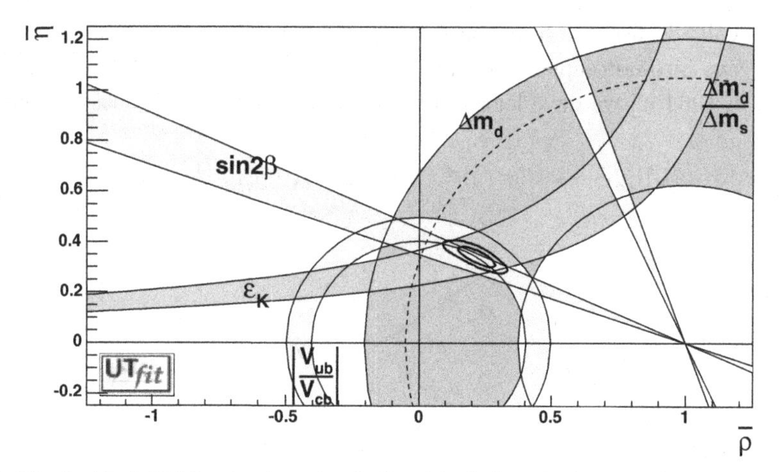

Fig. 4.2 The 'golden' CKM unitarity triangle fitted including the impacts from ϵ_K and ΔM_{B_s} from two other triangles [135].

4.5.4 *Consistent parametrization*

The SM produces at least the leading source of **CP** violation in $K_L \to 2\pi$ and B decays with good accuracy. Searching for ND we need even more precision and to measure the correlations with other FS. The 'landscape' of the CKM matrix is described *through* $\mathcal{O}(\lambda^6)$ *consistently* [136][34]:

$$V_{\rm CKM} \simeq \begin{pmatrix} 1-\frac{\lambda^2}{2}-\frac{\lambda^4}{8}+ & \lambda & \bar{h}\lambda^4 e^{-i\delta_{\rm QM}} \\ -\frac{\lambda^6}{16} & & \\ & 1-\frac{\lambda^2}{2}-\frac{\lambda^4}{8}(1+4f^2)+ & f\lambda^2+\bar{h}\lambda^3 e^{-i\delta_{\rm QM}} \\ -\lambda+\frac{\lambda^5}{2}f^2 & -f\bar{h}\lambda^5 e^{i\delta_{\rm QM}}+ & -\frac{\lambda^5}{2}\bar{h}e^{-i\delta_{\rm QM}} \\ & +\frac{\lambda^6}{16}(4f^2-4\bar{h}^2-1) & \\ f\lambda^3 & -f\lambda^2-\bar{h}\lambda^3 e^{i\delta_{\rm QM}} & 1-\frac{\lambda^4}{2}f^2+ \\ & +\frac{\lambda^4}{2}f+\frac{\lambda^6}{8}f & -f\bar{h}\lambda^5 e^{-i\delta_{\rm QM}}-\frac{\lambda^6}{2}\bar{h}^2 \end{pmatrix} \quad (4.58)$$

with $\bar{h} \simeq 1.35$, $f \simeq 0.75$ and $\delta_{\rm QM} \sim 90^o$ and only an expansion in $\lambda \simeq 0.223$. On the basis of this parametrization, we 'renovate' the unitarity triangles of the previous section:

$$\text{Triangle 1.1}: \; V_{ud}V_{us}^* \; [\mathcal{O}(\lambda)] + V_{cd}V_{cs}^* \; [\mathcal{O}(\lambda)] + V_{td}V_{ts}^* \; [\mathcal{O}(\lambda^{5\,\&\,6})] = 0 \quad (4.59)$$

$$\text{Triangle 1.2}: \; V_{ud}^*V_{cd} \; [\mathcal{O}(\lambda)] + V_{us}^*V_{cs} \; [\mathcal{O}(\lambda)] + V_{ub}^*V_{cb} \; [\mathcal{O}(\lambda^{6\,\&\,7})] = 0 \quad (4.60)$$

[34] We are not authors of this reference – i.e., we are not biased here.

$$\text{Triangle 2.1}: \; V_{us}V_{ub}^* \,[\mathcal{O}(\lambda^5)] + V_{cs}V_{cb}^* \,[\mathcal{O}(\lambda^{2\,\&\,3})] + V_{ts}V_{tb}^* \,[\mathcal{O}(\lambda^2)] = 0 \quad (4.61)$$

$$\text{Triangle 2.2}: \; V_{cd}^*V_{td} \,[\mathcal{O}(\lambda^4)] + V_{cs}^*V_{ts} \,[\mathcal{O}(\lambda^{2\,\&\,3})] + V_{cb}^*V_{tb} \,[\mathcal{O}(\lambda^{2\,\&\,3})] = 0 \quad (4.62)$$

$$\text{Triangle 3.1}: \; V_{ud}V_{ub}^* \,[\mathcal{O}(\lambda^4)] + V_{cd}V_{cb}^* \,[\mathcal{O}(\lambda^{3\,\&\,4})] + V_{td}V_{tb}^* \,[\mathcal{O}(\lambda^3)] = 0 \quad (4.63)$$

$$\text{Triangle 3.2}: \; V_{ud}^*V_{td} \,[\mathcal{O}(\lambda^3)] + V_{us}^*V_{ts} \,[\mathcal{O}(\lambda^{3\,\&\,4})] + V_{ub}^*V_{tb} \,[\mathcal{O}(\lambda^4)] = 0 \quad (4.64)$$

To investigate with accuracy flavor dynamics and **CP** violation in hadron decays, our community cannot focus only on the 'golden triangle'; we have to measure the *correlations* among triangles consistently [45, 137]. Let us emphasize some of the important points:

- One has to probe triangle 3.1 to attain precision in B^0 and B^+ transitions.
- Triangle 2.1 shows the impact on B_s^0 amplitudes and connections with B^0 and B^+ ones.
- Triangle 1.2 predicts small **CP** asymmetries in Singly Cabibbo Suppressed (SCS) D decays, but hardly in Doubly Cabibbo Suppressed (DCS) ones.
- Triangle 1.1 can be probed in tiny $K \to \pi\nu\bar{\nu}$ decays – there is a 'price'–with small *theoretical* uncertainties, namely with a 'prize'.

4.5.5 *Penguin operators vs. penguin diagrams*

Penguin diagrams were introduced by M. Shifman, A. Vainshtein and V. Zakharow in 1975 [110] to explain the measured $\Delta I(3/2) \ll \Delta I(1/2)$ amplitudes in kaon decays and later predicted the direct **CP** violation $\mathrm{Re}(\epsilon'/\epsilon_K) \neq 0$. These are based on *local* operators with two-body FS, although they come from loop diagrams.

The situation has significantly changed for the decays of beauty and charm hadrons, where FS consist mostly of many-body hadrons, not from two-body FS or even not two-body ones with narrow resonances. They show the 'road'; however, in general they cannot produce quantitative predictions.[35] Therefore we use the word of 'painting' the landscape of penguins– the situation is much more complex and also different for beauty and charm hadrons, and there is no reason why two-, three- and four-body FS should follow the same pattern.

One can add pairs of $\bar{q}q$ to "penguin" (or tree) diagrams and claims to produce the numbers of hadrons one wants. However, drawing diagrams is one thing, but describing the process is quite another thing.[36] The connections of penguin and

[35] Often 'penguin' contributions are referred to as 'penguin pollutions'. It is unfair to blame our lack of theoretical control on the water fowls rather than on the guilty party, namely 'us'.

[36] It was predicted 1500 years ago by Marinus, student of Proklos (known Neoplatonist philosopher influential on Western medieval philosophy as well as Islamic thought): "Only *being* good is one thing – but good *doing* it is the other one!"

tree diagrams with reality are often 'fuzzy' as pointed out in Refs. [44, 46]. The name of "penguin" diagrams is often used in broad sense: $Q\bar{q}_a \to q(\bar{q}_i q_i + \bar{q}_i q_i \bar{q}_j q_j + ...)\bar{q}_a$ following unitarity, where Q and q quarks carry the same charge with or without local operators. For heavy quarks – in particular for beauty quarks – one can compute the widths that depend on inclusive transitions, described by local operators. *Inclusive* decays of beauty hadrons show the impact of penguin diagrams in two classes of flavor changes – $b \Rightarrow s$ and $b \Rightarrow d$. However, *exclusive* transitions cannot claim, in general, to depend only on local operators: they depend significantly on strong long-distance dynamics.

Penguin diagrams give also imaginary parts that one needs for FSI [12,44]. Non-local penguin operator with internal charm lines can produce FSI: since $2m_c < m_b$, they can be *on*-shell and thus produce an imaginary part. However, they give us pictorials, but not much more. In the world of hadrons one measures rates and asymmetries, while in the world of quarks one produces amplitudes that have to include re-scattering, as pointed out in Eq. (2.60). That connection is far from trivial, and we are not yet at the end to understand the underlying dynamics.

4.5.6 *Transitions of hadrons with non-zero flavors*

Without a Higgs sector the SM is a self-consistent theory in the ultraviolet domain, albeit one with profoundly puzzling features like family replication. It is also out of touch with reality; all of its gauge bosons and fermions are massless. Coupling Higgs fields to gauge bosons in the required manner takes care of the former omission at the expense of the gauge hierarchy problem between the electro-weak, GUT and the Planck scales.

Obtaining a 'natural' solutions to the aforementioned hierarchy problem typically implies a scale not to exceed few TeV – or one comes with novel ideas as discussed below.

Here we focus on quarks. In the 'Yukawa-less' SM there are three families of $SU(2)_L$ quark doublets Q_L^i and right-handed quark singlets ($i = 1, 2, 3$); thus it is symmetric under the *global* group $SU(3)_q^3$:

$$SU(3)_q^3 \equiv SU(3)_Q \otimes SU(3)_U \otimes SU(3)_D \tag{4.65}$$

with the obvious transformation laws

$$Q_L^i \sim (3,1,1) \ , \ U_R^i \sim (1,3,1) \ , \ D_R^i \sim (1,1,3) \quad \text{under } SU(3)_q^3 \, . \tag{4.66}$$

'Switching on' the needed Yukawa couplings to the quark fields breaks the global $SU(3)_q^3$ massively – pun intended – and create non-diagonal flavor transitions. A priori ND can induce two types of transition operators, namely those with the same flavor and Lorentz structure as the SM and those with novel structures; the former can produce merely a re-calibration of the SM rates, while the latter can lead to qualitatively new effects, in particular new sources of **CP** violation beyond the CKM phase. The former scenario is labeled 'minimal flavor violation' (MFV). It is not a theory per se – it is a classification scheme. When one considers a specific model, one has to analyze whether this model represents a dynamical implementation of MFV and to which degree. For example, MFV is a very useful tool in the context of SUSY theories, where MFV arises naturally in minimal versions at least. It has little *predictive* power *per sé*; yet once deviations from the SM are observed, it provides a valuable diagnostic tool because, by comparing the pattern in different transitions, it allows to decide whether the emerging ND is of the MFV type or requires novel sources of flavor and in particular **CP** violations.

$SU(3)^3_q$ is broken by the SM Yukawa couplings; this symmetry can *formally* be recovered by introducing matrices Y_U and Y_D of dimensionless *auxiliary* fields that transform non-trivially under $SU(3)^3_q$:

$$Y_U \sim (3, \bar{3}, 1) \qquad Y_D \sim (3, 1, \bar{3}) \tag{4.67}$$

Y_U and Y_D are *not* physical fields; they are merely book-keeping devices convenient for tracking $SU(3)^3_q$ transformation properties of the interactions; thus they are also referred to as 'spurions'. With them one can express the SM Yukawa couplings contained in $\mathcal{L}_{\text{SM}}$ formally as $SU(3)^3_q$ invariant:

$$\mathcal{L}_Y = \bar{Q}_L Y_D D_R H + \bar{Q}_L Y_U U_R H_C + h.c. \tag{4.68}$$

with H and H_C describing the Higgs fields and its charge conjugate, respectively. Thus we can transform the matrices of the auxiliary fields using the (formal) $SU(3)^3_q$ symmetry:

$$Y_D \to \lambda_d \qquad Y_U \to V^\dagger \lambda_u \tag{4.69}$$

with λ denoting diagonal matrices of the Yukawa couplings and V the CKM matrix.

A theory satisfies MFV if all transition operators constructed from the SM fields and the "spurions" Y are invariant under $SU(3)^3_q$. In MFV the dynamics of flavor transitions is then completely controlled by the pattern of the Yukawa couplings; in particular all **CP** violation originates from the KM phase (ignoring the possibility of $\bar{\theta} \neq 0$ in QCD).

The effective transition operators for beauty, strange and charm decays get the leading terms arise from the dimension-6 level. Let us focus on $\Delta F = 2$ transitions for mesons made up from down-type quarks, namely **CP** odd contributions to $K^0 - \bar{K}^0$ and all aspects from $B^0_{(s)} - \bar{B}^0_{(s)}$ oscillations. Those are driven by four quark operators involving quark bilinears $\bar{Q}_L Y_U Y_U^\dagger Q_L$. For most practical purposes we can use an approximate expression for the Y. Noting that all SM Yukawa couplings are small except for top quarks, the leading non-diagonal element arises from the contraction of two Y_U, which transforms as (8,1,1) under $SU(3)^3_q$. The only sizable elements of the matrix $Y_U Y_U^\dagger$ are: $(\lambda_{FC})_{ij} = (Y_U Y_U^\dagger)_{ij} \simeq \lambda_t^2 V_{3i}^* V_{3j}$, $i \neq j$. ΔS and $\Delta B = 2$ operators are controlled by a single term. Since this effective operator has operator dimension six, we expect it appears in the OPE with a coefficient proportional to $1/\Lambda^2$.

From the measured values of ϵ_K, ΔM_{B^0} and $\Delta M_{B^0_s}$ one infers $\Lambda > 5$ TeV [95%] with similar bounds of a few TeV from other process like $B \to \gamma X_s, l^+l^- X_s$. With MFV SM-like effects can be due to relatively low scales of ND. For scenarios that are not of the MFV variety one can actually infer bounds higher by an order of magnitude, namely several 10 TeV.

In all these exercises one simplifying assumption has been made out of necessity, namely there is only a single ND contribution to observables; i.e., no allowance is made for cancellations between different possible contributions. Even with that caveat one cannot simply equate the scale Λ with the mass scale of new states. For the couplings g_X of such a state X to the quarks (or the SM fields in general) might be suppressed in a flavor universal way $1/\Lambda^2 \simeq g_X^2/M_X^2$ thus opening the door for even $M_X \ll \Lambda$.

4.6 *Beyond* the SM: Pontecorvo-Maki-Nakagawa-Sakata matrix

4.6.1 *Short history of the neutrino*

The history of neutrino dynamics is amazing with many lessons and unexpected results at different levels. In the twenties it was already known that the energy spectrum of electrons emitted in the transition of a nucleus into another one (with atomic number differing of ± 1 and the same mass number) – β decay–was continuous. This of course was puzzling, because the single emitted electron should have a well-defined energy while recoiling off a nucleus. Though it was discovered much later, we take as an example the simplest case of a free neutron. If we assume its decay is $n \to pe^-$, its e^- energy should have a fixed value, which is at variance with facts. Furthermore, it would break angular momentum conservation and violate fermion number as $\text{N}_{\text{fermion}} = 1 \to \text{N}_{\text{fermion}} = 2$.

The existence of neutrinos was hypothesized by W. Pauli as a "desperate remedy". By daring so he showed the power of symmetries: 'we' need the existence of very light neutral fermion in order to save energy and angular momentum conservation and describe the phenomenology of nuclear β-decay, where the missing energy and spin can be carried off by a neutral particle, which escapes direct detection.

At first the existence of neutrino was not suggested in papers or talks. On 4 December 1930, Pauli wrote a letter to the Physical Institute of the Federal Institute of Technology, Zürich, in which he proposed the electron neutrino as a potential solution to solve the problem of the continuous beta decay spectrum. An excerpt of the letter reads:

"Dear radioactive ladies and gentlemen,
As the bearer of these lines [...] will explain more exactly, considering [...] the continuous β-spectrum, I have hit upon a desperate remedy to save [...] the energy theorem. Namely [there is] the possibility that there could exist in the nuclei electrically neutral particles that I wish to call 'neutrons', which have spin 1/2 and obey the exclusion principle, and additionally differ from light quanta in that they

do not travel with the velocity of light [...] The continuous β-spectrum would then become understandable by the assumption that in β decay a 'neutron' is emitted together with the electron, in such a way that the sum of the energies of neutron and electron is constant. [...]
But I don't feel secure enough to publish anything about this idea, so I first turn confidently to you, dear radioactives, with a question as to the situation concerning experimental proof of such a 'neutron' [...] I admit that my remedy may appear to have a small a priori probability because 'neutrons', if they exist, would probably have long ago been seen. However, only those who wager can win, and the seriousness of the situation of the continuous β-spectrum can be made clear by the saying [...] 'One does best not to think about that at all, like the new taxes.' [...] So, dear radioactives, put it to test and set it right. [...]
With many greetings to you [...], your devoted servant, W. Pauli".[37]

Neutrons as part of nuclei have been discovered by Chadwick [138] only just in 1932, two years later. Of course, neutrons are heavy and feel strong forces, so they were immediately recognized as something different from the Pauli 'neutrons'. The word 'neutrinos' entered scientific vocabulary by 1932 due to the Italian culture used by Fermi and also Pauli; the word of "ino", in Italian language, is associated to smallness, and it seemed more appropriate for the proposed undetected particle.

In 1934 Enrico Fermi published his crucial contribution to the understanding of weak interactions and neutrino physics [139] in Italian in a relatively small journal and later in Germany [140], having his paper been rejected by the journal *Nature*. For a long time it was not understood the crucial impact of the 1934 paper of Fermi for the theory of β-decay and in general for weak forces. It is based on four-fermion interactions in the case of $n \to pe^-\bar{\nu}$ with thr coupling G_F. It is a very successful example of an effective quantum field theory, which captures most of neutrino physics at low energy when complemented with the V-A structure of the currents, which was done in the 50's. The interaction cross section of neutrinos with nuclei was estimated by Bethe and Peierls being extremely small ($\sigma < 10^{-44}$ cm^2) [141]. This number explained why the Pauli particle escaped detection until that time and put a formidable task to experimentalists. Present data lead to $G_F/(\hbar c)^3 = 1.1663787(6) \cdot 10^{-5}$ GeV^{-2}.

The features of neutrino dynamics are mostly obvious for 'fans' of invariances:

- Its mass must be very light at least, being the end point of the electron spectrum, very close to mass difference of neutron, proton and electron.
- It carries no electric charge.
- It is insensitive to strong forces and it has no color.
- It is sensitive for weak dynamics.[38]
- It is a fermion with spin $\frac{1}{2}$.

[37] There was a much more radical suggestion from the giants of quantum mechanics, namely to give up on conservation of energy and spin in $n \to pe^-$; later in time experiments closed that 'road'.
[38] Later we will shortly comment about sterile neutrinos.

For many years after Fermi's theory, the neutrino was considered as an undetectable particle. In 1946, the Italian physicist Bruno Pontecorvo,[39] a former student of Enrico Fermi, was the first to challenge this opinion [142]. Thanks to his inspiration, to the development of nuclear technology and to the invention of the delayed coincidence technique [143], F. Reines and C.L. Cowan [144] finally proved experimentally the existence of neutrinos (actually, electron anti-neutrinos). Pontecorvo in 1957 marks again the beginning of a new era in neutrino physics mentioning for the first time the possibility of neutrino oscillations [145],[40] although first as neutrino-anti-neutrino oscillations. At that time only one type of neutrino was known and the theory of a massless two component neutrino, according to which only a left-handed neutrino ν_L and right-handed antineutrino $\bar{\nu}_R$ exist,[41] with transitions between them forbidden by the conservation of the total angular momentum, had just received strong support by the measurement of neutrino helicity [146]. Pontecorvo realized that unusual transitions can happen only with two conditions at the same time: a neutrino has a non-zero mass, and lepton numbers are broken by two units, namely $\nu \rightleftharpoons \bar{\nu}$. To be more precise: neutrino flavor oscillations can happen; one can measure that *lepton flavor eigenstates* are produced, but then move forward as *mass eigenstate* or "to propagate" (in the vacuum).

In 1962 an experiment carried on by L.M. Lederman, M. Schwartz and J. Steinberger showed that ν_e and ν_μ are different states in quantum mechanics. Quite immediately, Z. Maki, M. Nakagawa and S. Sakata discussed the possibility that the two known flavors were a mixture of two neutrino mass eigenstates [147]. In 1967/8 the first phenomenological model for ν_e and ν_μ mixing and oscillations was worked out by Pontecorvo [148] and Gribov & Pontecorvo [149].

Now we know that we have three neutrino flavors with different masses thanks to a long effort started in the late 60s with solar [150],[42] atmospheric, reactor and accelerator neutrinos, which have finally proven the existence of oscillations [151–153], both in vacuum and in matter [154]. The large mixing angles (much bigger than those mixing the quark sector [155,156]) give a chance to find **CP** asymmetries in the lepton sector, as discussed below. This is similar to what happens in the quark sector, but the 3×3 matrix that connects flavor vs. matrix eigenstates of neutrinos is very different, as we will discuss, even if the neutrinos are 'only' Dirac fermions.[43] With Majorana neutrinos the landscape can be even richer, though not affecting neutrino oscillation phenomena.

[39] While both Cabibbo and Pontecorvo worked on 'mixing' – the first about quarks and the latter about leptons – they followed different 'roads': both started their careers in research in the center of Rome, but Cabibbo remained in Rome, while Pontecorvo did it in the Soviet Union. We should mention that Pontecorvo was about twenty years older than Cabibbo.

[40] A pioneering paper by Lee and Yang was published in 1956 to probe parity violation in weak forces [24], and in 1957 it was established [25], see **Sect. 2.8**; It was an amazing time both in the worlds of baryons and neutrinos, although that connection was not realized then.

[41] As said above, **CPT** invariance tells us that.

[42] In **Chapter 9** we will show the impact of the Homestake experiment due to R. Davier Jr.

[43] The analogous of the CKM matrix in the neutrino sector is the so-called PMNS (Pontecorvo, Maki, Nakagawa and Sakata) matrix. In the two matrices one 'pattern' is the same: the first person is Italian, while the others are Japanese; furthermore they did not meet in person.

Fermi got a well-deserved Nobel prize in 1938, the quoted reason was "for his demonstrations of the existence of new radioactive elements produced by neutron irradiation, and for his related discovery of nuclear reactions brought about by slow neutrons"; likewise W. Pauli in 1945 about "for the discovery of the Exclusion Principle, also called the Pauli Principle". Neither motivation mentioned neutrinos. Experimental achievements were awarded with Nobel awards:

- 1988: L.M. Lederman, M. Schwartz, J. Steinberger[44] "for the neutrino beam method and the demonstration of the doublet structure of the leptons through the discovery of the muon neutrino";
- 1995: F. Reines "for the detection of the neutrino" and "for pioneering experimental contributions to lepton physics";[45]
- 2002: R. Davis, Jr., M. Koshiba "for pioneering contributions to astrophysics, in particular for the detection of cosmic neutrinos";
- 2015: T. Kajita, A. B. McDonald "for the discovery of neutrino oscillations, which shows that neutrinos have mass".

4.6.2 *The Pontecorvo-Maki-Nakagawa-Sakata (PMNS) ansatz*

In the SM there are three neutrinos belonging to doublets of weak interaction: ν_i with $i = e, \mu, \tau$ are considered massless without good reason; again, 'par ordre du mufti'. The discovery of neutrino oscillations represented a powerful and elegant way to establish a non-vanishing mixing matrix between different weak eigenstates in the leptonic sector, namely $m_i^2 - m_j^2 \neq 0$ with $i \neq j$.[46]

There are two different classes of neutrinos with non-zero masses. If the neutrinos are Dirac fermions, the discussions of mixing in the lepton sector mimics the one in the quark sector *in principle*, see Eq. (4.36). The mixing is parametrized by an unitary mixing matrix – analogous to the CKM matrix in the quark sector – called the PMNS (Pontecorvo-Maki-Nakagawa-Sakata) matrix. The induced transformation

$$\begin{pmatrix} \nu_e \\ \nu_\mu \\ \nu_\tau \end{pmatrix} = \mathbf{V}_{\text{PMNS}} \begin{pmatrix} \nu_1 \\ \nu_2 \\ \nu_3 \end{pmatrix} \tag{4.70}$$

gives different massive stationary states with masses m_1, m_2 and m_3. Thus the counting of independent parameters of the PMNS matrix is analogous to the counting of independent parameters made for the CKM; therefore we are at liberty to employ the same formalism. The 3×3 PMNS matrix can be parameterized by three

[44] In 1949 Jack Steinberger was first refereed to as a theorist [96].
[45] Cowan-Reines neutrino experiment shows that scientific leaders need a long life time span to get an award; Cowan passed away in 1974, while their results were published in 1956.
[46] The value of $m_i^2 = m_j^2 \neq 0$ does not lead to oscillations.

Euler mixing angles θ_{12}, θ_{23} and θ_{13} and by a **CP** violating phase $\delta_{\mathbf{CP}}^{\text{lept}}$, which is an observable. It can be decomposed into three matrices:

$$\mathbf{V}_{\text{PMNS}} = \begin{pmatrix} 1 & 0 & 0 \\ 0 & c_{23} & s_{23} \\ 0 & -s_{23} & c_{23} \end{pmatrix} \begin{pmatrix} c_{13} & 0 & s_{13}e^{-i\delta_{\mathbf{CP}}^{\text{lept}}} \\ 0 & 1 & 0 \\ -s_{13}e^{i\delta_{\mathbf{CP}}^{\text{lept}}} & 0 & c_{13} \end{pmatrix} \begin{pmatrix} c_{12} & s_{12} & 0 \\ -s_{12} & c_{12} & 0 \\ 0 & 0 & 1 \end{pmatrix} \tag{4.71}$$

The relatively large difference between the two mass scales measured by neutrino oscillation experiments makes the three matrix shown in Eq. (4.71) quite directly connected to different class of experiments.[47] The angle θ_{12} and the squared mass difference Δm_{21}^2 are mostly measured by means of solar neutrino data, θ_{23} and Δm_{32}^2 by means of atmospheric neutrinos and long baseline neutrino experiments, while θ_{13} was obtained by short distance ($\approx$1 km) reactor experiments and T2K experiment. Recent analysis of many experiments are done with a full three flavor analysis, so we can expect that most next generation experiments will be sensitive to the whole PMNS matrix, although with different and complementary sensitivity, depending on energy, neutrino type, detection technique and source-detector base distance. We will discuss it further in **Chapter 9**, where we will also introduce the *Mikheyev-Smirnev-Wolfenstein effect.* Let us remark that neutrino oscillations happen not only in vacuum, but also in matter. There are crucial examples, obtained integrating the source's position inside the Sun's core or the distances from the reactors, chiefly from the KamLAND experiment.

If the neutrinos are Majorana fermions, the 'situations' are very different. The 1937 paper by Ettore[48] Majorana [157] was basically ignored first; much later his work had large impact on neutrino physics.[49] Now his name has entered the scientific literature as a well-known one – actually as a 'hot' item.

One can add a Majorana mass term to the Lagrangian of the form

$$\mathcal{L}_{ML} = -\frac{1}{2}\,\overline{\hat{\nu}_R^C}\; m_{ML}\; \hat{\nu}_L + h.c. \tag{4.72}$$

where m_{ML} is a 3×3 matrix and $\hat{\nu} \equiv (\hat{\nu}_e, \hat{\nu}_\mu, \hat{\nu}_\tau)$ are the neutrino states whose left-handed components enter the weak Lagrangian; the overline C indicates the conjugate state. The states that diagonalize the mass terms in the lepton sector are obtained by a field redefinition, that for the left-handed neutrino corresponds to

$$\widetilde{\nu}_L = U_L^{\nu\,\dagger}\hat{\nu}_L \tag{4.73}$$

[47]The scale dominating solar neutrino and long distance reactor experiments is usually called $\Delta m_{21}^2 = m_2^2 - m_1^2$, being ν_1 and ν_2 the two mass eigenstates with mass $m_1 < m_2$ by definition. A significantly larger mass difference dominates atmospheric neutrinos and long baseline accelerator experiments, usually called $\Delta m_{32}^2 = m_3^2 - m_2^2$. Data say that $\Delta m_{21}^2 \simeq 7.5 \cdot 10^{-5}$ eV^2 and $|\Delta m_{32}^2| \sim 2 \cdot 10^{-3}$ eV^2. The fact that they are so different has been extremely beneficial for neutrino physics, allowing two-flavor analysis to be sufficient until recently.

[48]The Italian name 'Ettore' of 'Hector' in Greek mythology might be a bad choice for a first name.

[49]Majorana was a complex person. His colleagues in Rome gave him the nickname 'the Grand Inquisitor'. One can – and should – read a book with the title "The Pope of Physics" [158].

where $\tilde{\nu} \equiv (\nu_1, \nu_2, \nu_3)$ are the Majorana mass eigenstates. In analogy to the quark case, the leptonic weak charged current contains the PMNS unitary matrix

$$\mathbf{U}_{\text{PMNS}} = U_L^{l\,\dagger}\, U_L^{\nu} \tag{4.74}$$

that connects mass eigenstates and weakly interacting left-handed states

$$\nu_L = \mathbf{U}_{\text{PMNS}}\, \tilde{\nu}_L \tag{4.75}$$

where $\nu \equiv (\nu_e, \nu_\mu, \nu_\tau)$ are the states participating to the weak interactions after the diagonalization of the mass terms.

The counting of the phases of the PMNS matrix changes. As said above, in the Dirac case for a complex $n \times n$ unitary matrix, we have $n\,(n+1)/2$ phase parameters. One may eliminate $2\,n-1$ parameters by a re-phasing of the quark fields, and obtain $(n-1)\,(n-2)/2$ independent phases (that is, one independent phase for three families). This counting includes the total lepton number, which remains a symmetry of the massive theory and thus cannot be used to reduce the number of physical parameters in the mass matrix. In the Majorana case, there is no independent right-handed neutrino field, nor is lepton number a good symmetry. With other words: Eq. (4.72) is *not* invariant under a global $U(1)$ transformation. Therefore, one can eliminate only n parameters, while $(n^2 - n)/2$ independent phases remain. If one has two families of quarks and Dirac leptons, then no **CP** asymmetries can happen. On the other hand, **CP** violation could be produced even with only two families if neutrinos are of Majorana type [157].

With three families of quarks and Dirac leptons, we get a single source of **CP** violation for quarks and another one for leptons. In a world of Majorana neutrinos we get three phases, so the **CP** violating landscape is richer. However, it should be noted that the connections of the two Majorana phases with neutrino observables are subtle. Most experiments, and chiefly oscillation experiments, do not distinguish Dirac and Majorana neutrinos. Other details will be discussed in **Chapter 9**.

A standard parametrization of the mixing matrix for Majorana neutrinos $\mathbf{U}_{\text{PMNS}}$ is given by the parameterization in terms of the product of two matrices: one with the usual CKM parameterization (Dirac case) $\mathbf{V}_{\text{PMNS}}$ and the other one containing additional phases α_1, α_2 and α_3:

$$\mathbf{U}_{\text{PMNS}} = \mathbf{V}_{\text{PMNS}}(\theta_{12}, \theta_{13}, \theta_{23}, \delta_{\mathbf{CP}}^{\text{lept}}) \begin{pmatrix} e^{i\alpha_1} & 0 & 0 \\ 0 & e^{i\alpha_2} & 0 \\ 0 & 0 & e^{i\alpha_3} \end{pmatrix} \tag{4.76}$$

The phase of $\delta_{\mathbf{CP}}^{\text{lept}}$ can be measured and will be discussed in **Chapter 9**. Only the differences between phases α_i can be probed; thus one phase can be set to zero without loss of generality (like $\alpha_1 = 0$). Only the two $\alpha_{2,3}$ phases – referred to as Majorana phases – are physical observables.[50]

[50]The diagonal phase matrix is nonphysical in the charged-current interaction because it can be absorbed into the charged lepton fields.

4.6.3 CP *violation*

There are three important comments one can make on neutrino time evolution:

- The weak interaction eigenstates ν_e, ν_μ and ν_τ are non-trivial linear combinations of the mass eigenstates ν_i and their time dependence are governed by

$$|\nu_\alpha(t)\rangle = \sum_{i=1}^{3} \mathbf{U}_{\alpha i} e^{-iE_i t}|\nu_i\rangle \qquad \alpha = e, \mu, \tau \tag{4.77}$$

- The transition probabilities for $\nu_\alpha \to \nu_\beta$ are obtained by evaluating $|\langle\nu_\beta|\nu_\alpha(t)\rangle|^2$. It is best to translate 'time' into 'space'; i.e., to express these probabilities as a function of the *distance* L from the production point rather than time t, since one can control L directly. For $m_i \ll E$ one gets:

$$\begin{aligned} P(\nu_\alpha \to \nu_\beta; L) = \delta_{\alpha\beta} - 4\sum_{i>j} \sin^2\left(\frac{m_i^2 - m_j^2}{4} \cdot \frac{L}{E}\right) \cdot \mathrm{Re}[V_{\alpha i} V^*_{\alpha j} V^*_{\beta i} V_{\beta j}] \\ +2\sum_{i>j} \sin\left(\frac{m_i^2 - m_j^2}{4} \cdot \frac{L}{E}\right) \cdot \mathrm{Im}[V_{\alpha i} V^*_{\alpha j} V^*_{\beta i} V_{\beta j}] \,. \end{aligned} \tag{4.78}$$

  Mostly we focus on $\mathbf{V}_{\mathrm{PMNS}}$, not $\mathbf{U}_{\mathrm{PMNS}}$; we will explain soon why.
  The probability depends on the magnitude of $\Delta m_{ij}^2 \equiv m_i^2 - m_j^2$ and the PMNS parameters. Indeed, for ultra-relativistic neutrinos, when $m_i \ll E$, it is useful to make the following approximation $E_k - E_j \simeq (m_k^2 - m_j^2)/2E$, where $E = p$ is the neutrino energy, neglecting the mass contribution. In an experiment one determines L/E independently (or varies it in a controlled way).
- There are three classes of transitions one can probe:
  (a) "Appearance" reactions:

$$P(\nu_\alpha \to \nu_\beta; L) > 0 \;\text{ for }\; \alpha \neq \beta \,. \tag{4.79}$$

  (b) "Disappearance" reactions:

$$P(\nu_\alpha \to \nu_\alpha; L) < 1 \,. \tag{4.80}$$

  (c) There is another 'road' for the 'disappearance': that could be caused by their decays. This would require the intervention of ND as well; one can distinguish these two scenarios by their dependence on L (or t).

When one probes neutrinos oscillations, it does not matter whether neutrinos have Dirac or Majorana nature; they follow

$$\begin{aligned} P(\nu_\alpha \to \nu_\beta) = |\langle\nu_\beta|\nu_\alpha(t)\rangle|^2 &= \left|\sum_{k=1}^{3} \mathbf{V}_{\beta\mathbf{k}}\, \mathbf{e}^{-\mathbf{i}\mathbf{E}_\mathbf{k}\mathbf{t}}\, \mathbf{V}^*_{\alpha\mathbf{k}}\right|^2 \\ &= \left|\sum_{k=1}^{3} \mathbf{U}_{\beta\mathbf{k}}\, \mathbf{e}^{-\mathbf{i}\mathbf{E}_\mathbf{k}\mathbf{t}}\, \mathbf{U}^*_{\alpha\mathbf{k}}\right|^2 \;; \end{aligned} \tag{4.81}$$

likewise for antineutrino transition probabilities: $P(\bar{\nu}_\alpha \to \bar{\nu}_\beta) = |\langle \bar{\nu}_\beta | \bar{\nu}_\alpha(t) \rangle|^2$. Majorana phases play no role in neutrino oscillations.

For three families of Dirac leptons, the PMNS matrix mimics the CKM matrix describing quark mixing: the independent parameters can be expressed as three leptonic angles and one leptonic **CP** violating phase. We know that the values of the PMNS matrix are very different from the CKM matrix; one is not surprised, since in the SM these values are arbitrary.

The probability transition and, as a consequence, the **CP** asymmetries, are independent from Majorana phases, being affected only by possible irreducible complex phases in the **V** matrix.

The simplest PMNS measure of **CP** violation – which is equivalent to **T** violation if **CPT** is conserved – would be the difference of oscillation probabilities between neutrinos and anti-neutrinos:

$$\Delta^{\mathbf{CP}}_{\alpha\beta} \equiv P(\nu_\alpha \to \nu_\beta) - P(\bar{\nu}_\alpha \to \bar{\nu}_\beta) = |\langle \nu_\beta | \nu_\alpha(t) \rangle|^2 - |\langle \bar{\nu}_\beta | \bar{\nu}_\alpha(t) \rangle|^2 \,. \tag{4.82}$$

A **CP** asymmetry can be written as

$$\begin{aligned} \Delta^{\mathbf{CP}}_{\alpha\beta} &= 4 \sum_{k>j} \mathrm{Im}\,(\mathbf{V}^*_{\alpha\mathbf{k}}\, \mathbf{V}_{\beta\mathbf{k}}\, \mathbf{V}_{\alpha\mathbf{j}}\, \mathbf{V}^*_{\beta\mathbf{j}}) \sin\left(\frac{\Delta m^2_{kj}}{2} \cdot \frac{L}{E}\right) = \\ &= 4 \sum_{k>j} J^{jk}_{\alpha\beta} \sin\left(\frac{\Delta m^2_{kj}}{2} \cdot \frac{L}{E}\right) . \end{aligned} \tag{4.83}$$

The quantity $J^{jk}_{\alpha\beta}$ presents several properties:

- the antisymmetric relations

$$J^{jk}_{\alpha\beta} = -J^{jk}_{\beta\alpha} \quad , \quad J^{jk}_{\alpha\beta} = -J^{kj}_{\alpha\beta} \tag{4.84}$$

  hold for every α, β and for every k, j, as one can immediately verify;
- From the unitary of the matrix **V** we have

$$\sum_j J^{jk}_{\alpha\beta} = \sum_\alpha J^{jk}_{\alpha\beta} = 0 \,. \tag{4.85}$$

  The index β runs over the 3 flavors e, μ, τ.[51]
- The cyclic relations

$$J^{jk}_{e\mu} = J^{jk}_{\mu\tau} = J^{jk}_{\tau e} \tag{4.86}$$

  are obtained by using the previous relations.

All $J^{jk}_{\alpha\beta}$ are equal in absolute value and can differ by a sign. Let us focus on the case of three massive neutrinos and set (by convention):

$$J \equiv J^{12}_{e\mu} \tag{4.87}$$

[51] If the number of massive neutrinos is bigger than the number of flavor neutrinos as a consequence, e.g., of a flavor neutrino - sterile neutrino mixing, the index β is bound to include other indexes, the ones relative to the sterile state.

For all non-zero elements $J^{jk}_{\alpha\beta}$ the following relation holds

$$J^{jk}_{\alpha\beta} = \pm J \tag{4.88}$$

where the sign depends on the indexes. The quantity J is analogous to what is called the Jarlskog invariant in the quark sector [132]: again, J is independent by changes of arbitrary phases and parametrization in the $\mathbf{V}_{\rm PMNS}$ matrix. Following Eqs. (4.38,4.39,4.71) used for the CKM matrix one can write J for the PMNS matrix:

$$J = c_{12}c_{23}c^2_{13}s_{12}s_{23}s_{13}\sin(\delta_{\rm lept}) = \frac{1}{8}\sin(2\theta_{12})\sin(2\theta_{13})\sin(2\theta_{23})\cos(\theta_{13})\sin(\delta^{\rm lept}_{\mathbf{CP}}) \,. \tag{4.89}$$

Of course, the situation is very different between quarks and leptons. Yet, the patterns are not so different, as long as we discuss Dirac leptons.

Using Eq. (4.83) we get

$$\Delta^{\mathbf{CP}}_{\alpha\beta} = \pm 4J\left[-\sin\left(\frac{\Delta m^2_{21}}{2}\cdot\frac{L}{E}\right)+\sin\left(\frac{\Delta m^2_{31}}{2}\cdot\frac{L}{E}\right)-\sin\left(\frac{\Delta m^2_{32}}{2}\cdot\frac{L}{E}\right)\right] \tag{4.90}$$

where the $\pm$ sign depends on the indexes α and β. In particular, one has

$$\begin{aligned}\Delta^{\mathbf{CP}}_{e\mu} &= \Delta^{\mathbf{CP}}_{\mu\tau} = -\Delta^{\mathbf{CP}}_{e\tau}\\ &= 4J\left[+\sin\left(\frac{\Delta m^2_{21}}{2}\cdot\frac{L}{E}\right)+\sin\left(\frac{\Delta m^2_{32}}{2}\cdot\frac{L}{E}\right)+\sin\left(\frac{\Delta m^2_{13}}{2}\cdot\frac{L}{E}\right)\right]\\ &= -16J\left[\sin\left(\frac{\Delta m^2_{21}}{4}\cdot\frac{L}{E}\right)\cdot\sin\left(\frac{\Delta m^2_{32}}{4}\cdot\frac{L}{E}\right)\cdot\sin\left(\frac{\Delta m^2_{13}}{4}\cdot\frac{L}{E}\right)\right]\,.\end{aligned} \tag{4.91}$$

The last equality comes from the identity $\sin 2\theta+\sin 2\alpha+\sin 2\beta = -4\sin\theta\sin\alpha\sin\beta$, which holds when $\theta+\alpha+\beta = 0$, as it can easily be verified by using the trigonometric formulas of addition. It is clear that the parameter J controls the magnitude of **CP** violation effects in neutrino oscillations in the case of three-neutrino mixing.

4.6.4 *Two generation case*

If the neutrinos are Dirac ones in the two generation case, the PMNS matrix is described by a Euler-type angle θ, and its pattern is analogous to the Cabibbo matrix for the quarks, but not quantitatively, and even semi-quantitatively. One gets with $\alpha \neq \beta$:

$$\begin{aligned}P(\nu_\alpha\to\nu_\beta) &= \frac{1}{2}\sin^2(2\theta)\left[1-\cos\left(\frac{\Delta m^2}{2}\cdot\frac{L}{E}\right)\right]\\ &= \sin^2(2\theta)\sin^2\left(\frac{\Delta m^2}{4}\cdot\frac{L}{E}\right) = P(\bar\nu_\alpha\to\bar\nu_\beta)\end{aligned} \tag{4.92}$$

The survival probability is easily obtained here:

$$P(\nu_\alpha\to\nu_\alpha) = 1-P(\nu_\alpha\to\nu_\beta) = 1-\sin^2 2\theta\sin^2\left(\frac{\Delta m^2}{4}\cdot\frac{L}{E}\right) = P(\bar\nu_\alpha\to\bar\nu_\alpha)$$

CP asymmetry cannot happen for Dirac neutrinos with two generation case. In principle, the situation would be quite different with Majorana neutrinos, allowing the possibility of non-zero **CP** asymmetry – but how could one find it? It would be a true challenge.

4.6.5 *Three generations case*

CP violation may indeed occur, and it is actually expected in the three generation case. We now know that all three mixing angles in the PMNS matrix are not zero and are actually much larger that those of the CKM matrix, so **CP** violation in the lepton sector is indeed possible and will likely be discovered in the next decade.

The basic formula for three generations is Eq. (4.91). It shows that **CP** violation depends on J (Eq. (4.89)); i.e., it is relevant only when all angles are not too small and depends on three factors of the type $\sin[(\Delta m^2_{ij}/4) \cdot (L/E)]$. This means that the effect is measurable only if the neutrino energy and the oscillation distance are chosen properly, so that none of these kinematic factors is too small. Practically, this is possible only by means of long baseline neutrino beams with appropriate energy and oscillation distance. The only neutrino beams available so far are muon neutrino beams, so the only transitions which are sensitive to **CP** violation effects are in practice $P(\nu_\mu \longrightarrow \nu_e)$ and $P(\bar{\nu}_\mu \longrightarrow \bar{\nu}_e)$. It must be stressed, however, that matter effects, i.e. resonant matter induced neutrino oscillations due to the propagation of the neutrino in electronic matter, make the analysis and the extraction of **CP** violation subtle. This matter will be discussed in **Chapter 9**.

4.6.6 *Majorana fermions in particle and nuclear physics*

Single beta decays of nuclei $A^N_Z \to A^{N-1}_{Z+1} + \beta^- + \bar{\nu}$ have been measured for a long time. In many cases those are energetically forbidden for elements with *even* atomic number Z and *even* neutron number N. Still double beta decays happen

$$A^N_Z \to A^{N-2}_{Z+2} + 2\beta^- + 2\bar{\nu} \,, \tag{4.93}$$

but their rates are greatly inhibited. It is a true achievement to measure them. These do *not* tell us, whether the neutrinos are Dirac or Majorana ones.

The next step is to probe neutrino-*less* double decays

$$A^N_Z \to A^{N-2}_{Z+2} + 2\beta^- \,. \tag{4.94}$$

The existence of these decays would mean that neutrinos are Majorana fields [157] – leading to *effective* Majorana neutrino masses $\langle m_{\beta\beta} \rangle$. None has been found yet. It is a true challenge, but our community has to continue this search[52]; we will come back to it in **Sect. 9.5.1**.

Often physics can connect different regions that one might not think about before. A good example was written down as a 2015 'Colloquium: Majorana fermions in nuclear, particle, and solid-state physics' [161]. There is a crucial question: are neutrinos "Dirac" or "Majorana" ones? Its influence does not stop with particle physics, even though that was the original consideration. The formalism developed has found many uses in cosmology, HEP and Nuclear Physics – and solid-state physics, including superconducting. The equations that were derived also arise in

[52] There is an important review article from 2008 [159]. See also Ref. [160].

solid-state physics to describe electronic states in materials with superconducting order, in particular in one and two dimensions at zero energy, called Majorana zero modes. They are endowed with remarkable physical properties that might lead to quantum computing. They are investigated experimentally. Our main comments on Majorana fermions in particle dynamics: (a) The backgrounds in HEP are very different from solid-matter dynamics. (b) A possible impact of Majorana particles and anti-particles are basically the same. Majorana zero modes could emerge in a class of systems called "topological superconductors". It is not easy to follow the arguments in different regions, but it is interesting.[53]

4.6.7 *Short summary of neutrino dynamics*

Let us summarize the main lessons of this **Section**:

- Data have told us that the masses of three neutrinos are *not* degenerate and their mixing angles are *not* zero. Thus neutrino physics is a hunting region for ND!
- **CP** violation in neutrino oscillations can*not* provide information whether they are of Dirac or Majorana natures, see Eq. (4.81).
- **CP** violation cannot show up in "disappearance" experiments, where one flavor converts to the same one, but only in "appearance" experiments, where one flavor oscillates into another one and the detector is flavor sensitive.
- Neutrinos represent another 'road' for probing the impact of Dark Matter, that we will discuss in **Sect. 15.1**.

By drawing on analogies with $K^0-\bar{K}^0$ and $B^0_{(s)}-\bar{B}^0_{(s)}$ oscillations, we can formulate, in principle, the essence of neutrino oscillations, although the situations are very different. In the SM we describe the $\Delta L = 1$ current $J_\mu = \bar{\ell}\gamma_\mu(1-\gamma_5)\nu_\ell$ with $\ell = \tau, \mu, e$ and massless ν_ℓ. That would hold also for 'massive' neutrinos, in principle, as long these neutrinos have the same mass. In our Universe, where the ν_ℓ have different masses, one can describe the coupling of ν_ℓ with a charged lepton ℓ as

$$J_\mu = \bar{\ell}^{\rm flav}\gamma_\mu(1-\gamma_5)\nu_\ell^{\rm flav} = \bar{\ell}^{\rm m}\gamma_\mu(1-\gamma_5)\,\mathbf{U}_{\rm PMNS}\,\nu_\ell^{\rm m}\,. \tag{4.95}$$

4.7 History of Higgs dynamics and its future

Crucial papers by Higgs and by Brout and Englert were published in 1964; they pointed out the 'road' to complete the SM based on $SU(3)_C\times SU(2)_L\times U(1)_Y$ gauge symmetry. The 'Higgs' boson has been found by ATLAS and CMS in 2012 [4].[54]

[53] A 2016 article [162] is also interesting; maybe its statements are not truly novel, but they show one has to be careful on how to apply to **CPT**, **CP** and **C** symmetries to fermions and to probe that with external fields. Ref. [162] gives a good list of references that will help a reader to get a deeper understanding about this situation.

[54] To be precise: these collaborations found a spin-zero boson that is at least mostly a scalar one, while a pseudo-scalar one is at best a non-leading one.

P. Higgs and F. Englert got the Nobel prize on Dec. 10, 2013; Englert's co-researcher R. Brout had passed away in 2011. Many people had sizably contributed to the evolution of Higgs dynamics, but one can say there are three other pioneers: Guralnik, Hagen and Kibble [163].[55]

Actors in Higgs dynamics play in different stages, namely when gauge symmetries write the story, Yukawa dynamics connects Higgs and fermion fields. It has been suggested in the 1960s: "Higgs" dynamics can do the 'job', never mind whether it is a piece of art or not. Establishing the existence of a boson with spin zero has been a great achievement on the experimental side.

Theoretical engineering allows the neutral Higgs field ϕ to acquire a vacuum expectation value $\langle\phi\rangle \simeq$ (VEV). Minimizing an Higgs potential of the type $V_H = M_\phi^2|\phi|^2 + \lambda|\phi|^4$ with $M_\phi^2 < 0$ leads to $\langle\phi\rangle = \sqrt{-M_\phi^2/2\lambda}$. From the measured Fermi constant G_F one infers $\langle\phi\rangle \simeq 174$ GeV for the neutral component of the $SU(2)_L$ doublet; therefore one expects $|M_\phi| \sim \mathcal{O}(100)$ GeV.

In QFT one has to include loop corrections for observables. Most of them have logarithmic dependence on their scales, and we know how to deal with that. In some cases the situation is quite different and we talk about *scalar* masses, that depend on the *square* of the UV cut-off $\Lambda_{\rm UV}$:

$$\Delta M_\phi^2 \propto \Lambda_{\rm UV}^2 + ... ; \tag{4.96}$$

That introduces a quadratic infinity in the scalar mass. It can be removed through the renormalization process. Mathematically this can be arranged for by introducing counter-terms into the Lagrangian. Yet physicists tend to view such a situation as very contrived requiring extreme *fine tuning.* In their view $\Lambda_{\rm UV}^2$ is *not* merely a mathematical 'place holder' for an infinity. It parametrizes the sensitivity of an observable – in this case the scalar mass – to the dynamics being relevant at (very) high energy scales. It implies that the dynamics being effective at the electroweak scale of about 100 GeV had to be highly cognizant of whether new layers of dynamics enter at much higher scales, close to Planck scale $\sim \mathcal{O}(10^{19})$ GeV or 'at least' to GUT scales $\sim \mathcal{O}(10^{15})$ GeV, to achieve the needed cancellation. Often this is referred to as the 'gauge hierarchy problem' of the SM. Indeed, it is a problem, at least for many theorists: the SM "Higgs" is 'un-natural'.

In **Chapter 14** we will discuss about several strengths of SUSY – one of them is that it fixes the naturalness problem of the Higgs mass in the Standard Model. This happens through cancellations of $\Lambda_{\rm UV}^2$ terms done by the bosons, SUSY partners of the fermions, in the scalar mass quantum corrections due to loops. Never mind

[55]It could be called the 'Englert-Brout-Higgs-Guralnik-Hagen-Kibble' mechanism; however, that is too long even for physicists with ambition for literature.

there is no sign that SUSY has been found in the data so far. With the quadratic Higgs mass renormalization removed by SUSY, electroweak symmetry breaking in SUSY can be induced *radiatively*, if the top quark is sufficiently massive; then its Yukawa coupling dominates the renormalization of the Higgs mass: $\partial M_\phi^2/\partial \log\mu = 3(g_t^Y)^2 m_t^2/8\pi^2 + ...$ This happens around $m_t > 160$ GeV. This feature had been noted in 1982 when such a mass for a top quark was perceived as extravagantly high [164]. In 1986 it was pointed out that top quarks might be 'ultra-heavy', with limits $m_t > 100 - 125$ GeV [51]; since strong forces scale as $\sim \mathcal{O}(1$ GeV) or, in time units, as $\sim 10^{-23}$ seconds, top quarks (semi-)weakly decay *before* they can produce top hadrons or $\bar{t}t$ quarkonium bound states.

We cannot see a good reason why the SM Higgs dynamics or close to it should be the end of our 'adventure' on spin-zero forces; we should reflect on how this item fits into our understanding of our Universe or – even better – leads to a new puzzle in our Universe.

4.8 Resume so far

We have learnt that the landscapes of flavor dynamics of quarks and leptons are 'complex' in different ways and directions. We need tools to analyze the transitions both on the experimental and theoretical sides with accuracy or even better with precision. Actually, our community has very good tool boxes; however, to apply them it needs time, connections among transitions and more thinking. We give a few more comments

- Is it enough to describe with three 'families' – or we have then to go beyond the item 'families'?
- In the world of quarks and gluons the measured rates show the impact of strong dynamics on the weak transitions. The goal is to establish the impact of ND and its features. One has to probe many-body FS with regional transitions – QCD cannot give only the background.
- We have mentioned that top quarks cannot produce top hadrons or a $[\bar{t}t]$ quarkonium. Yet top quarks are not 'free', since they carry unbroken color QN 3 or $\bar{3}$; thus they have to find another source of color QN. We will discuss the dynamics of top quark in **Chapter 8**.
- The situation is very different in the world of leptons – in particular for neutrinos. Are the neutrinos members of the Dirac family or the Majorana one? Most theorists think that neutrinos belong to the latter.
- Neutrino oscillations are seen as a wonderful hunting region for the existence of ND and its features, if we have enough data and control the background. There is 'only' a practical question: our detectors do not work in 'vacuum'– the background is "matter" (but not "anti-matter").

Our understanding of fundamental dynamics is based on symmetries in many ways – although often that is not obvious. We can learn from paintings (or other arts). Some artists can have 'visions', not just describe the landscape: when one talks about triangles, quadrangles etc. in the different flavor dynamics of quarks and leptons, the paintings of the Russian artist Kandinsky come to mind, see **Fig. 4.3**.

Fig. 4.3 Composition VIII by Vassily Kandinsky (1923) [Guggenheim Museum, New York (Collection Online)].

Chapter 5

Rare decays of flavor hadrons and leptons

While the main title of this book is "New Era for **CP** Asymmetries", we have also a sub-title "... and Rare Decays of Hadrons and Leptons". The goal is to find hints of new dynamics (ND) and possibly even its features there.

Rare decays provide a rich landscape for new lessons about fundamental dynamics. Of course, 'we' need much more data, but also more refined tools and analyses of backgrounds. Furthermore, it is crucial to think about the connections between different transitions. We can distinguish two classes of rare decays of flavor hadrons and leptons:

- **Class I** is that of the exactly forbidden decays, where the SM gives zero value for the corresponding decay amplitude. The final states of this class of decays usually include two different charged leptons. Many experiments are devoted to the search of these decays, also because the clear lepton signature often offers a small or sometimes completely negligible background. Furthermore, it is usually easy to draw Feynman diagrams for such ND and provide specific models for the new amplitude.
- **Class II** is made of those decays that are not totally forbidden by the SM, but are highly suppressed. Also this class of events is interesting because the ND might give larger values than the expected ones from the SM or give sizable interference effects in different kinematic regions. We already know that the SM is incomplete at least in lepton dynamics, as established with neutrino oscillations. It would not be a miracle to find other footprints in lepton transitions.

5.1 Rare decays of strange, beauty and charm hadrons

5.1.1 Class I

Some decays are not allowed in the SM, but in some models of ND they can happen. The candidates of **Class I** can be divided into two main groups.

The first group includes :

$$B^0/B_s^0 \to \mu^\pm\tau^\mp/e^\pm\tau^\mp/\pi^+\pi^-e^\pm\tau^\mp \tag{5.1}$$
$$B^+ \to \pi^+\mu^\pm\tau^\mp/K^+\mu^\pm\tau^\mp \tag{5.2}$$
$$\Lambda_b^0 \to p\pi^-\mu^\pm\tau^\mp/pK^-\mu^\pm\tau^\mp \tag{5.3}$$

All these decays explicitly violate family lepton number, which we already know to be broken in neutrino oscillations. Some models with non-minimal Higgs dynamics, for example, produce rates that might be within the sensitivity of ATLAS, CMS, LHCb and Belle-II experiments in the forthcoming years.

The second group includes:

$$D^0 \to \mu^\pm e^\mp/\pi^+\pi^-\mu^\pm e^\mp/K^+K^-\mu^\pm e^\mp \tag{5.4}$$
$$D^+/D_s^+ \to \pi^+\mu^\pm e^\mp/\pi^+\pi^+\pi^-\mu^\pm e^\mp \tag{5.5}$$
$$\Lambda_c^+ \to p\mu^\pm e^\mp \tag{5.6}$$

Models of ND introducing large effects in these decays are less 'natural', since the SM background comes from Cabibbo favored transitions, not suppressed CKM ones as for $\Delta B \neq 0$. However, ATLAS, CMS, LHCb and Belle II experiments will look for these decays with unprecedented sensitivity in the near future.

So far there are no 'golden' channels for finding the impact of ND in rare decays of charm and beauty hadrons (including baryons).

5.1.2 Class II

The 'landscape' is more complex for rare decays in the **Class II**: the SM does contribute to these transitions, thus one has to be sure about the expected SM values. One wants to establish the impact of ND and, to be realistic, such impact should enhance rates, not suppress them.

We have a 'golden' goal in $\Delta S \neq 0$ dynamics, namely the truly tiny rates for $K^+ \to \pi^+\bar{\nu}\nu$ with a non-zero value from the SM; for example, the prediction from the 'Buras School' gives $\mathrm{BR}(K^+ \to \pi^+\bar{\nu}\nu)|_{\mathrm{SM}} = (0.839 \pm 0.030) \cdot 10^{-10}$ [165].

Present data on $\mathrm{BR}(K^+ \to \pi^+\bar{\nu}\nu) = (1.7 \pm 1.1) \cdot 10^{-10}$ [8] tell us that it probably happens, although its uncertainty is still large; on the other hand, there is room for a branching ratio larger that the one SM predicts. NA62 experiment at CERN is studying $K^+ \to \pi^+\bar{\nu}\nu$ decay with a large statistical set. The result published in 2019 [166], obtained with just 2% of available statistics yields $\mathrm{BR}(K^+ \to \pi^+\bar{\nu}\nu) < 14 \times 10^{-10}$. An improved results published in 2020 [167], obtained with about 20% of the collected statistics, has yielded two candidate events with an expected background of 1.5 events, from which the collaboration derived, combining all available data, the upper limit $\mathrm{BR}(K^+ \to \pi^+\bar{\nu}\nu) < 1.78 \times 10^{-10}$. The final sensitivity of NA62 with full statistics is expected to yield a measurement of the SM branching ratio to 10% precision and be able, therefore, to probe ND at that level. We have another 'golden' mode in $\Delta S \neq 0$, namely $K_L \to \pi^0\bar{\nu}\nu$ decay; the rate for $K_L \to \pi^0\bar{\nu}\nu$ is basically based on **CP** violation. So far, the data have been giving only limits, and those are much larger than the SM predictions. The KOTO experiment at J-PARC should make a significant progress in measuring the branching ratio for $\mathrm{BR}(K_L \to \pi^0 \bar{\nu}\nu)$ [168]. We will discuss both 'golden' modes in **Sect. 6.5.10**.

Rare decays of D mesons like $D^0 \to \gamma\gamma/\mu^+\mu^-$ and $D \to X_u l^+ l^- \ /X_u \bar{\nu}\nu$ [169] are also of **Class II**. In these decays short distance dynamics of the SM give non-zero values, but very tiny; long distance dynamics suggest bigger values, but still tiny. Experimental limits are larger by orders of ten compared to SM predictions, but future results from LHCb and Belle-II might bring the experimental limits close to predictions and make these decays an interesting probe of ND.

5.2 Rare decays of charged leptons μ and τ beyond the SM

The SM does *not*[1] allow $\mu^- \to e^-\gamma$ or $e^-e^+e^-$; PDG2019 tells us:

$$\mathrm{BR}(\mu^- \to e^-\gamma) \leq 4.2 \cdot 10^{-13} \ , \ \mathrm{BR}(\mu^- \to e^-e^+e^-) \ \leq 1.0 \cdot 10^{-12} \ . \tag{5.7}$$

The future perspectives are intriguing and may lead to surprises. A next generation experiment at PSI (Switzerland), namely MEG II (MEG: Muon to Electron and Gamma experiment) [170], should reach a sensitivity of $\sim 4 \cdot 10^{-14}$ on $\mathrm{BR}(\mu^- \to e^-\gamma)$ in the near future, improving of an order of magnitude the final result of the MEG experiment, that is an upper limit of $\sim 4.2 \cdot 10^{-13}$ at 90%CL.

Also one can probe $\mu \Rightarrow e$ conversion in the nucleus electric field, namely by $\mu^- N \to e^- N$. Up to now we have a limit on a branching ratio of few$\cdot 10^{-13}$ from the SINDRUM II experiment [171]. However, a new Mu2e experiment at Fermilab could improve the current limit by four orders of magnitude, measuring the muon-to-electron conversion with a single event sensitivity of few$\cdot 10^{-17}$ [172]!

Rare decays of the τ lepton may or may not include light hadrons [8]:

$$\mathrm{BR}(\tau^- \to e^-\gamma) \leq 3.3 \cdot 10^{-8} \ , \ \mathrm{BR}(\tau^- \to \mu^-\gamma) \leq 4.4 \cdot 10^{-8} \tag{5.8}$$

$$\mathrm{BR}(\tau^- \to e^-e^+e^-) \leq 2.7 \cdot 10^{-8} \ , \ \mathrm{BR}(\tau^- \to \mu^-\mu^+\mu^-) \leq 2.1 \cdot 10^{-8} \tag{5.9}$$

$$\mathrm{BR}(\tau^- \to e^-\rho^0) \leq 1.8 \cdot 10^{-8} \ , \ \mathrm{BR}(\tau^- \to \mu^-\rho^0) \leq 1.2 \cdot 10^{-8} \ . \tag{5.10}$$

The ATLAS and CMS experiments will open a new era for probing τ decays with their huge future statistics.

One expects that the ratio of $\Delta S = 0$ vs. $\Delta S = 1$ is typically of $\mathcal{O}(10)$. Indeed, we find [8]:

$$\mathrm{BR}(\tau^- \to \nu\pi^-) = (10.82 \pm 0.05) \cdot 10^{-2} \ \text{vs.} \ \mathrm{BR}(\tau^- \to \nu K^-) = (6.96 \pm 0.10) \cdot 10^{-3} \ ;$$

However, present data [8] also shows :

$$\mathrm{BR}(\tau^- \to \nu\phi\pi^-) \simeq (3.4 \pm 0.6) \cdot 10^{-5} \ \text{vs.} \ \mathrm{BR}(\tau^- \to \nu\phi K^-) = (4.4 \pm 1.6) \cdot 10^{-5} \ .$$

Do we have less control of strong forces close to threshold in case of $\tau^- \to \nu\phi K^-/\nu K^- K^+ K^-$ – or is it a sign of the impact of ND?

[1] In the SM with massive mixed neutrinos this is not formally true. Neutrino mass term may induce such a decay, but the SM value for that is completely negligible, so the statement in the text is correct to any practical purpose. Any signal of $\mu^- \to e^-\gamma$ decay would be a signal of ND.

5.3 $|V_{us}|$ from the decays of strange hadrons and the τ lepton

The comparison of rare and **CP** violating decays of the kaon, hyperons and the τ lepton might teach us something important. Particularly, it is interesting to compare the value of $|V_{us}|$ extracted from semi-leptonic decays of strange hadrons and from the suppressed semi-hadronic decays of the τ lepton.

In order to do so, a precise determination of $|V_{ud}|$ is crucial. A recent survey has presented about 20 super-allowed $0^+ \to 0^+$ nuclear β decays [173]: it has led to $V_{ud} = 0.97417 \pm 0.00021$ – i.e., uncertainty around $2 \cdot 10^{-4}$. It sets the scale of precision (including also some theoretical assumptions). It leads to an unitarity test of the CKM matrix $|V_{ud}|^2 + |V_{us}|^2 + |V_{ub}|^2 = 0.99978 \pm 0.00055$, when one uses PDG 'recommended' value for $|V_{us}|$. The value of $|V_{ub}|^2 \sim 2 \cdot 10^{-5}$ has no impact.

The values of $|V_{us}|$ can be measured directly in the weak decays of strange hadrons and the charged lepton τ. For $|V_{us}|$ we have a 'Pope', namely the measured $|V_{us}|$ from semi-leptonic decays of kaons:

$$|V_{us}|_K = 0.2248 \pm 0.0006 \,, \tag{5.11}$$

as well as 'Cardinals' from hyperon decays:

$$|V_{us}|_\Lambda = 0.2250 \pm 0.0027 \,. \tag{5.12}$$

Not surprisingly they agree. The situation is not so clear about the 'Protestant' camp from τ decays:

$$|V_{us}|_\tau = 0.2202 \pm 0.0015 \,. \tag{5.13}$$

These 'actors' have to learn about their 'plays' – or is there a 'cryptic' lesson?

Chapter 6

Oscillations with neutral mesons and neutrons

Here we mostly discuss neutral mesons and anti-mesons that are distinguished by an internal quantum number F. In these systems we can observe oscillations, namely transitions with $\Delta F = 2$ units.[1]

Oscillations depend crucially on "time".[2] They may occur even in absence of **CP** asymmetries, while *indirect* **CP** violation cannot happen without them. Oscillations with **CP** violation have been established experimentally for *strange* and *beauty* mesons in the previous century. $D^0 - \bar{D}^0$ oscillations have recently been established as well.

Another interesting phenomena are neutron $\rightleftharpoons$ anti-neutron oscillations[3]; in that case baryon number violation is also involved, and one must use tools that are quite different from those for mesons; we discuss it in **Sect. 6.8**.

We do not contemplate violation of electric charge conservation and, therefore, we analyze only electrically neutral states.

First a few general comments about neutral hadrons with non-zero flavors:

- One can distinguish between indirect vs. direct **CP** violation. In indirect one we observe the features of *neutral initial* states based on oscillations.[4] To probe them one mostly focuses on two-body FS, for practical reasons. For *direct* **CP** violation in flavor hadron decays we emphasize the impact of *final* states, in particular for charm and beauty mesons with many-body FS.
- For *Down*-type quarks d, s and b we have three candidates for *indirect* **CP** violation: K^0, B^0 and B^0_s. For the first two mesons *indirect* **CP** violation has been established with accuracy. Fast experimental progress has been driven in the recent past by Tevatron experiments and is now being continued at LHC.

---

[1] Mostly F means 'flavor', but sometimes F can refer to neutral baryons; we will come back to it.

[2] We are talking about "proper time" as defined in the special theory of relativity – i.e., a Lorentz scalar.

[3] We use a different symbol, namely "$\rightleftharpoons$".

[4] Remember **Sect. 2.5** about *mixing* vs. *oscillations*.

For *Up*-type quarks u, c and t we have only one candidate for oscillations: D^0. In 2019 **CP** asymmetry has been established [174], but only *direct* one; we will discuss it in **Sect. 7.5**. Top quarks decay *before* they can hadronize [51]; thus oscillations cannot happen with them.

- Hadronization – and non-perturbative dynamics in general – is usually viewed as a complication (if not an outright nuisance) for our description of fundamental forces. Indeed, based on present uncertainties, we cannot rule out that sizable fractions of observables as ΔM_K, ΔM_{B^0}, $\Delta M_{B^0_s}/\Delta M_{B^0}$ and ϵ_K come from ND. Such a view misses the deeper truth: with*out* hadronization neutral mesons could not form bound states of quarks and anti-quarks. *With* hadronization $P^0 - \bar{P}^0$ oscillations happen in very different landscapes: we have $K^0 - \bar{K}^0$, $B^0 - \bar{B}^0$, $B^0_s - \bar{B}^0_s$ and $D^0 - \bar{D}^0$ oscillations. Oscillations involve quantum mechanical coherence over macroscopic distances and allow to measure truly tiny mass differences, as $\mathrm{Im} M^K_{12} \sim \mathcal{O}(10^{-8})$ eV, that drives indirect **CP** in K_L decays, or tiny mass differences as $\Delta M_{B^0} \sim \mathcal{O}(10^{-4})$ eV. One needs *coherent* states $|P^0\rangle$ and $|\bar{P}^0\rangle$ and interference of transition amplitudes $T(P^0 \to f)$ and $T(P^0 \to \bar{P}^0 \to f)$ to produce oscillations and **CP** asymmetries. Hadronization provides the portals for this to happen: it acts as a 'cooling' process enhancing the coherence of amplitudes. Its intrinsic strength is of great advantage in this context. Of course, these statements do not make it easier to produce quantitative predictions or analyses with accuracy.
- There is an example of non-local connections of QM, pointed out first by Einstein, Podolsky and Rosen (EPR) in 1935 [20]. EPR correlations, which represent some of the most puzzling features of quantum mechanics, serve as an essential precision tool, which is routinely used in B-factories for measurements like

$$e^+e^- \to \Upsilon(4S) \to \bar{B}^0 B^0 \to [l^\pm X^\mp]_B [\psi K_S]_B \tag{6.1}$$

Due to EPR correlations, once one of the beauty mesons is tagged – namely a B^0 by its semileptonic decay into a positively charged lepton – we know that its pair produced partner has to be a $\bar{B}^0$ at that time, and that its amplitude for decaying later into a **CP** eigenstate depends only on the time elapsed since the first decay. While we might not have an intuitive grasp of EPR correlations, we should appreciate them as a gift from Nature rather than as a paradox: *EPR correlations are a blessing in disguise!*

6.1 Master equations of time evolution

For neutral mesons we have two definitions of eigenstates, namely flavor eigenstates – $|P^0\rangle$ and $|\bar{P}^0\rangle$ – and mass eigenstates – $|P_1\rangle$ and $|P_2\rangle$. **CPT** invariance tells us:

$$|P_1\rangle = p|P^0\rangle + q|\bar{P}^0\rangle \quad , \quad |P_2\rangle = p|P^0\rangle - q|\bar{P}^0\rangle \tag{6.2}$$

They describe 'mass' eigenvectors of a non-Hermitian Hamiltonian matrix $\mathbf{H} = \mathbf{M} - \frac{i}{2}\mathbf{\Gamma}$, while $\mathbf{M}$ and $\mathbf{\Gamma}$ are Hermitian. The complex eigenvalues of $\mathbf{H}$ are[5]:

$$M_1 - \frac{i}{2}\Gamma_1 = M_{11} - \frac{i}{2}\Gamma_{11} + \frac{q}{p}\left(M_{12} - \frac{i}{2}\Gamma_{12}\right) \tag{6.3}$$

$$M_2 - \frac{i}{2}\Gamma_2 = M_{11} - \frac{i}{2}\Gamma_{11} - \frac{q}{p}\left(M_{12} - \frac{i}{2}\Gamma_{12}\right) \tag{6.4}$$

$$\left(\frac{q}{p}\right)^2 = \frac{M_{12}^* - \frac{i}{2}\Gamma_{12}^*}{M_{12} - \frac{i}{2}\Gamma_{12}} \tag{6.5}$$

Two solutions exists:

$$\frac{q}{p} = \pm\sqrt{\frac{M_{12}^* - \frac{i}{2}\Gamma_{12}^*}{M_{12} - \frac{i}{2}\Gamma_{12}}}\ ; \tag{6.6}$$

choosing the negative rather than the positive sign in Eq. (6.6) is equivalent to interchanging the labels $1 \leftrightarrow 2$ of the mass eigenstates. The two off-diagonal elements would be zero through flavor conservation, as would happen if only electromagnetic or strong interaction were involved. Yet they are different from *zero*, since weak interactions can produce $\Delta F = 2$ transitions. There are two parts: M_{12} mainly quantifies short-distance dynamics from *off*-scale intermediate states, while Γ_{12} for *on*-scale ones. There is another ambiguity: anti-particles are defined up to a phase only. Going from $|\bar{P}^0\rangle$ to $e^{i\xi}|\bar{P}^0\rangle$ will modify the off-diagonal elements of the $\mathbf{M} - \frac{i}{2}\mathbf{\Gamma}$ matrix:

$$M_{12} - \frac{i}{2}\Gamma_{12} \Longrightarrow e^{i\xi}\left(M_{12} - \frac{i}{2}\Gamma_{12}\right)\ ,\ \frac{q}{p} \Longrightarrow e^{-i\xi}\frac{q}{p}\ , \tag{6.7}$$

yet it leaves their product invariant:

$$\frac{q}{p}\left(M_{12} - \frac{i}{2}\Gamma_{12}\right) \Longrightarrow \frac{q}{p}\left(M_{12} - \frac{i}{2}\Gamma_{12}\right)\ ! \tag{6.8}$$

The masses and widths of $P_{1,2}$ being observables are insensitive to the arbitrary phase:

$$M_2 - M_1 \equiv \Delta M = -2\,\mathrm{Re}\left(\frac{q}{p}(M_{12} - \frac{i}{2}\Gamma_{12})\right) \tag{6.9}$$

$$\Gamma_2 - \Gamma_1 \equiv \Delta\Gamma = +4\,\mathrm{Im}\left(\frac{q}{p}(M_{12} - \frac{i}{2}\Gamma_{12})\right)\ . \tag{6.10}$$

The time evolution of states starting as P^0 or $\bar{P}^0$ is[6]:

$$|P^0(t)\rangle = f_+(t)|P^0\rangle + \frac{q}{p}f_-(t)|\bar{P}^0\rangle\ ,\ |\bar{P}^0(t)\rangle = f_+(t)|\bar{P}^0\rangle + \frac{p}{q}f_-(t)|P^0\rangle \tag{6.11}$$

[5] $M_{22} = M_{11}$ and $\Gamma_{22} = \Gamma_{11}$ based on **CPT** invariance.
[6] The reader can find the demonstration in Ref. [12].

with

$$f_\pm(t) = \frac{1}{2}e^{-iM_1 t}e^{-\frac{1}{2}\Gamma_1 t}\left[1 \pm e^{\frac{1}{2}\Delta\Gamma t}e^{-i\Delta M t}\right] . \tag{6.12}$$

First one looks at a f being a **CP** eigenstate; we denote the amplitudes for $P^0 \to f$ and $\bar{P}^0 \to f$ transitions as A_f and $\bar{A}_f$, respectively, and their ratios: $\bar{\rho}_f = \bar{A}_f/A_f = 1/\rho_f$. The widths are described by

$$\frac{\Gamma(P^0(t) \to f)}{|A_f|^2} \propto e^{-\Gamma_1 t}\left[K_+(t) + K_-(t)\left|\frac{q}{p}\right|^2 |\bar{\rho}_f|^2 + 2\mathrm{Re}\left[L^*(t)\left(\frac{q}{p}\right)\bar{\rho}_f\right]\right]$$

$$\frac{\Gamma(\bar{P}^0(t) \to f)}{|\bar{A}_f|^2} \propto e^{-\Gamma_1 t}\left[K_+(t) + K_-(t)\left|\frac{p}{q}\right|^2 |\rho_f|^2 + 2\mathrm{Re}\left[L^*(t)\left(\frac{p}{q}\right)\rho_f\right]\right] \tag{6.13}$$

where

$$|f_\pm(t)|^2 = \frac{1}{4}e^{-\Gamma_1 t}K_\pm(t) \, , \; f_-(t)f_+^*(t) = \frac{1}{4}e^{-\Gamma_1 t}L^*(t)$$

$$K_\pm(t) = 1 + e^{\Delta\Gamma t} \pm 2e^{\frac{1}{2}\Delta\Gamma t}\cos\Delta M t \, , \; L^*(t) = 1 - e^{\Delta\Gamma t} + 2ie^{\frac{1}{2}\Delta\Gamma t}\sin\Delta M t \, .$$

Oscillations thus provide for the presence of different amplitudes contributing *coherently*; their relative weights vary with the time of decay. There are two decay chains, $P^0 \to f$ and $P^0 \to \bar{P}^0 \to f$, which interfere according to the principles of QM.

In general, we can conveniently separate out the main exponential time dependence of the decay rate in both channels:

$$\Gamma(P^0(t) \to f) \propto \frac{1}{2}e^{-\Gamma_1 t} \cdot G_f(t) \, ,$$

$$G_f(t) = a + be^{\Delta\Gamma t} + ce^{\Delta\Gamma t/2}\cos\Delta M t + de^{\Delta\Gamma t/2}\sin\Delta M t \tag{6.14}$$

$$a = |A_f|^2\left[\frac{1}{2}\left(1 + \left|\frac{q}{p}\bar{\rho}_f\right|^2\right) + \mathrm{Re}\left(\frac{q}{p}\bar{\rho}_f\right)\right] \tag{6.15}$$

$$b = |A_f|^2\left[\frac{1}{2}\left(1 + \left|\frac{q}{p}\bar{\rho}_f\right|^2\right) - \mathrm{Re}\left(\frac{q}{p}\bar{\rho}_f\right)\right] \tag{6.16}$$

$$c = |A_f|^2\left[1 - \left|\frac{q}{p}\bar{\rho}_f\right|^2\right] \tag{6.17}$$

$$d = -2|A_f|^2\mathrm{Im}\left(\frac{q}{p}\bar{\rho}_f\right) \tag{6.18}$$

$$\Gamma(\bar{P}^0(t) \to \bar{f}) \propto \frac{1}{2}e^{-\Gamma_1 t} \cdot \bar{G}_{\bar{f}}(t) \ ,$$

$$\bar{G}_{\bar{f}}(t) = \bar{a} + \bar{b}e^{\Delta\Gamma\, t} + \bar{c}e^{\Delta\Gamma\, t/2}\cos\Delta M\, t + \bar{d}e^{\Delta\Gamma\, t/2}\sin\Delta M\, t \quad (6.19)$$

$$\bar{a} = |\bar{A}_{\bar{f}}|^2 \left[\frac{1}{2}\left(1 + \left|\frac{p}{q}\rho_{\bar{f}}\right|^2\right) + \mathrm{Re}\left(\frac{p}{q}\rho_{\bar{f}}\right)\right] \quad (6.20)$$

$$\bar{b} = |\bar{A}_{\bar{f}}|^2 \left[\frac{1}{2}\left(1 + \left|\frac{p}{q}\rho_{\bar{f}}\right|^2\right) - \mathrm{Re}\left(\frac{p}{q}\rho_{\bar{f}}\right)\right] \quad (6.21)$$

$$\bar{c} = |\bar{A}_{\bar{f}}|^2 \left[1 - \left|\frac{p}{q}\rho_{\bar{f}}\right|^2\right] \quad (6.22)$$

$$\bar{d} = -2|\bar{A}_{\bar{f}}|^2 \mathrm{Im}\left(\frac{p}{q}\rho_{\bar{f}}\right) \quad (6.23)$$

Obviously, **CP** violation occurs when

$$G_f(t) \neq \bar{G}_{\bar{f}}(t) \quad (6.24)$$

Above we have 'painted' the possible 'landscapes'. Now we focus on special cases.

6.1.1 Flavor-specific *decays*

Here we discuss final states that come from either P^0 or $\bar{P}^0$, but not both:

$$P^0 \to f \not\leftarrow \bar{P}^0 \quad \text{or} \quad P^0 \not\to f \leftarrow \bar{P}^0 \quad (6.25)$$

Prominent flavor-specific channels for neutral mesons (K^0, D^0, B^0) are provided by semi-leptonic decays with $X^\pm$ as hadronic FS:

$$P^0 \to l^+ + \nu_l + X^- \not\leftarrow \bar{P}^0 \quad , \quad P^0 \not\to l^- + \bar{\nu}_l + X^+ \leftarrow \bar{P}^0 \quad (6.26)$$

$$\bar{A}_{l^+} = 0 = A_{l^-} \quad (6.27)$$

By comparing with Eq. (6.13), it leads to

$$\Gamma(P^0(t) \to l^+ + \nu_l + X^-) \propto e^{-\Gamma_1 t}K_+(t)|A_{l^+}|^2$$

$$\Gamma(P^0(t) \to l^- + \bar{\nu}_l + X^+) \propto e^{-\Gamma_1 t}K_-(t)\left|\frac{q}{p}\right|^2 |\bar{A}_{l^-}|^2$$

$$\Gamma(\bar{P}^0(t) \to l^- + \bar{\nu}_l + X^+) \propto e^{-\Gamma_1 t}K_+(t)|\bar{A}_{l^-}|^2$$

$$\Gamma(\bar{P}^0(t) \to l^+ + \nu_l + X^-) \propto e^{-\Gamma_1 t}K_-(t)\left|\frac{p}{q}\right|^2 |A_{l^+}|^2 \quad (6.28)$$

Furthermore **CPT** invariance leads to the hadronic $|A_{l^+}|^2 = |\bar{A}_{l^-}|^2$. Thus one probes indirect **CP** violation in semi-leptonic decays with the asymmetry:

$$A_{SL} \equiv \frac{\Gamma(P^0(t) \to l^-\bar{\nu}_l X^+) - \Gamma(\bar{P}^0(t) \to l^+\nu_l X^-)}{\Gamma(P^0(t) \to l^-\bar{\nu}_l X^+) + \Gamma(\bar{P}^0(t) \to l^+\nu_l X^-)} = \frac{|q/p|^2 - |p/q|^2}{|q/p|^2 + |p/q|^2} \ ; \quad (6.29)$$

its value is independent of time t, while it needs oscillations. Here we are talking about *indirect* **CP** violation that comes with oscillations; i.e., it depends on the *initial* states with non-zero value of flavor.

6.1.2 Flavor-non-specific *decays*

In flavor-non-specific decays, final states are fed both by P^0 and $\bar{P}^0$ decays, not necessarily with the same rate:

$$P^0 \to f \leftarrow \bar{P}^0 \,. \tag{6.30}$$

CP eigenstates $f_\pm$ fall into this category.

- There is a special case, when we have[7]

$$|\bar{A}_f| = |A_f| \,, \qquad \frac{q}{p} A_f \bar{A}_f^* \simeq \left(\frac{p}{q} \bar{A}_f A_f^* \right)^* \tag{6.31}$$

  – assuming $\Delta\Gamma \simeq 0$, this leads to

$$\Gamma(P^0(t) \to f) \propto e^{-\Gamma_1 t} \left[|A_f|^2 - \mathrm{Im}\left(\frac{q}{p} A_f \bar{A}_f^* \right) \cdot \sin\Delta M t \right]$$
$$\Gamma(\bar{P}^0(t) \to f) \propto e^{-\Gamma_1 t} \left[|A_f|^2 + \mathrm{Im}\left(\frac{q}{p} A_f \bar{A}_f^* \right) \cdot \sin\Delta M t \right] \tag{6.32}$$

  with indirect **CP** violation described by:

$$\frac{\Gamma(P^0(t) \to f) - \Gamma(\bar{P}^0(t) \to f)}{\Gamma(P^0(t) \to f) + \Gamma(\bar{P}^0(t) \to f)} = -\frac{\mathrm{Im}\left(\frac{q}{p} A_f \bar{A}_f^* \right)}{|A_f|^2} \cdot \sin\Delta M t \tag{6.33}$$

  The 'well-known' example is $B^0 \to J/\psi K_S$ vs. $\bar{B}^0 \to J/\psi K_S$ [134].[8]

  – one can also probe FS with $\Delta\Gamma \neq 0$. Eqs. (6.14,6.19) yield

$$e^{\Gamma_1 t}\Gamma(P^0(t) \to f) \propto \left[|A_f|^2 (1 + e^{\Delta\Gamma t}) + \mathrm{Re}\left(\frac{q}{p} \bar{A}_f A_f^* \right) \left[1 - e^{\Delta\Gamma t}\right] \right.$$
$$\left. - 2\mathrm{Im}\left(\frac{q}{p} \bar{A}_f A_f^* \right) e^{\frac{1}{2}\Delta\Gamma t} \sin\Delta M t \right] \tag{6.34}$$

$$e^{\Gamma_1 t}\Gamma(\bar{P}^0(t) \to f) \propto \left[|A_f|^2 (1 + e^{\Delta\Gamma t}) + \mathrm{Re}\left(\frac{q}{p} \bar{A}_f A_f^* \right) \left[1 - e^{\Delta\Gamma t}\right] \right.$$
$$\left. + 2\mathrm{Im}\left(\frac{q}{p} \bar{A}_f A_f^* \right) e^{\frac{1}{2}\Delta\Gamma t} \sin\Delta M t \right] \tag{6.35}$$

  with indirect **CP** violation being expressed in this case by:

$$\frac{\Gamma(P^0(t) \to f) - \Gamma(\bar{P}^0(t) \to f)}{\Gamma(P^0(t) \to f) + \Gamma(\bar{P}^0(t) \to f)} = -\frac{2\mathrm{Im}\left(\frac{q}{p} \bar{A}_f A_f^* \right)}{|A_f|^2 (1 + e^{\Delta\Gamma t}) + \mathrm{Re}\left(\frac{q}{p} \bar{A}_f A_f^* \right)(1 - e^{\Delta\Gamma t})} \tag{6.36}$$

  An obvious example are $B_s^0 \to [J/\psi]\phi/[J/\psi]K^+K^-$ decays, where one can use the reasonably precise value $\Delta\Gamma(B_s^0) = (0.085 \pm 0.004) \cdot 10^{12}\ \mathrm{s}^{-1}$ [8].

Thus **CP** asymmetry arises if two conditions are satisfied:

(a) $\Delta M \neq 0$ due to $P^0 - \bar{P}^0$ oscillations

[7]Here we have assumed $|p|^2 \simeq |q|^2$, which is realistic for beauty mesons.

[8]It is not fair to use the word 'well-known'; it means not giving a reference to this original paper.

(b) weak phases like $\mathrm{Im}\left(\frac{q}{p}\bar{A}_f A_f^*\right) = |A_f|^2 \mathrm{Im}\left(\frac{q}{p}\bar{\rho}_f\right)$ exist.

- There is another special case, when the FS gives

$$\bar{\rho}_f = \bar{A}_f / A_f = 1/\rho_f \neq 1 \tag{6.37}$$

Eqs. (6.14,6.19) give[9]

$$\frac{\Gamma(P^0(t) \to f)}{e^{-\Gamma_1 t}|A_f|^2} \propto \left[1 + |\bar{\rho}_f|^2 + (1 - |\bar{\rho}_f|^2)\cos\Delta M\, t - 2\mathrm{Im}\left(\frac{q}{p}\bar{\rho}_f\right)\sin\Delta M\, t\right]$$
$$\frac{\Gamma(\bar{P}^0(t) \to f)}{e^{-\Gamma_1 t}|\bar{A}_f|^2} \propto \left[1 + |\rho_f|^2 + (1 - |\rho_f|^2)\cos\Delta M\, t - 2\mathrm{Im}\left(\frac{p}{q}\rho_f\right)\sin\Delta M\, t\right]$$

The **CP** asymmetry can be expressed as

$$\begin{aligned}\frac{\Gamma(P^0(t) \to f) - \Gamma(\bar{P}^0(t) \to f)}{\Gamma(P^0(t) \to f) + \Gamma(\bar{P}^0(t) \to f)} &= C_f \cos\Delta M\, t - S_f \sin\Delta M\, t \\ C_f = \frac{1 - |\bar{\rho}_f|^2}{1 + |\bar{\rho}_f|^2} \quad &, \quad S_f = \frac{2\mathrm{Im}((q/p)\bar{\rho}_f)}{1 + |\bar{\rho}_f|^2} \,.\end{aligned} \tag{6.38}$$

It has two sources, which can be clearly separated by the time dependence, namely C_f and S_f. The former requires $|\bar{\rho}(f)| \neq 1$, i.e. *direct* **CP** violation residing in $H_{\Delta F=1}$, which we discuss in **Chapter 7**. A consistence check for data is provided by $C_f^2 + S_f^2 \leq 1$ [12].

6.2 Time-integrated rates

Not all systems of neutral P^0 are equivalent when we observe the time dependence of the decay rates. The tracks of short-lived particles such as B mesons, which live for only about 1.5 ps, are more difficult to measure than those of the K mesons, which live at least two order of magnitude more.

One can show the impact of oscillations on time-integrated rates from Eq. (6.13):

$$\langle K_\pm \rangle = \int_0^\infty dt\, e^{-\Gamma_1 t} K_\pm(t) = \frac{2}{\Gamma}\left[\frac{1}{1-y^2} \pm \frac{1}{1+x^2}\right] \tag{6.39}$$

$$\langle L^* \rangle = \int_0^\infty dt\, e^{-\Gamma_1 t} L^*(t) = \frac{2}{\Gamma}\left[-\frac{y}{1-y^2} + i\frac{x}{1+x^2}\right] \tag{6.40}$$

$$x \equiv \frac{\Delta M}{\Gamma} \,,\; y \equiv \frac{\Delta\Gamma}{2\Gamma} \,,\; \Gamma \equiv \frac{1}{2}(\Gamma_1 + \Gamma_2) \tag{6.41}$$

We observe that the oscillations are calibrated by the decay rates.

[9] Mostly assumed $|q| \simeq |p|$.

For flavor specific decays, as semi-leptonic decays, from Eq. (6.28) one finds

$$\int_0^\infty dt\,\Gamma(P^0(t) \to l^+ + \nu_l + X^-) \propto \frac{2+x^2-y^2}{(1-y^2)(1+x^2)}|A_{l^+}|^2 \tag{6.42}$$

$$\int_0^\infty dt\,\Gamma(P^0(t) \to l^- + \bar{\nu}_l + X^+) \propto \frac{x^2+y^2}{(1-y^2)(1+x^2)}\left|\frac{q}{p}\right|^2 |\bar{A}_{l^-}|^2 \tag{6.43}$$

$$\int_0^\infty dt\,\Gamma(\bar{P}^0(t) \to l^- + \bar{\nu}_l + X^+) \propto \frac{2+x^2-y^2}{(1-y^2)(1+x^2)}|\bar{A}_{l^-}|^2 \tag{6.44}$$

$$\int_0^\infty dt\,\Gamma(\bar{P}^0(t) \to l^+ + \nu_l + X^-) \propto \frac{x^2+y^2}{(1-y^2)(1+x^2)}\left|\frac{p}{q}\right|^2 |A_{l^+}|^2 \tag{6.45}$$

CPT invariance leads to $|\bar{A}_{l^-}|^2 = |A_{l^+}|^2$; thus one recovers Eq. (6.29)

$$A_{SL} \equiv \frac{\Gamma(P^0(t) \to l^-\bar{\nu}_l X^+) - \Gamma(\bar{P}^0(t) \to l^+\nu_l X^-)}{\Gamma(P^0(t) \to l^-\bar{\nu}_l X^+) + \Gamma(\bar{P}^0(t) \to l^+\nu_l X^-)} = \frac{|q/p|^2 - |p/q|^2}{|q/p|^2 + |p/q|^2}, \tag{6.46}$$

which describes indirect **CP** violation in semi-leptonic decays. This **CP** violation is not time dependent.

The situations for non-leptonic transitions are very 'complex', which is not surprising. First one focuses on simple cases, namely assuming $|\bar{\rho}_f| = 1 = |q/p|$ and $y = 0$, see Eq. (6.32). For time-integrated rates one gets:

$$\int_0^\infty dt\,\Gamma(P^0(t) \to f) \propto |A_f|^2 \cdot \left(1 - \frac{x}{1+x^2}\mathrm{Im}\frac{q}{p}\bar{\rho}_f\right) \tag{6.47}$$

$$\int_0^\infty dt\,\Gamma(\bar{P}^0(t) \to f) \propto |A_f|^2 \cdot \left(1 + \frac{x}{1+x^2}\mathrm{Im}\frac{q}{p}\bar{\rho}_f\right); \tag{6.48}$$

thus

$$A_{\mathbf{CP}}(P^0 \to f) = -\frac{x}{1+x^2}\,\mathrm{Im}\,\frac{q}{p}\,\bar{\rho}_f\,. \tag{6.49}$$

The observable **CP** violation depends on the ratio $x = \Delta M/\Gamma$ rather than on ΔM only, as expected: the asymmetry is due to the interference between the amplitudes of decay and oscillation – it is the *relative* weight of the latter that counts.

In the beginning of the 1980's it was pointed out that the SM predicts sizable or even large **CP** asymmetries in the weak decays of beauty mesons [134] and that one may get much more information about the underlying dynamics by probing time evolution.

The best way to study time evolution is to produce coherent B meson pairs with a B-factory, i.e. an e^+e^- collider machine with a center of mass energy corresponding to $\Upsilon(4S)$, the lowest energy bottomonium state which can decay in two B mesons. However, in order to exploit such time dependence, the experimental apparatus must be able to clearly separate the two mesons so that flavor, position and momentum can be precisely determined. The historical symmetric collider machines could not do that, because the two B mesons originating from $\Upsilon(4S)$ decays are too close and can not be unambiguously separated. In the 1990's it was realized that this problem would have been solved by an asymmetric e^+e^- collider, in which the $\Upsilon(4S)$ is produced with a substantial boost and the two B mesons are well separated. The BaBar and Belle collaborations at SLAC (California) and KEK (Japan) did it with great successes which we will describe below. Belle II has started data taking and will continue testing $\Upsilon(4S) \to B^0\bar{B}^0$ and $\Upsilon(5S) \to B_s^0\bar{B}_s^0$ processes with high precision.

6.3 Re-generation

In **Sect. 6.1** we have discussed how a beam made up of P^0 states evolves in time while traveling in vacuum. Now we talk about a beam passing through nuclear matter along the z axis; z and t are related by $z = \gamma\beta t$, $\gamma = 1/\sqrt{1-\beta^2}$, and β denotes the velocity of P^0.

Let us define $|\psi(t)\rangle = a(t)|P^0\rangle + b(t)|\bar{P}^0\rangle$ and consider a medium that can be described in terms of indices of refraction n and $\bar{n}$ [175]. The evolution equation through the medium then reads as follows:

$$\frac{d|\psi(t)\rangle}{dt} = -\left[\left(\frac{1}{2}\mathbf{\Gamma} + i\mathbf{M}\right) - ik\beta\gamma\begin{pmatrix} n-1 & 0 \\ 0 & \bar{n}-1 \end{pmatrix}\right]|\psi(t)\rangle\,, \tag{6.50}$$

where k is the momentum of the particle. Refraction effectively adds a term $-k\beta\gamma(n-1)$ to M_{11} and $-k\beta\gamma(\bar{n}-1)$ to M_{22}, while keeping the *off*-diagonal matrix element unchanged:

$$\frac{d|\psi(t)\rangle}{dt} = -\left[\left(\frac{1}{2}\mathbf{\Gamma} + i\tilde{\mathbf{M}}\right)\right]|\psi(t)\rangle \tag{6.51}$$

$$\tilde{\mathbf{M}} = \begin{pmatrix} M_{11} - k\beta\gamma(n-1) & M_{12} \\ M_{12}^* & M_{22} - k\beta\gamma(\bar{n}-1) \end{pmatrix}\,. \tag{6.52}$$

The eigenvectors and eigenvalues of $\tilde{\mathbf{M}}$ differ from those of $\mathbf{M}$. In particular there is one important *qualitative* distinction: since

$$\tilde{M}_{11} \neq \tilde{M}_{22} \tag{6.53}$$

we are dealing with a situation where **CPT** invariance is effectively broken: P^0 and $\bar{P}^0$ propagate through matter rather than anti-matter; thus it *mimics* **CPT** violation. This observation can be expressed in general terms: an observed difference in two conjugate transition rates can be caused by *an asymmetry in the prevailing boundary conditions* rather than *a difference in the fundamental dynamics*. By studying the evolution on the states, one can easily found that P_2 (P_1) is re-generated from an initially pure P_1 (P_2) beam when traversing matter [12]. These observations are of practical interest for actual tests of **CPT** invariance. Real experiments are *not* undertaken in a perfect 'vacuum', but in an environment that is dominated by matter. In particular, a detector is *not* **CPT** invariant; this places inherent limits on **CPT** tests.

6.4 Resume on $P^0 - \bar{P}^0$ oscillations

There are four classes of **CP** asymmetries in the transitions of flavor states[10]:

(1) One class unambiguously reflects $\Delta F = 1$ dynamics and it is characterized by

$$A_f \neq \bar{A}_{\bar{f}}. \tag{6.54}$$

Such an asymmetry can occur also in the decays of charged mesons, baryons and leptons. It is referred to as *direct* **CP** violation (or **CP** violation in decay).

(2) One can probe lepton asymmetry, see Eq. (6.46):

$$|q/p|^2 \neq |p/q|^2 . \tag{6.55}$$

This is doubtless driven by the dynamics in the $\Delta F = 2$ sector. It represents *indirect* **CP** violation in $P^0 - \bar{P}^0$ oscillations.[11] Its impact depends on time, as seen in Eq. (6.28); however, the lepton asymmetry is time independent.

(3) Another class reflects combined effects of $\Delta F = 2$ and $\Delta F = 1$ transitions and can be referred to as **CP** violation *involving* $P^0 - \bar{P}^0$ *oscillations*[12]:

$$\text{Im}\left[\frac{q}{p}\bar{\rho}_f\right] = \left|\frac{q}{p}\bar{\rho}_f\right| \sin(\arg(q/p) + \arg(\bar{\rho}_f)) \neq 0 . \tag{6.56}$$

As long as such an asymmetry is found for a *single* FS f only (or in a single pair of **CP** conjugate states f and $\bar{f}$), it is meaningless to differentiate between *indirect* and *direct* **CP** violation. For a change in the phase convention will shift the weight between arg (q/p) and $\arg(\bar{\rho}_f)$. Changing $\mathbf{CP}|P^0\rangle \equiv |\bar{P}^0\rangle$ to the equivalent definition $\mathbf{CP}|P^0\rangle \equiv e^{i\xi}|\bar{P}^0\rangle$ with real ξ, will obviously have no effect on $|\bar{\rho}_f|$ or $|q/p|$. Yet q/p and $\bar{\rho}_f$ are affected by it, but their product remains invariant: $(q/p)\bar{\rho}_f \Rightarrow (q/p)\bar{\rho}_f$. Thus the sum of $\arg(q/p) + \arg(\bar{\rho}(f))$ is *not* sensitive to changes in the phase convention for the anti-particle and thus qualifies as an observable (see Eq. (6.8)).

(4) Once one has found a **CP** asymmetry in two different FS f_1 and f_2 (or in two conjugate pairs), one can establish *direct* **CP** violation by observing:

$$\sin(\arg(q/p) + \arg(\bar{\rho}_{f_1})) \neq \sin(\arg(q/p) + \arg(\bar{\rho}_{f_2})) , \tag{6.57}$$

since $\arg(q/p)$ depends only on P^0.

Very soon after **CP** violation was discovered in kaon decays in 1964 [2], Wolfenstein suggested to ascribe it to ND with a 'superweak' interaction [176]. The impact of this ND is only for indirect **CP** violation based on $\langle K^0|H^{\text{odd}}_{SW}|\bar{K}^0\rangle \neq 0$ with $[\Delta S = 2] \neq 0$, while $[\Delta S = 1] = 0$.[13] It is a classification, not a model.

[10] Here flavor is strange, charm or beauty hadrons, but not top quarks.

[11] It is also referred to as **CP** violation in mixing.

[12] An alternative name is **CP** violation in interference between a decay with and without mixing.

[13] A general classification of dynamical models was defined using a general decomposition of the Hamiltonian into **CP** even and odd parts. The simplest model can be built by adding four quark coupling with the hierarchy $0 < H_{SW} \ll H_W$ (for example, $H^{\text{odd}}_{SW} \sim 10^{-8}\, H_W$). In the general case it was suggested that it would lead to $[\Delta F = 2] \neq 0$ and $[\Delta F = 1] = 0$.

Direct **CP** violation was established in the 1990's for K^0 decays, in the early years of this century for B^+, B^0 and B^0_s ones and quite recently, in 2019, for a D^0 transition. Details for all these phenomena are reported in **Chapter 7**.

6.5 $K^0 - \bar{K}^0$ oscillations

Some people (like us) think that one can learn from the history: in the October 1946 Rochester and Butler exposed cloud chambers (the state-of-the-art tracking chambers of that period) to *cosmic* rays and recorded two charged particles coming from the same location. They concluded they had observed the decay of an unobserved neutral particle[14] with a mass of (435 ± 100) MeV into two charged particles with a mass around 100 MeV; in today's language: $K^{\rm neut} \to \pi^+\pi^-$.[15] This experiment had witnessed the first 'strange' particle.[16] Its main peculiarity: its production rate greatly exceeds its decay rate. This apparent paradox was resolved by A. Pais in 1952 through the concept of *associated production* in fundamental physics [177]: a new quantum number – not surprisingly called "strangeness" (S) – was introduced; it is conserved by strong, but not weak forces. It was the first example of flavour F. Such particles can be produced pairwise with strong forces – like $pp \to K\bar{K} + X$ or $\pi p \to \Lambda K + X$; then they decay weakly: $K \to \pi\pi$ or $\Lambda \to p\pi$.

Since kaons carry non-zero flavor S, they can*not* be their anti-particles – unlike the π^0 (or η and η'); thus K^0 and $\bar{K}^0$ had to exist that are differing by two units of S. How can one establish their separate existence? This challenge was successfully taken by Gell-Mann and Pais [37]. Their analysis, introducing mixing through careful quantum mechanical reasoning, has yielded some of the glorious pages in the theoretical development.

It appeared 'natural' to assume **CP** invariance for the analysis of kaons – partly for its simplicity and partly because illustrious theorists had made strong-worded pronouncements why our Nature had better obey that symmetry.

Let us turn off weak forces: K^0 and $\bar{K}^0$ can neither decay nor transform into each other. For the single particle system the Hamilton operator H is a mass matrix that is diagonal and real:

$$H = \begin{pmatrix} M_K & 0 \\ 0 & M_K \end{pmatrix}, \tag{6.58}$$

where **CPT** symmetry enforces the equality of the two diagonal elements; thus K^0 and $\bar{K}^0$ are two *degenerate* mass eigenstates. The situation changes already *qualitatively* once weak forces are included, namely decay processes with $|\Delta S| = 1$ take place: $K^0 \to \pi\pi$ and $\bar{K}^0 \to \pi\pi$.

[14]Tracking chambers, being based on ionization, cannot record the passage of a neutral particle.
[15]For the record: charged pions were found in 1947 and the neutral one officially in 1950.
[16]'We' had also learnt that K and π carry spin zero.

Weak forces mix K^0 and $\bar{K}^0$ through the chain $K^0 \to \pi^+\pi^- \to \bar{K}^0$; the mixing term is included in the Hamiltonian as δ:

$$H = \begin{pmatrix} M_K & \delta \\ \delta & M_K \end{pmatrix} . \tag{6.59}$$

Since δ is second order in weak interactions, it is tiny compared to M_K. Still the Hamiltonian is no more diagonal; the new eigenstates are two **CP** eigenstates:

$$|K_1\rangle = \frac{1}{\sqrt{2}}(|K^0\rangle + |\bar{K}^0\rangle) \;\; , \;\; |K_2\rangle = \frac{1}{\sqrt{2}}(|K^0\rangle - |\bar{K}^0\rangle) \tag{6.60}$$

with different masses and lifetimes. To be clear:

$$\mathbf{CP}|K^0\rangle = |\bar{K}^0\rangle \;\; , \;\; \mathbf{CP}|K_{1/2}\rangle = \pm|K_{1/2}\rangle \;\; , \;\; \mathbf{CP}|\pi\pi\rangle = +|\pi\pi\rangle \; . \tag{6.61}$$

For non-leptonic decays **CP** symmetry leads to[17]:

$$K_1 \to 2\pi \;\; , \;\; K_2 \not\to 2\pi \;\; , \;\; K_2 \to 3\pi \tag{6.62}$$

The phase space for $K_2 \to 3\pi$ is very restricted and it does not come as a surprise that the **CP** *odd* K_2 lives much longer than the **CP** *even* K_1:

$$\tau(K_2) \sim 5 \cdot 10^{-8}\,\mathrm{s} \quad \text{vs.} \quad \tau(K_1) \sim 10^{-10}\,\mathrm{s}\,. \tag{6.63}$$

The numerically large difference of lifetimes reflects a dynamical 'accident'.[18] Much lighter pions would lead to $\tau(K_2) \sim \tau(K_1)$; that being the case, it is likely that **CP** violation would not have been established in the previous century. Therefore our community is 'very lucky' to have hierarchies as $2M_\pi < 3M_\pi < M_K < 4M_\pi$.

One renames $K_1 \Rightarrow K_S$ and $K_2 \Rightarrow K_L$. The ratio of oscillations and weak decays for strange mesons is very unusual:

$$\frac{2\Delta M_K}{\Gamma(K_L) + \Gamma(K_S)} \simeq \frac{2\Delta M_K}{\Gamma(K_S)} \simeq 0.98\,. \tag{6.64}$$

QM tells us that both $|\Delta M_K|$ and the sign of ΔM_K are observables. It turns out that:

$$M(K_L) > M(K_S) \tag{6.65}$$

leading to a nice mnemonic for K_L vs. K_S for English readers: L means both *larger* mass and *longer* lifetimes, whereas S denotes *smaller* mass and *shorter* lifetimes.

In 1964 a 'storm' happened, see Ref. [2]:

$$\frac{\Gamma(K_L \to \pi^+\pi^-)}{\Gamma(K_S \to \pi^+\pi^-)} = [(2.0 \pm 0.4) \cdot 10^{-3}]^2\;! \tag{6.66}$$

Our community can be forgiven for not worrying about the numerically tiny effect of Eq. (6.66), when calculations of $\Gamma(K_L \to \mu^+\mu^-)$ etc. yielded infinities in the absence of a renormalizable theory for weak interactions. The (somewhat) renormalizable Glashow-Salam-Weinberg theory [127] had been formulated in the 1960s and scored excellent success in predicting neutral currents. In the 1960s the present SM had

[17] $\mathrm{BR}(K_1 \to \pi^+\pi^-\pi^0) \sim \text{few}\times 10^{-7}$.

[18] Or is there a deeper reason which we do not understand?

gauge bosons and Higgs dynamics plus u, d and s quarks, although one was not sure whether they were physical fields. In 1970 it was pointed out that the SM is a real (i.e. renormalizable) theory, when one adds a fourth quark with $m_c \leq 2$ GeV; it also explained the absence of S changing neutral currents [131]. However, a general skepticism about the existence of the charm quark may have been the reason for most to be content with a 'superweak' **CP** violation. In 1974 a novel era opened with the observation of the super-narrow resonance J/ψ with its mass $\simeq 3.1$ GeV.

Another novel era opened, joining with the first one, when a crucial paper by M. Kobayashi and T. Maskawa was published in 1973 [128]. It listed three possible sources of **CP** violation in QFT, namely non-minimal Higgs dynamics, right-handed $W_R^{\pm}$ – and (at least) three families of quarks. The third option is what has been established so far: (a) The SM does not produce "superweak" dynamics, which means that we have both $[\Delta F = 2] \neq 0 \neq [\Delta F = 1]$. (b) So far, the established **CP** asymmetries with hadrons are consistent with the SM predictions.

6.5.1 *The 'year of 1964'*

The shock that 'our' community felt about **CP** violation [2] shows by looking at the alternatives with **CP** invariance:

- Right away it was suggested that a so far unobserved very light neutral meson U with $\mathbf{CP}|U\rangle = -|U\rangle$ restores **CP** invariance: $K_L \to K_S + U \to (\pi\pi)_{K_S} + U$. It was mimicking Pauli's brilliant introduction of the neutrino to reconcile the continuous lepton electron spectrum observed in β decays with conservation of energy and momentum (see **Sect. 4.6**). More careful studies revealed that the decay rate evolution of $K^0 \to \pi^+\pi^-$ was *not* described by a simple sum of $K_S \to \pi^+\pi^-$ and $K_L \to \pi^+\pi^-$: it also exhibited an *interference* region between the two transitions; interference requires *indistinguishable* FS. Observing interference between $K_L \to \pi\pi$ and $K_S \to \pi\pi$ with K_S being coherently regenerated from a K_L beam ruled out this scenario[19] [34].
- Observing $K_L \to \pi\pi$ to occur does not establish **CP** violation by itself. What is needed for this conclusion is to have two mass *non*-degenerate[20] states K_L and K_S, with $K_S \to \pi\pi$ and $K_L \to \pi\pi\pi$, $\pi^+\pi^-$ – within the framework of *linear* quantum mechanics with its superposition principle! It was suggested to introduce a judiciously chosen *non*-linear term into the Schrödinger equation to induce $K_L \to \pi^+\pi^-$ with **CP** *conserving* dynamics [34]. This would violate the superposition principle – yet that had not been tested to the required sensitivity before. This option could be proved false, since it predicted very different oscillation rates in the neutral kaons – and it was quickly.

[19]This story represents a modern example of the ancient Roman saying: "Quod licet Jovi, non licet bovi"; i.e., "What is permitted to Jove, is not permitted to a bull". That means that we mere mortals can*not* get away with speculation like 'Jove = Jupiter = Pauli'.
[20]It does not matter at this point how tiny ΔM_K is.

- The observed **CP** asymmetry might be an *environmental* effect due to the preponderance of matter vs. anti-matter in our corner of our Universe. One advocated possibility was the existence of a novel long-range force of cosmological origin. It was an old history (by our scale), when most people did not think about Dark Matter and/or Dark Energy.

These escape roads show how *unprepared* the theoretical community was to accept **CP** violation – except for Lev Okun [1]! On the other hand, our community came quickly and decisively around to accept **CP** violation as a fact.

6.5.2 $\Delta S = 2$ *amplitudes*

The SM has been very successful in describing oscillations of $K^0 - \bar{K}^0$. It is a 'complex' landscape: proper treatment of **CP** phenomenology needs the full machinery of theoretical tools available in the QFT and appreciation of its subtleties.

With*out* an *elementary* $\Delta S = 2$ interaction in the SM (i.e., without Ref. [176]), one gets the effective $\Delta S = 2$ coupling by iterating the basic $\Delta S = 1$ one:

$$\mathcal{L}_{\text{eff}}(\Delta S = 2) = \mathcal{L}_{\text{eff}}(\Delta S = 1) \otimes \mathcal{L}_{\text{eff}}(\Delta S = 1) \,. \tag{6.67}$$

Thus one gets the celebrated box diagrams, namely (re-fined) one-loop diagrams. We have three families of quarks – $[u, d]$, $[c, s]$ and $[t, b]$ – and apply the general GIM mechanism; integrating over the internal lines yields a *convergent* result [178]:

$$\begin{aligned}\mathcal{L}^{\text{box}}_{\text{eff}}(\Delta S = 2; \mu) = \left(\frac{G_F}{4\pi}\right)^2 M_W^2 \cdot \big(&\eta_{\text{cc}}(\mu)\lambda_c^2 E(x_c;\mu) + \eta_{\text{tt}}(\mu)\lambda_t^2 E(x_t;\mu) + \\ &+ 2\eta_{\text{ct}}(\mu)\lambda_c\lambda_t E(x_c, x_t;\mu)\big) \left[\bar{d}\gamma_\mu(1-\gamma_5)s\right]^2\Big|_{(\mu)}\end{aligned} \tag{6.68}$$

with λ_i denoting combinations of CKM parameters $\lambda_i = \mathbf{V}_{is}\mathbf{V}^*_{id}$ $(i = c, t)$, while $E(x_i;\mu)$ and $E(x_c, x_t;\mu)$ reflect box loops with the internal c and t quarks:

$$E(x_i;\mu) = x_i^2\left(\frac{1}{4} + \frac{9}{4(1-x_i)} - \frac{3}{2(1-x_i)^2}\right) - \frac{3}{2}\left(\frac{x_i}{1-x_i}\right)^3 \log x_i$$

$$E(x_c, x_t;\mu) = x_c x_t\left[\left(\frac{1}{4} + \frac{3}{2}\frac{1}{1-x_t} - \frac{3}{4}\frac{1}{(1-x_t)^2}\right)\frac{\log x_t}{x_t - x_c} + (x_c \leftrightarrow x_t)\right. \tag{6.69}$$

$$\left. - \frac{3}{4}\frac{1}{(1-x_c)(1-x_t)}\right] \;, \; x_i = \frac{m_i}{M_W} \;; \tag{6.70}$$

$\eta_{qq'}$ contain QCD radiative corrections evolving the effective Lagrangian from M_W down below m_c[21]:

$$\eta_{cc}(\mu) \simeq 1.38 \pm 0.20 \;\;,\;\; \eta_{tt}(\mu) \simeq 0.57 \pm 0.01 \;\;,\;\; \eta_{ct}(\mu) \simeq 0.47 \pm 0.04 \tag{6.71}$$

with $\mu \sim 0.5 - 1$ GeV. On-shell matrix elements of local $\Delta S = 2$ operators are parametrized as

$$\langle K^0|(\bar{d}\gamma_\mu(1-\gamma_5)s)(\bar{d}\gamma^\mu(1-\gamma_5)s)|\bar{K}^0\rangle = -\frac{4}{3}B_K f_K^2 M_K \tag{6.72}$$

[21] The items of scale dependence and other uncertainties are discussed in Refs. [179], [180].

where $f_K \simeq 156$ MeV and B_K is called the 'bag' factor for historical reasons. In 2009 a value $B_K^{\rm theor} \simeq 0.8 \pm 0.2$, based on chiral symmetry, $1/N_C$ and QCD sum rules, was generally used [12]. The stage has changed with the arrival of Lattice QCD (LQCD) determinations: recent results are quite a progress [181][22]:

$$\widehat{B}_K^{\rm LQCD} = 0.7625 \pm 0.0097 \,. \tag{6.73}$$

6.5.3 ΔM_K

The Glashow-Illiopolous-Maiani (GIM) mechanism [131] (published in 1970) pointed out there are *no* flavor-changing neutral currents if there is a fourth quark c – "charm" – with $M_K < m_c \leq 2$ GeV in addition to u, d and s.[23] However, this statement did not convince the skeptics: they needed a 'Damascus' experience to change from 'Saulus' into 'Paulus' – i.e., from disbelievers into believers. It happened very quickly in the so called 'November revolution' in 1974; an unusually narrow resonance at 3096 MeV/c^2 (named J/ψ) was found by two experiments – one at Stanford Linear Accelerator on the west coast of the US and the other one at Brookhaven National Laboratory ($p\,Be$ collisions) on its east coast. It was soon established as a bound state $[\bar{c}c]$, especially after the observation at SLAC of its radial excited state ψ' at at 3686 MeV/c^2, which gave no room for other interpretations.

Given the values of the CKM factors, the charm produces the dominant contributions to the kaon mass difference (see Eq. (6.68)):

$$\begin{aligned} M_{K_L} - M_{K_S} &= \Delta M_K = 2\langle \bar{K}^0 | \mathcal{L}_{\rm eff}^{\rm box}(\Delta S = 2; \mu) | K^0 \rangle \\ &\simeq 2\left(\frac{G_F}{4\pi}\right)^2 M_W^2 \eta_{cc}(\mu) \lambda_c^2 E(x_c;\mu)\, \langle \bar{K}^0 | \left[\bar{d}\gamma_\mu(1-\gamma_5)s\right]\Big|^2_{(\mu)} |K^0\rangle \\ &\sim\ 3 \cdot 10^{-12}\,\text{MeV} \,. \end{aligned} \tag{6.74}$$

The approximate numerical result is obtained after substituting the matrix element as in Eq. (6.72). K_L is (a bit) heavier than K_S, and the SM gives at least the right order of ΔM_K. These conclusions are highly non-trivial. Furthermore, analyzing the kaon decay rate evolution in time one can extract the experimental value $\Delta M_K|_{\rm exp.} \simeq 3.48 \cdot 10^{-12}$ MeV and conclude: $\Delta M_K \sim 0.6 \cdot \Delta M_K|_{\rm exp.}$. It is an achievement for short-distance dynamics. We can assign the deficit in $\Delta M_K|_{\rm exp.}$ to long distance physics corresponding to non-local operators. Thus we have $\Delta M_K|_{LD} \sim 0.4 \cdot \Delta M_K|_{\rm exp.}$, where the long-distance dynamics can come from hadron intermediate states, that is: $K^0 \to \pi, \eta, \eta', 2\pi, ... \to \bar{K}^0$.

6.5.4 *Indirect* CP *violation in* K^0 *transitions*

The two pion FS in the weak decays of kaons can be classified by their isospin I. Due to Bose statistics two S wave pions in neutral $K \to \pi\pi$ can form an $I = 0$ or 2,

[22] One can link B_K to a renormalization group independent B-parameter $\widehat{B}_K$. See 'FLAG Review 2019' [181], in particular **Sect. 6**.

[23] 'Charm' has 'magic power' to prevent bad luck – like 'charming' a venomous snake.

yet not a $I = 1$ configuration. Thus Watson's theorem, see **Sect. 2.10.1**, tells us that the amplitude for K^0 or $\bar{K}^0$ decaying into two pions with total isospin I can be expressed as follows:

$$\langle (2\pi)_I | H_W | K^0 \rangle = \mathcal{A}_I e^{i\delta_I} \tag{6.75}$$

$$\langle (2\pi)_I | H_W | \bar{K}^0 \rangle = \bar{\mathcal{A}}_I e^{i\delta_I} \; ; \tag{6.76}$$

δ_I denote the S wave $(\pi\pi)_I$ phase shift at the energy M_K. If H_W conserves **CP** invariance, both $\mathcal{A}_I$ and $\bar{\mathcal{A}}_I$ are real with $\mathcal{A}_I = \bar{\mathcal{A}}_I$. In that case the phase of the decay amplitude is determined by strong (and electric) FSI, since they follow both **CP** and **T** invariance. When there is **CP** violation, an additional phase referred to as *weak* phase will emerge in $\mathcal{A}_I$ and $\bar{\mathcal{A}}_I$. In **Sect. 6.1** we discussed oscillations in general, here we talk about the special case of neutral kaons; expressing the two-pion state in terms of $(2\pi)_{I=0,2}$, we have

$$\begin{aligned}
\langle \pi^+\pi^- | H_W | K_L \rangle &= +\sqrt{\frac{2}{3}} e^{i\delta_0}\, p\, \mathcal{A}_0 \left[\Delta_0 + \frac{1}{\sqrt{2}} e^{i(\delta_2 - \delta_0)} \omega \Delta_2 \right] \\
\langle \pi^0\pi^0 | H_W | K_L \rangle &= -\sqrt{\frac{2}{3}}\, p\, \mathcal{A}_0 \left[\Delta_0 - \sqrt{2} e^{i(\delta_2 - \delta_0)} \omega \Delta_2 \right] \\
\langle \pi^+\pi^- | H_W | K_S \rangle &= +\frac{2}{\sqrt{3}}\, e^{i\delta_0} \left(1 + \frac{1}{\sqrt{2}}\, \omega\, e^{i(\delta_2 - \delta_0)} \right) \\
\langle \pi^0\pi^0 | H_W | K_S \rangle &= -\frac{2}{\sqrt{3}}\, e^{i\delta_0} \left(1 - \sqrt{2}\, \omega\, e^{i(\delta_2 - \delta_0)} \right)
\end{aligned} \tag{6.77}$$

with $\omega \equiv A_2/A_0$ and $\Delta_I = 1 - q/p\, \bar{\mathcal{A}}_I/\mathcal{A}_I$. Since $\omega \simeq 1/20$, we can ignore terms of $\mathcal{O}(\omega^2)$. One can set $q/p \equiv e^{i\phi}$, where ϕ is complex in general; **CPT** invariance implies $\bar{\mathcal{A}}_I = \mathcal{A}_I^*$. It leads to

$$\Delta_I = 1 - \frac{q}{p} \frac{\bar{\mathcal{A}}_I}{\mathcal{A}_I} = 1 - \exp[i(\phi - 2 \arg \mathcal{A}_I)] \simeq -i(\phi - 2 \arg \mathcal{A}_I) \tag{6.78}$$

since **CP** violation is small in weak kaon decays. Then one finds[24]:

$$\phi \simeq \frac{-\mathrm{Im} M_{12} + i\, \mathrm{Im}\,(\Gamma_{12}/2)}{+\mathrm{Re} M_{12} - i\, \mathrm{Re}(\Gamma_{12}/2)} \, . \tag{6.79}$$

As mentioned before, one gets the effective $\Delta S = 2$ coupling by iterating the basic $\Delta S = 1$; moreover, in order to build Γ_{12}, one inserts a set of the intermediate on-shell states, which are basically $(\pi\pi)_{I=0}$, $(\pi\pi)_{I=2}$, $\pi^+\pi^-\pi^0$, $3\pi^0$ and $\pi l \nu$. According to phenomenology and the $\Delta = 1/2$ rule, one can ignore all but the $(\pi\pi)_{I=0}$ state, which yields $\Gamma_{12} \sim 2\pi\, \rho_{2\pi} \mathcal{A}_0^* \bar{\mathcal{A}}_0$, where $\rho_{2\pi}$ is a phase space factor. By naming ξ_i the phase of the amplitude A_i, we can write $\Gamma_{12} \sim 2\pi\, \rho_{2\pi} e^{-i\xi_0} |\mathcal{A}_0|^2$. Retaining **CP** violation only to the first order, one can set $\langle 2\pi | H_W | K^0 \rangle = \langle 2\pi | H_W | K_S \rangle / \sqrt{2}$ leading to

$$\Gamma_{12} \sim \frac{1}{2} e^{-2i\xi_0}\, \Gamma_S \simeq \frac{1}{2} e^{-2i\xi_0}\, \Delta\Gamma_K \, . \tag{6.80}$$

[24] For a derivation, see **Sect. 7.2** in Ref. [12].

One can use the ratios of non-leptonic amplitudes for $K_L \to \pi\pi$ vs. $K_S \to \pi\pi$ and define:

$$\eta_{+-} \equiv \frac{\langle \pi^+\pi^-|H_W|K_L\rangle}{\langle \pi^+\pi^-|H_W|K_S\rangle} \equiv \epsilon_K + \epsilon' \quad , \quad \eta_{00} \equiv \frac{\langle \pi^0\pi^0|H_W|K_L\rangle}{\langle \pi^0\pi^0|H_W|K_S\rangle} \equiv \epsilon_K - 2\epsilon' \ . \quad (6.81)$$

In the case of *indirect* **CP** violation only, there is not a difference between $\eta_{+-} = \epsilon_K = \eta_{00}$. Using the above relations, included Eq. (6.79), one arrives at

$$\epsilon_K \simeq -i \left(\frac{\operatorname{Im} M_{12} - \frac{i}{2}\operatorname{Im}\Gamma_{12}}{\Delta M_K + \frac{i}{2}\Delta\Gamma_K} + \xi_0 \right) , \quad (6.82)$$

where the denominator of the first term is well approximated in the kaon case by $(M_S - M_L) - \frac{i}{2}(\Gamma_S - \Gamma_L)$, the difference of the two eigenvalues of the $\mathbf{M} - \frac{i}{2}\mathbf{\Gamma}$ matrix, see Eqs. (6.9,6.10). Furthermore, using Eq. (6.80), one obtains

$$\epsilon_K \simeq \frac{1}{\sqrt{1 + (\frac{\Delta\Gamma_K}{2\Delta M_K})^2}} e^{i\phi_{SW}} \left(-\frac{\operatorname{Im}M_{12}}{\Delta M_K} + \xi_0 \right) \simeq \frac{1}{\sqrt{2}} e^{i\phi_{SW}} \left(-\frac{\operatorname{Im}M_{12}}{\Delta M_K} + \xi_0 \right)$$

where

$$e^{i\phi_{SW}} \equiv \operatorname{tg}^{-1}\frac{2\Delta M_K}{\Delta\Gamma_K} \ ; \quad (6.83)$$

$\Delta\Gamma_K/(2\Delta M_K)|_{\text{exp.}} \simeq 1$ and $\phi_{SW} = (43.51 \pm 0.05)^o$ [8] are basically independent of **CP** violation. The ξ_0 term comes from long-distance dynamics. The expression for ϵ'/ϵ_K and the experimental result of $\epsilon' \ll \epsilon_K$ show that long-distance can*not* give sizable contributions to ϵ_K by themselves–we will discuss that in **Sect. 7.2**. By itself ϵ_K is *not* an observable, but $|\epsilon_K|$ is. From Eq. (6.77) one finds:

$$|\epsilon_K| \simeq \frac{1}{2}\Delta_0 = \frac{1}{2}\left(1 - \frac{q}{p}\frac{\bar{\mathcal{A}}_0}{\mathcal{A}_0}\right) . \quad (6.84)$$

If also weak dynamics would follow **CP** and **T** *invariance*, it would lead to $(q\,\bar{\mathcal{A}}_0/p\,\mathcal{A}_0) = 1$; thus $|\epsilon_K| = 0$. We observe that by a change of phase $q/p \to q/p\, e^{i\phi}$ and $\bar{\mathcal{A}}_0/\mathcal{A}_0 \to e^{-i\phi}\,\bar{\mathcal{A}}_0/\mathcal{A}_0$, but $(q/p)(\bar{\mathcal{A}}_0/\mathcal{A}_0) \Rightarrow (q/p)(\bar{\mathcal{A}}_0/\mathcal{A}_0)$; i.e., this product is phase independent as it should. The formalism we have adopted allows us to discuss **CP** violation without choosing a phase convention.

Indirect **CP** violation in kaons has been established [8]:

$$|\epsilon_K| = \left| \frac{2}{3}\eta_{+-} + \frac{1}{3}\eta_{00} \right| = (2.228 \pm 0.011) \cdot 10^{-3}|_{\text{exp}} \ . \quad (6.85)$$

Within the KM ansatz [128], the interplay of all three families is essential. Since $m_t \gg m_c > M_K$, the relevant effective operator is the local $\mathcal{L}_{\text{eff}}^{\text{box}}(\Delta S = 2)$, discussed in **Sect. 6.5.2**. By using it to evaluate $\operatorname{Im}M_{12}$, one finds:

$$\begin{aligned} |\epsilon_K|_{\text{KM}} &\simeq |\epsilon_K|_{\text{KM}}^{\text{box}} \simeq \frac{G_F^2}{8\pi^2} M_W^2 \frac{|\langle K^0|(\bar{d}\gamma_\mu(1-\gamma_5)s)(\bar{d}\gamma_\mu(1-\gamma_5)s)|\bar{K}^0\rangle|_{(\mu)}}{\sqrt{2}\,\Delta M_K} \\ &\cdot \left[\operatorname{Im}(\lambda_t^2)E(x_t)\eta_{tt}(\mu) + \operatorname{Im}(\lambda_c^2)E(x_c)\eta_{cc}(\mu) + 2\operatorname{Im}(\lambda_c\lambda_t)E(x_c,x_t)\eta_{ct}(\mu)\right] \\ &\simeq 2.23 \cdot 10^{-3} \ . \end{aligned} \quad (6.86)$$

The numerical result is obtained by using Eq. (6.72) and other approximate values given above. In the SM indirect **CP** violation proceeds from box diagrams; it represents a scenario where second order interactions are accessible at experiment, allowing a good control of ND. Before top quarks had been established directly by the CDF and D0 collaborations in the middle of the 1990's, it had already been stated that the SM needs them with a mass well beyond 100 GeV to produce the measured $|\epsilon_K|$ (and the large $B^0 - \bar{B}^0$ oscillations).

6.5.5 *Bell-Steinberger inequality*

Allow us to go back to history about $K^0 - \bar{K}^0$ oscillations. With K_1 and K_2 denoting the two mass eigenstates of the $K^0 - \bar{K}^0$ complex one gets[25]

$$\langle K_1|\frac{1}{2}\mathbf{\Gamma} + i\mathbf{M}|K_2\rangle = \left(\frac{1}{2}\Gamma_2 + iM_2\right)\langle K_1|K_2\rangle \tag{6.87}$$

$$\langle K_2|\frac{1}{2}\mathbf{\Gamma} + i\mathbf{M}|K_1\rangle = \left(\frac{1}{2}\Gamma_1 + iM_1\right)\langle K_2|K_1\rangle \,. \tag{6.88}$$

Taking the second equation and adding the complex conjugate of the first one yields:

$$\langle K_2|\mathbf{\Gamma}|K_1\rangle = \left[\frac{1}{2}(\Gamma_1 + \Gamma_2) + i(M_1 - M_2)\right]\langle K_2|K_1\rangle \,. \tag{6.89}$$

Using the relation

$$\langle K_2|\mathbf{\Gamma}|K_1\rangle = 2\pi\sum_f \delta(M_P - M_f)\,\langle f|H|K_2\rangle^\star\,\langle f|H|K_1\rangle \tag{6.90}$$

as well as the Schwartz inequality

$$|\langle K_2|\mathbf{\Gamma}|K_1\rangle|^2 \leq \sum_f \Gamma_1^f\Gamma_2^f \,, \tag{6.91}$$

one arrives to[26]:

$$|\langle K_2|K_1\rangle| \leq \sqrt{\frac{4\sum_f \Gamma_1^f\Gamma_2^f}{(\Gamma_1+\Gamma_2)^2 + 4(M_1 - M_2)^2}} \,, \tag{6.92}$$

which is called the Bell-Steinberger inequality. It was first derived for neutral kaons, but holds in general for neutral mesons. The two mass eigenstates are no longer orthogonal to each other if **CP** (or in principle **CPT**) invariance is violated. This inequality means that the amount of that non-orthogonality is constrained by a sum over exclusive K_1 and K_2 widths. It is numerically quite relevant for kaons. Using the measured $\Gamma_L \ll \Gamma_S \simeq 2\,\Delta M_K$, one gets

$$|\langle K_L|K_S\rangle| \leq \sqrt{\frac{4\sum_f \Gamma_L^f\Gamma_S^f}{(\Gamma_L+\Gamma_S)^2 + 4(\Delta M_K)^2}} \simeq \sqrt{\frac{\sum_f 2\Gamma_L^f\Gamma_S^f}{\Gamma_S^2}} \leq \sqrt{\frac{2\Gamma_L}{\Gamma_S}} \simeq 0.06 \tag{6.93}$$

[25]The Eqs. (6.87,6.88) hold in general for neutral mesons.

[26]See the Proceedings of the 1965 Oxford Conference [182].

where one uses $\Gamma_L \Gamma_S = \sum_{f_L, f_S} \Gamma_L^{f_L} \Gamma_S^{f_S} \geq \sum_f \Gamma_L^f \Gamma_S^f$. Since **CP** invariance requires $\langle K_L | K_S \rangle = 0$, the following message is conveyed: *unitarity* as expressed through the Bell-Steinberger relation together with the experimental findings $\Gamma_L \ll \Gamma_S \simeq 2\Delta M_K$ already imposes a near-orthogonality of K_L and K_S, *irrespective* of **CP** violation! This inequality responded to a comment from Pais reported in **Sect. 2.9**: "... **CP** invariance as a near miss, made the news even harder to digest.".[27]

6.5.6 CP *violation in* K_L *semi-leptonic decays*

CP violation was found first in 1964 in non-leptonic transition $K_L \to \pi^+\pi^-$, but it has to be probed also in semi-leptonic decays of K_L, which are flavor specific:

$$\begin{aligned} \langle l^+ \nu_l \pi^- | H_W | K_L \rangle &= + \, p_K \, \langle l^+ \nu_l \pi^- | H_W | K^0 \rangle \\ \langle l^- \bar{\nu}_l \pi^+ | H_W | K_L \rangle &= - \, q_K \, \langle l^- \bar{\nu}_l \pi^+ | H_W | \bar{K}^0 \rangle \, . \end{aligned} \tag{6.94}$$

One can compare with two semi-leptonic decays of K_L (see Eq. (6.46)):

$$\begin{aligned} A_{\mathbf{CP},SL} \equiv& \frac{\Gamma(K_L \to l^+ \nu_l \pi^-) - \Gamma(K_L \to l^- \bar{\nu}_l \pi^+)}{\Gamma(K_L \to l^+ \nu_l \pi^-) + \Gamma(K_L \to l^- \bar{\nu}_l \pi^+)} \simeq \frac{|p_K|^2 - |q_K|^2}{|p_K|^2 + |q_K|^2} \simeq \\ \simeq& \, 2 \, \frac{\mathrm{Re}\,(\epsilon_K)}{1 + |\epsilon_K|^2} \simeq 2 \, \mathrm{Re}\,(\epsilon_K) \simeq \sqrt{2} \, |\epsilon_K| \, . \end{aligned} \tag{6.95}$$

The asymmetry $A_{\mathbf{CP},SL}$ is based on $K^0 - \bar{K}^0$ oscillations, but it is time independent. Its present value:

$$A_{\mathbf{CP},SL} = (3.32 \pm 0.06) \cdot 10^{-3}|_{\mathrm{PDG2020}} \, , \tag{6.96}$$

is consistent with Eq. (6.95).

6.5.7 *Kabir test*

$K^0 - \bar{K}^0$ oscillations exhibit a measured **T** violation . The most direct test of it is given by the "Kabir Test" proposed in 1970 [183]. It was performed by the CPLEAR collaboration at CERN with a low energy $\bar{p}$ beam stopped by an hydrogen target. The two processes $\bar{p}p|_{\text{close to rest}} \to K^0 K^- \pi^+$ vs. $\bar{K}^0 K^+ \pi^-$ were studied; actually one compares $K^0 K^-$ vs. $\bar{K}^0 K^+$ from their *strong productions* and later from their *weak decays*. The "strangeness" of the neutral kaon at the time of production is tagged by the charged kaon, namely $K^{neutr} \to \pi^+\pi^-$ and $K^+ \to \mu^+\nu, \pi^+\pi^0/K^- \to \mu^-\bar{\nu}, \pi^-\pi^0$; likewise later. Again, oscillations tell us that rate$(\bar{K}^0 \Rightarrow K^0) \neq 0 \neq$ rate$(K^0 \Rightarrow \bar{K}^0)$. The measured asymmetry was [15]:

$$A_{\mathbf{T}} = \frac{\Gamma(K^0 \Rightarrow \bar{K}^0) - \Gamma(\bar{K}^0 \Rightarrow K^0)}{\Gamma(K^0 \Rightarrow \bar{K}^0) + \Gamma(\bar{K}^0 \Rightarrow K^0)} = (6.6 \pm 1.3|_{\mathrm{stat}} \pm 1.0|_{\mathrm{syst}}) \cdot 10^{-3} \, . \tag{6.97}$$

It is different from zero by 4σ uncertainty. The present SM tells us:

$$A_{\mathbf{T},\mathbf{SM}} = 2\sqrt{2} \cdot |\epsilon_K| \simeq (6.3 \pm 0.03) \cdot 10^{-3} \, . \tag{6.98}$$

In order to assess **T** violation , setting (or not) **CPT** symmetry , some assumptions are needed [184].

[27]The situation is very different for beauty and charm mesons, where many-body FS are crucial.

6.5.8 CP *violation in* $K_{L(S)} \to \pi^+\pi^-\gamma$

The modes $K_{L,S} \to \pi^+\pi^-\gamma$ have been measured [8]:

$$\mathrm{BR}(K_L \to \pi^+\pi^-\gamma) = (4.15{\pm}0.15)\cdot 10^{-5} \ , \ \mathrm{BR}(K_S \to \pi^+\pi^-\gamma) = (1.79{\pm}0.05)\cdot 10^{-3} \ .$$

These are not pure **CP** eigenstates; yet these can produce **CP** violation by an interference between K_S and K_L. The photon can be emitted in two ways: by QED internal bremsstrahlung (*IB*) and by emission from one of the pions in the $\pi^+\pi^-$ state or from the weak decay vertex (direct emission *DE*). The photon emission can proceed though an electric (E_l) or magnetic (M_l) multi-pole. Electric and magnetic multi-poles have opposite transformation properties: **CP** eigenvalues of $\pi\pi\gamma$ states are $(-1)^{l+1}$ for E_l and $(-1)^l$ for M_l.

We can focus on a special aspect: interference of E_1 and M_1 amplitudes, which yields a circularly polarized photon. It leads to a triple correlation between the pion momenta and the photon polarization: $P_\perp^\gamma = \langle \vec{\epsilon}_\gamma \cdot (\vec{p}_{\pi^+} \times \vec{p}_{\pi^-})\rangle$, which is **CP** *odd*; its leading contribution is proportional to $|\eta_{+-}|$ entering in the E_1 amplitude.

Bremsstrahlung emission is essentially an electric dipole transition. *DE* contributions are generally smaller, with strong dominance of lowest multipolarity terms. *DE* terms $E_1(M_1)$ in $K_L(K_S) \to \pi^+\pi^-\gamma$ are also suppressed. The photon energy spectra for the two types of emission (*IB* and *DE*) are different, therefore allowing the measurement of their relative importance in kaon decays.

In analogy to ϵ_K one can define the ratio of E_1 amplitudes

$$\eta_{+-\gamma} = \frac{T(K_L \to \pi^+\pi^-\gamma, E_1)}{T(K_S \to \pi^+\pi^-\gamma, E_1)} \tag{6.99}$$

that measure **CP** violation. One gets $|\eta_{+-\gamma}| \simeq |\eta_{+-}|$ and compare the decay rate evolution of $K_L \to \pi^+\pi^-\gamma$ and $K_S \to \pi^+\pi^-\gamma$ as a function of the time of decay. It has been demonstrated that $K_L - K_S$ interference occurs in $K \to \pi^+\pi^-\gamma$, and it is very well consistent with those of $K_L \to \pi^+\pi^-$: $|\eta_{+-\gamma}| = (2.35 \pm 0.07)\cdot 10^{-3}$ vs. $|\eta_{+-}| = (2.228 \pm 0.011)\cdot 10^{-3}$. Yet it is not the end of the story – see next.

6.5.9 CP *asymmetry in* $K_L \to \pi^+\pi^-e^+e^-$

Direct measurement of the polarization of a real photon in HEP experiments is a difficult task,[28] as talked about just above in **Sect. 6.5.8**. The decays $K_{L,S} \to \pi^+\pi^-e^+e^-$ proceed through an intermediate state of two pions plus a *off-shell* photon followed by its conversion into a lepton pair $K_L \to \pi^+\pi^-\gamma^* \Rightarrow \pi^+\pi^-e^+e^-$. It has been pointed out first by Sehgal *et al.* [185] that the angle Φ between the e^+e^- and $\pi^+\pi^-$ planes in the K_L rest frame can be measured:

$$\Phi \equiv \angle(\vec{n}_l, \vec{n}_\pi)$$
$$\vec{n}_l \equiv \vec{p}_{e^+} \times \vec{p}_{e^-}/|\vec{p}_{e^+} \times \vec{p}_{e^-}| \ , \ \vec{n}_\pi \equiv \vec{p}_{\pi^+} \times \vec{p}_{\pi^-}/|\vec{p}_{\pi^+} \times \vec{p}_{\pi^-}| \ ; \tag{6.100}$$

[28] A truly beautiful exception to this statement is the historical measurement of neutrino helicity made by Goldhaber, Grodzins and Sunyar in 1957 [146].

one can analyze the decay rate as a function of Φ:

$$\frac{\mathrm{d}\Gamma}{\mathrm{d}\Phi} \equiv \Gamma_1 \cos^2\Phi + \Gamma_2 \sin^2\Phi + \Gamma_3 \cos\Phi \sin\Phi\,. \tag{6.101}$$

With $\cos\Phi\sin\Phi = (\vec{n}_l \cdot \vec{n}_\pi)(\vec{n}_l \times \vec{n}_\pi)\cdot(\vec{p}_{\pi^+} + \vec{p}_{\pi^-})/|\vec{p}_{\pi^+} + \vec{p}_{\pi^-}|$ one notes:

$$\cos\Phi\sin\Phi \overset{\mathbf{T,CP}}{\Longrightarrow} -\cos\Phi\sin\Phi \tag{6.102}$$

under both **T** and **CP** transformations; i.e., the observable Γ_3 represents **T** and **CP** odd correlations. It can be compared the Φ distribution integrated over two quadrants:

$$\langle A_{\mathbf{CP}}\rangle = \frac{\int_0^{\pi/2} \mathrm{d}\Phi\, \frac{d\Gamma}{d\Phi} - \int_{\pi/2}^{\pi} \mathrm{d}\Phi\, \frac{d\Gamma}{d\Phi}}{\int_0^{\pi} \mathrm{d}\Phi\, \frac{d\Gamma}{d\Phi}} = \frac{2\Gamma_3}{\pi(\Gamma_1+\Gamma_2)} \tag{6.103}$$

The amplitude for $K_L \to \pi^+\pi^- e^+ e^-$ can be written as (see Ref. [185]):

$$\begin{aligned} T(K_L \to \pi^+\pi^- e^+ e^-) &= e|T(K_S \to \pi^+\pi^-)| \\ &\cdot \left[\frac{g_{BR}}{M_K^4}\left(\frac{p_+^\mu}{p_+\cdot k} - \frac{p_-^\mu}{p_-\cdot k}\right) + \frac{g_{M1}}{M_K^4}\epsilon_{\mu\nu\alpha\beta}k^\nu p_+^\alpha p_-^\beta\right]\frac{e}{k^2}\bar{u}(k_-)\gamma^\mu v(k_+) \end{aligned} \tag{6.104}$$

(with $k = k_+ + k_-$) in terms of two couplings g_{BR} and g_{M1} for the bremsstrahlung and M_1 transitions. From observed rates one infers $g_{BR} = \eta_{+-}e^{i\delta_0(M_K^2)}$ and $g_{M1} = 0.76\, i\, e^{i\,\delta_1(s_\pi)}$, where $\delta_{0,1}$ denote the S- and P-wave $\pi\pi$ phase shifts. Thus one gets:

$$\langle A_{\mathbf{CP}}\rangle \simeq (15\cos\Theta_1)\% + \left(38\left|\frac{g_{E1}}{g_{M1}}\right|\cos\Theta_2\right)\% \tag{6.105}$$

with $\Theta_1 = \phi_{+-} + \delta_0 - \delta_1 - \frac{\pi}{2}$ (mod π) and $\Theta_2 = \phi_{+-} + \delta_0 - \frac{\pi}{2}$ (mod π); g_{E1} is a coupling for the E_1 transition and δ_1 the P-wave $\pi\pi$ phase shift averaged over the kinematical region. For $\phi_{+-} = \phi_{\mathrm{SM}}$ and experimental values $\delta_0 - \delta_1 \sim 30^o$ and $g_{E1}/g_{M1} \simeq 0.05$, one gets [185]:

$$\langle A_{\mathbf{CP}}\rangle = (14.3 \pm 1.3)\%\,. \tag{6.106}$$

It was a true predictions from theorists before they had the data.

The main theoretical uncertainty resides in what one assumes for the hadronic form factors. In the above analysis a phenomenological ansatz was employed; a later chiral perturbative theory yields a similar number [186].

The KTeV experiment at FNAL and subsequently NA48 at CERN measured the rates and asymmetry for this mode, see Ref. [187] (and [8]):

$$\langle A_{\mathbf{CP}}\rangle = (13.7 \pm 1.5)\%\,. \tag{6.107}$$

The discovery of such a spectacularly large **CP** asymmetry is a very significant result. One should note that this $\langle A_{\mathbf{CP}}\rangle$ is driven by $|\epsilon_K| \simeq 0.223\%$ entering the analysis, not the prediction of a specific model or theory. This **CP** asymmetry is so large because the **CP** violating amplitude is enhanced by kinematic bremsstrahlung factors. It is an excellent example that one can (and has to) use very small branching ratios to get deeper understanding underlying dynamics. There is a price vs. prize,[29] just compare:

$$\mathrm{BR}(K_L \to \pi^+\pi^- e^+ e^-) \simeq 3.11 \cdot 10^{-7} \quad \text{vs.} \quad \langle A_{\mathbf{CP}} \rangle \simeq 0.14\,! \tag{6.108}$$

Analyses of many-body FS should *not* been seen as a background to the information one can get from two-body ones. In **Sect. 7.1.2** we will discuss **T**-odd moments and even one-dimensional observables for D and B mesons.

6.5.10 $K^+ \to \pi^+ \bar{\nu}\nu$ *and* $K_L \to \pi^0 \bar{\nu}\nu$

There are different classes of very rare decays of K mesons[30]: we focus on the ultra-rare $K^+ \to \pi^+ \bar{\nu}\nu$ and $K_L \to \pi^0 \bar{\nu}\nu$. The SM predicts non-zero values for their branching ratios, but tiny ones with very small uncertainties. Not surprisingly, they are referred to as two 'golden modes'. The measurements of these branching ratios test experimentalists' expertise and patience. One can go back to ancient history: according to the Greek mythology, there were two wars over Troja.[31] In a similar vein: we talked about the heroic campaign over $K^0 - \bar{K}^0$ oscillations – ΔM_K and ϵ_K – as the first one, to be followed by a likewise epic struggle over $K^+ \to \pi^+ \bar{\nu}\nu$ and $K_L \to \pi^0 \bar{\nu}\nu$.

Let us first examine the theoretical side. The SM with three families of quarks has six triangles with very different features, but the same non-zero areas, see **Sect. 4.5.1** (and **Sects. 4.5.3, 4.5.4**). The $K \to \pi \bar{\nu}\nu$ decays are driven by loop diagrams in the SM. Let us consider the unitarity triangle I.1 in Eq. (4.59):

$$V_{ud}V_{us}^* \; [\mathcal{O}(\lambda)] + V_{cd}V_{cs}^* \; [\mathcal{O}(\lambda)] + V_{td}V_{ts}^* \; [\mathcal{O}(\lambda^5)] = 0\,. \tag{6.109}$$

The three vectors $V_{id}V_{is}^*$ converge to form an elongated triangle in the complex plane. We have that

- the side $V_{ud}V_{us}^*$ is connected to $\mathrm{BR}(K^+ \to \pi^0 l^+ \nu)$ decay
- the side $V_{td}V_{ts}^*$ can be described by $\mathrm{BR}(K^+ \to \pi^+ \bar{\nu}\nu)$
- the height is connected with $\mathrm{BR}(K_L \to \pi^0 \bar{\nu}\nu)$.

The $K \to \pi \bar{\nu}\nu$ decays are sensitive to the magnitude and imaginary part of V_{td}. The rate of $K_L \to \pi^0 \bar{\nu}\nu$ basically depends on **CP** violation that has been measured in $K_L \to \pi\pi$.[32]

The matrix element for $K \to \pi \bar{\nu}\nu$ transitions is given by [178]:

$$\langle \pi \bar{\nu}\nu | H(d\bar{s} \to \bar{\nu}\nu) | K \rangle = \frac{G_F}{\sqrt{2}} \frac{\alpha}{2\pi \sin^2\theta_W} \cdot \sum_{i=e,\mu,\tau} D \, \langle \pi | (\bar{s}d)_{V-A} | K \rangle (\bar{\nu}\nu)_{V-A} \tag{6.110}$$

[29] For the record: so far one can reproduce η_{+-} with **T** *invariant* dynamics through *fine-tuning* **CPT** breaking [188]; these authors did not like that at all.

[30] There is a good and clear '2020 Review of Particle Physics, 74. Rare Kaon Decays' [8].

[31] The Trojan War described in Homer's Iliad was the second one.

[32] The SM assumes massless neutrinos.

where[33]

$$D = \mathbf{V}_{td}\mathbf{V}^*_{ts}\, Z(x_t) + \mathbf{V}_{cd}\mathbf{V}^*_{cs}\, Z(x_c) \tag{6.111}$$

$$Z(x_i) = \frac{x_i}{8}\left[\frac{3(x_i-2)}{(1-x_i)^2}\,\mathrm{log}x_i + \frac{x_i+2}{x_i-1}\right]\ ,\ x_i = \frac{m_i^2}{M_W^2}\ . \tag{6.112}$$

Here we have ignored QCD corrections as well as terms of $\mathcal{O}(m_\tau^2/M_W^2)$. For these corrections and details of the computation we refer to Ref. [189].

In evaluating the hadronic matrix element $\langle\pi|J_\mu^{\rm had}|K\rangle$ one cannot ignore long-distance dynamics. Isospin invariance 'saves the day': it implies the equality of $\langle\pi^+|J_\mu^{\rm had,neut}|K^+\rangle$ and $\langle\pi^0|J_\mu^{\rm had,,neut}|K^0\rangle$ with $\langle\pi^0|J_\mu^{\rm had,charg}|K^+\rangle$; the latter has been extracted by the ordinary semi-leptonic $K^+\to\pi^0 l^+\nu$ decay. In particular, we have

$$\frac{\mathrm{BR}(K_L\to\pi^0\bar\nu\nu)}{\mathrm{BR}(K^+\to\pi^0 e^+\nu_e)} \simeq \frac{3}{2}\frac{\tau(K_L)}{\tau(K^+)}\frac{\alpha^2}{\pi^2\sin^4\theta_W}\left|\frac{\mathrm{Im}(V_{td}V_{ts}^*)}{V_{us}}Z(x_t)\right|^2\ . \tag{6.113}$$

The predictions from the SM are well controlled on the theoretical side. After hard work theorists are entitled to show their achievements [165, 190]:

$$\mathrm{BR}(K^+\to\pi^+\bar\nu\nu)|_{\rm SM} = (8.39\pm0.30)\cdot10^{-11}\left[\frac{|V_{cb}|}{40.7\cdot10^{-3}}\right]^{2.8}\left[\frac{\phi_3/\gamma}{73.2^0}\right]^{0.74} \tag{6.114}$$

$$\mathrm{BR}(K_L\to\pi^0\bar\nu\nu)|_{\rm SM} = (3.36\pm0.05)\cdot10^{-11}\left[\frac{|V_{cb}|}{40.7\cdot10^{-3}}\right]^{2}\left[\frac{\phi_3/\gamma}{73.2^0}\right]^{2} \tag{6.115}$$

Parametric uncertainty in the CKM angles can result in numbers with different central values, that with the actual precision can reach 10%. These predictions are connected in the SM. There is a (nearly) model independent upper bound [191]:

$$\mathrm{BR}(K_L\to\pi^0\nu\bar\nu) \le 1.1\cdot10^{-8}\ . \tag{6.116}$$

Using the BNL 787/949 90%CL bound on $K^+\to\pi^+\bar\nu\nu$ decay rate [192], this limit leads to $\mathrm{BR}(K_L\to\pi^0\bar\nu\nu)\le 1.46\cdot10^{-9}$.[34] Using instead the NA62 result in [193], the upper bound predicts $\mathrm{BR}(K_L\to\pi^0\bar\nu\nu)\le 8.14\cdot10^{-10}$. The PDG2020 gives

$$\mathrm{BR}(K^+\to\pi^+\bar\nu\nu)|_{\rm PDG2020} = (17\pm11)\cdot10^{-11} \tag{6.117}$$

$$\mathrm{BR}(K_L\to\pi^0\bar\nu\nu)|_{\rm PDG2020} < 300\cdot10^{-11}\ [\,=3\cdot10^{-9}\,]\ . \tag{6.118}$$

The PDG2020 limit is based on the 2016–2018 run from the KOTO collaboration [194]. At the ICHEP 2020 conference it was given an improved result [194]:

$$\mathrm{BR}(K_L\to\pi^0\bar\nu\nu) < 71\cdot10^{-11}\ . \tag{6.119}$$

However, the KOTO collaboration in 2021 has published an updated upper limit [195]:

$$\mathrm{BR}(K_L\to\pi^0\nu\bar\nu) < 490\cdot10^{-11}\ ,$$

[33] A reader might be confused by expressions as $1/(x_i-1)^2$ and $1/(x_i-1)$, since one gets infinities for $x_i\simeq1$: $-\log x/(x-1)^2+1/(x-1)$. However, one can write $\log x = (x-1)+\frac{1}{2}(x-1)^2-\frac{1}{3}(x-1)^3+\ldots = -1+1-\frac{1}{2}+\frac{1}{3}(x-1)$ etc.; the formula is consistent.

[34] In models with lepton flavor violation, the final state is not necessarily a **CP** eigenstate; then **CP** conserving contributions could dominate the decay rate.

which is worse than that presented at ICHEP 2020 because a possible new background source was identified.

- In 1997 the first data of $K^+ \to \pi^+ \bar{\nu}\nu$ came from the Brookhaven National Laboratory with kaon decays 'at rest' with a few candidates.
- The NA62 experiment at CERN proposed to measure the branching ratio of $K^+ \to \pi^+ \bar{\nu}\nu$ with 10% uncertainty. It has been running in 2016 - 2018; its 2016 data set has given one candidate [166]; a sample of its 2017 data set has given two candidates [167]. The NA62 collaboration has published a new result in 2021 that provides the strongest evidence yet for the existence of this truly rare process at 3.4σ significance [196]:

$$\mathrm{BR}(K^+ \to \pi^+ \nu\bar{\nu}) = (10.6^{+4.0}_{-3.4}|_{\mathrm{stat}} \pm 0.9|_{\mathrm{syst}}) \cdot 10^{-11} \tag{6.120}$$

  where the first error is statistical and the second one systematic. It is compatible with the SM prediction within one standard deviation.
- By $\sim$ 2024 the KOTO experiment might have the sensitivity to reach the SM prediction for $K_L \to \pi^0 \bar{\nu}\nu$.
- Experimenters are always thinking about novel tools to enhance our understanding of fundamental forces. For example: the 'NA62/KLEVER project' at CERN can measure $K_L \to \pi^0 \bar{\nu}\nu$ in the future [197].
- We 'suggest' that 'our' community needs about 1000 events of these modes to extract the value of $\mathbf{V}_{td}$ and/or identify likely the impact of ND.

Our colleagues from the theory side did an excellent job about the predictions from the SM and possible ND. Our community has the right to brag about the progress achieved in the previous decade and our current better understanding of flavor dynamics. Again, there is a price for a prize. The load is now on the shoulders of our experimental colleagues.

Actually, that is not the end of the story for the 21st century in a different 'dimension'; we have another player on the stage: LQCD. As we have said before, the differences between experimenters and theorists are not so clear when one talks about LQCD: to make progress our colleagues need more lattice data and refined analyses. A very recent example is an exploratory LQCD study of the long-distance contribution to the $K^+ \to \pi^+ \bar{\nu}\nu$ decay amplitude [198]. The long-distance contributions are expected to be of $O(5\%)$ in the branching ratio. Up to now the non-perturbative hadronic matrix element of the local four-fermion operator has been determined by using connections (isospin rotation) with the experimental measurement of ordinary semi-leptonic $K^+ \to \pi^0 l^+ \nu$, decay. The LQCD community would like to probe $K^+ \to \pi^+ \bar{\nu}\nu$ from first-principle calculations. Present studies use *un*physical quark masses. It is actually premature to draw conclusions; future progress can be expected in the next decade.

6.6 B^0 and B_s^0 systems

The data of the (experimental) birth of b quark physics can be placed on 1977: the collaboration of the E288 experiment at Fermilab, led by Leon Lederman, observed a narrow resonance at an energy of about 9.5 GeV in the reaction $p+N \to \mu^+\mu^- + X$ that is the di-muon production in proton-nucleon collision [199]. This resonance was named Υ.[35] Subsequently it was identified as one of the states of $[\bar{b}b]$ systems; it is often referred to as 'bottomonium'.[36] The first fully reconstructed B mesons were reported in 1983 by the CLEO collaboration [201], but already in 1980 [134] it had been pointed out that the SM predicts sizable **CP** asymmetries in the transitions of beauty mesons. Predictions of **CP** asymmetries in the transitions of B^0 and B_s^0 (and D^0) neutral mesons had also been given as early as 1987 [44].

The first evidence for *long* beauty lifetimes was given by the MAC collaboration [202] at the PEP ring of SLAC, quickly followed by the MARK-II collaboration, also at PEP-II [203]. The total width of the beauty quark can be easily guessed by scaling the expression for the muon width and considering that the decay of the b quark into the c quark involves a larger number of decay channels:

$$\Gamma_\mu \sim \mathcal{O}\left(\frac{G_F^2}{192\pi^3} m_\mu^5\right) \Rightarrow \Gamma_b \sim \mathcal{O}\left(\frac{G_F^2}{192\pi^3} m_b^5 |V_{cb}|^2 \cdot (2 \cdot 3 + 3)\right), \tag{6.121}$$

In the leading tree diagram, the virtual photon couples the pair (c, s) with 2 pairs of flavors $((u, d)$ and $(c, s))$, coming in 3 colors, or 3 pairs of charged leptons (e, μ and τ) with the respective neutrinos. Before data were available, it was guessed that $|V_{cb}| \sim \sin\theta_C$, but we have since learnt that $|V_{cb}| \sim (\sin\theta_C)^2 \sim 0.05$, i.e. that the lifetimes of beauty hadrons are 'long': $\tau_b \sim 1.5 \cdot 10^{-12}$ s.

6.6.1 $\bar{B}^0 - B^0$ *and* $\bar{B}_s^0 - B_s^0$ *oscillations*

$\bar{B}^0 - B^0$ oscillations were first discovered in 1987 when it was observed the existence of same-sign di-lepton events in the decay channel $e^+e^- \to \Upsilon(4S) \to B^0\bar{B}^0 \to$

[35] One should read this article in the CERN Courier [200]. It tells us about HEP results in the 1970's: (a) The 'improved' E288 experiment began taking data on 15 May 1977; after one week of taking data, a 'bump' appeared at 9.5 GeV. (b) On 21 May 1977 fire broke out in a device that measures current in a magnet, and the fire spread to the wiring etc. threatening the future of E288. Lederman was on the phone searching for a salvage expert. He found a Dutchman who lived in Spain. The expert agreed to come, but needed 10 days to get a US visa. Lederman called the US embassy, asking for an exception. Not possible, said the embassy official. Lederman mentioned that he was a Columbia University professor; the official was a Columbia graduate. The salvage expert was at Fermilab two days later. Collaborators used the expert's 'secret formulas' to treat electronic circuit boards, and E288 was back online by 27 May. (c) By 15 June, the collaborators had collected enough data to prove the existence of a 'bump' at 9.5 GeV: evidence for a new particle, the upsilon (Υ). On 1 July the collaborators submitted a paper to Physics Review Letters; it was published with*out* review [199]. (d) The 'old' E288 had shown several events with a mass $\sim$ 6 GeV in 1976. It has found a new particle, named 'upsilon'. Later it was realized it was a fluctuations, and this 'upsilon' became known as 'oopsLeon'.

[36] However, we prefer to call b quarks the 'beauty' ones.

$\mu^\pm \mu^\pm + X$ [204]. Without oscillations the two leptons from B^0 decays should carry opposite charges.

In contrast to the neutral kaon system, there is little distinction between the two neutral B meson mass eigenstates, $M_{B^0_H}$ and $M_{B^0_L}$[37]: we define $\Delta M_{B^0} \equiv M_{B^0_H} - M_{B^0_L} > 0$ and $\Delta M_{B^0_s} \equiv M_{B^0_{s,H}} - M_{B^0_{s,L}} > 0$. Also the difference in their widths can be measured. Based on the SM one can give qualitative expectations:

$$\Delta B = 2 : \Delta M_{B^0} < \Gamma_{B^0} \; , \; \Delta\Gamma_{B^0} \ll \Gamma_{B^0} \quad , \; \Delta\Gamma_{B^0} \ll \Delta M_{B^0} \; ;$$
$$\Delta M_{B^0_s} > \Gamma_{B^0_s} \; , \; \Delta\Gamma_{B^0_s} \sim \mathcal{O}(\Gamma_{B^0_s}) \; , \; \Delta\Gamma_{B^0_s} \ll \Delta M_{B^0_s} \; . \quad (6.122)$$

The large oscillations of neutral B mesons follow from box diagrams, sporting two W boson and two internal quark exchanges. These are basic diagrams contributing to $\Delta F = 2$ transitions between a neutral flavored particle and its antiparticle. In the case of B mesons, the main contribution to $\Delta M_{B^0_{(s)}}$ comes from virtual top quarks. The mass difference is proportional to the squared mass of the quark being exchanged; in this case it is basically $\sim m_t^2/M_W^2$. Studies of B oscillations have suggested that top quarks must be very heavy before its *direct* discovery. Our community can 'brag' about success with a true *pre*dictions of $m_t > 100$ GeV, based on $x_d = \Delta M_{B^0}/\Gamma_{B^0} \sim 0.8$. The different sizes of the CKM couplings in the box diagrams are coherent with the observed $\Delta M_{B^0} \ll \Delta M_{B^0_s}$. On general grounds one can predict that $\Delta\Gamma_{B^0_{(s)}} \ll \Delta M_{B^0_{(s)}}$: the weak decays of b quarks basically lead to c (or $c\,\bar{c}\,c$) on-shell states. The charm quark is considerably lighter than the $B^0_{(s)}$ mass; it is then the latter that sets the scale for $\Delta\Gamma_{B^0_{(s)}}$, which is in turn much smaller than the top mass. The relative width differences in the B^0 and B^0_s systems is also expected in the SM, being originated by different strength of the couplings in the FS to which both neutral B mesons can decay. Such decays, responsible of $\Delta\Gamma \neq 0$, involve $b \to c\bar{c}q$ quark level transitions, which are Cabibbo suppressed if $q = d$ and Cabibbo favored if $q = s$. It leads to a sizable $\Delta\Gamma_{B^0_s}$. Finally, the pattern of $\Delta B = 2$ oscillation is 'natural': $M_{B_H} > M_{B_L}$, while $\Gamma_{B_L} > \Gamma_{B_H}$.

Now we list 2019 data about oscillations:

$$\Delta M_{B^0} = (3.334 \pm 0.013) \cdot 10^{-13} \text{ GeV} \; , \; x_d = \frac{\Delta M_{B^0}}{\Gamma_{B^0}} = 0.769 \pm 0.004$$
$$\Delta M_{B^0_s} = (1.1688 \pm 0.0014) \cdot 10^{-11} \text{ GeV} \; , \; x_s = \frac{\Delta M_{B^0_s}}{\Gamma_{B^0_s}} = 26.81 \pm 0.08$$
$$\Delta\Gamma_{B^0_s} = (0.090 \pm 0.005) \cdot 10^{12} \text{ s}^{-1} \; , \; y_s = \frac{\Delta\Gamma_{B^0_s}}{2\Gamma_{B^0_s}} = 0.068 \pm 0.004$$

The SM also predicts $y_{B^0} = \Delta\Gamma_{B^0}/2\Gamma_{B^0_d} \sim 0.005$, although it is beyond present data. Since $x = \frac{\Delta M_B}{\Gamma_B} = \Delta M_B \cdot \tau_B$, this parameter represents the ratio of lifetime and oscillation time, and it is important for the observability of oscillations. It should not be very small, with $x \sim 1$ being optimal.

[37] H means that *measured* mass is higher, while L is lower; it is a definition.

6.6.2 *Indirect* CP *violation in* B^0 *and* B_s^0 *transitions*

Indirect **CP** violation was found in semi-leptonic decays of K_L, but the approach used for kaons did not continue for neutral B mesons, which is not surprising[38]:

$$\frac{\mathrm{Re}(\epsilon_{B^0})}{(1+|\epsilon_{B^0})|^2)} = (-0.5 \pm 0.4)\cdot 10^{-3}, \frac{\mathrm{Re}(\epsilon_{B_s^0})}{(1+|\epsilon_{B_s^0})|^2)} = (-0.15 \pm 0.70)\cdot 10^{-3} \quad (6.123)$$

Indirect **CP** violation has been established in $B^0 \to J/\psi K_S$ (and $B^0 \to J/\psi K_L$):

$$S_{J/\psi(nS)K^0}|_{\mathrm{PDG2020}} = 0.701 \pm 0.017\ ,\ C_{J/\psi(nS)K^0}|_{\mathrm{PDG2020}} = 0.005 \pm 0.020\ .$$

These are golden modes for analyzing **CP** violation; they have an accessible branching ratios ($\sim$ several$\cdot 10^{-4}$) and a very strong signature from $J/\psi \to \mu^+\mu^-$ decay.

They arise from quark transitions $b \to c\bar{c}s$ at the tree level with a CKM form factor $V_{cb}^* V_{cs}$. A second amplitude is generated by one-loop QCD penguin diagrams $b \to sg^* \to s\,\bar{c}c$; i.e., an *off*-shell gluon can produce an *on*-shell pair $\bar{c}c$. These contributions have different dependence on the CKM parameters: $V_{cb}^* V_{cs}$ for the tree diagram, $V_{tb}^* V_{ts}$ and $V_{ub}^* V_{us}$ for the penguin diagram with a virtual top quark and up quark, respectively. Thus one can use the unitarity relation $V_{tb}^* V_{ts} + V_{cb}^* V_{cs} + V_{ub}^* V_{us} = 0$ to connect 'tree' and penguin diagrams. The sides of this UT triangle have different sizes, namely $\mathcal{O}(\lambda^2)$ vs. $\mathcal{O}(\lambda^4)$ (see **Sects. 4.5.3, 4.5.4**). The weak phases of the tree diagram and the top-quark penguin are the same to leading order in λ, thus the presence of this penguin does not change the asymmetry. It is different for the remaining penguin – yet this is suppressed, by a ratio $\sim \lambda^2 \sim 0.05$. Thus we have an amplitude that effectively has only a single CKM coefficient and hence one overall weak phase. This then ensues $|\bar{A}/A| = 1$; i.e., no direct **CP** violation there. One can go beyond simple hand-waving arguments: the authors of Refs. [134] and [12] predicted $C_{J/\psi(nS)K^0} \sim \mathcal{O}(10^{-3})$, while others predicted 1–2%, which is still covered by experimental uncertainties.

The situation is quite different for indirect **CP** violation in B_s^0 transitions: so far it has not been found there. While the SM tells us that B_s^0 oscillations are stronger than for B^0 ones, its indirect **CP** violation is very small. Still it is a good hunting region for ND. Major candidates are: $B_s^0 \to J/\psi\,\phi$ and $B_s^0 \to J/\psi\, f_0(980)$.[39] These are considered golden modes with relatively large branching ratios $\sim 10^{-4} - 10^{-3}$ and suppressed penguin contributions .

To summarize: indirect **CP** violation needs oscillations – in this case $\Delta B = 2$ one – and weak phases. It shows the impact of the initial states of neutral B mesons. Furthermore, one focuses on $b \to c\bar{c}s$ transitions that produce basically a single non-leptonic FS; thus the strong phase is not an observable.

(a) The SM can produce a very large indirect **CP** violation in $B^0 \to J/\psi\, K^0$ that is consistent with the present data.

[38] Remember that in the kaon formalism described above, for an arbitrary phase convention, $2\mathrm{Re}(\epsilon)/(1+|\epsilon|^2) = (1-|q/p|^2)/(1+|q/p|^2)$ holds.

[39] Or in the views of theorists: $B_s^0 \to J/\psi\,\eta/\eta'$.

(b) PDG2020 gives only limits on indirect **CP** violation in B_s^0 transitions, namely a weak phase ϕ_s between $\sim (-0.08$ and $+0.04)$, while the SM predicts ~ 0.001. With more data and re-fined analyses, this becomes an interesting 'hunting' region of ND.

6.7 D^0 oscillations and its indirect CP violation

There is one member of Up-type mesons which undergo oscillations, namely $D^0 = [c\bar{u}]$. It is not easy to predict its oscillations and indirect **CP** violation. As we have just seen, oscillations in the B^0 and B_s^0 systems can be described by an effective operator governed by short-distance box diagrams. In $\Delta C = 2$ transitions one can use an analogous four-quark local operator describing the transition $c\bar{u} \Rightarrow u\bar{c}$, driven by box quark diagrams with *internal* $\bar{b}b$ fields. Yet its short-distance coefficient is highly CKM suppressed: $|V_{cb}^* V_{ub}|^2 \sim \mathcal{O}(\lambda^{10})$. For internal light fields, that is for transitions $c\bar{u} \to s\bar{s}/d\bar{d} \to u\bar{c}$, the CKM factor of the effective operators are of order $|V_{cs(d)}^* V_{us(d)}|^2 \sim \mathcal{O}(\lambda^2)$, but the price to pay is GIM-suppression with a higher power, namely m_s^4/m_c^4. This suggests that we cannot describe their impact with a local operator, and possible major contributions of intrinsically non-perturbative origin.

6.7.1 $D^0 - \bar{D}^0$ *oscillation*

The first evidence for $D^0 - \bar{D}^0$ oscillations was obtained in 2007 by Belle [205] and BABAR [206], confirmed the same year by CDF [207] and, six years later, by LHCb [208]. The oscillation of the D^0 system are described by

$$x_D \equiv \frac{\Delta M_D}{\bar{\Gamma}_D} \quad , \quad y_D \equiv \frac{\Delta \Gamma_D}{2\bar{\Gamma}_D} \quad ; \tag{6.124}$$

Non-leptonic rates are affected by strong forces; it is sometimes convenient to describe oscillations in terms of

$$x'_D \equiv x_D \cos\delta_f + y_D \sin\delta_f \quad , \quad y'_D \equiv -x_D \sin\delta_f + y_D \cos\delta_f \; , \tag{6.125}$$

where δ_f describes relative strong phase. We have $x_D^2 + y_D^2 = (x'_D)^2 + (y'_D)^2 \neq 0$.

Overall oscillation strength provides the scale and *conservative* bounds:

$$\left.\frac{\Delta M_D}{\bar{\Gamma}_D}\right|_{\rm SM} , \left.\frac{\Delta \Gamma_D}{\bar{\Gamma}_D}\right|_{\rm SM} \sim [SU(3)_F \; breaking] \times \sin^2\theta_C < \text{few} \times 0.01 \; .$$

The dependence on $\sin^2\theta_C$ follows from the Cabibbo suppression mentioned in **Sect. 6.7**. Due to the GIM mechanism one has $\Delta M_D = \Delta\Gamma_D = 0$ in the limit of flavour symmetry; $D^0 - \bar{D}^0$ oscillations are driven by $SU(3)_F$ breaking characterized by $m_s^2 \neq m_d^2$. A semi-quantitative description of $SU(3)_F$ *breaking* is a central challenge. It was discussed in Ref. [209]. Its findings are based on the analysis of D^0 box diagrams: (a) Within the SM the most likely values are $x_D \sim \mathcal{O}(10^{-3}) \sim y_D$; the order of magnitude is similar, but the underlying dynamics is different. (b) However, remaining within the SM, one cannot rule out the possibility of larger values, up to 10^{-2}.[40] Recent results from the LHCb experiment [210] give:

$$x_D = \frac{\Delta M_D}{\Gamma_D} = (+3.97 \pm 0.46 \pm 0.29) \cdot 10^{-3}, \; y_D = \frac{\Delta \Gamma_D}{2\Gamma_D} \simeq (+4.59 \pm 1.20 \pm 0.85) \cdot 10^{-3}. \tag{6.126}$$

D^0 oscillation has been established, although its features – x_D vs. y_D – are not clear yet.

Finding two different lifetimes of neutral D mesons give an unequivocal manifestation of oscillations. Then the mass eigenstates of neutral D mesons have to be **CP** eigenstates as well. First one assumes **CP** symmetry for $D^0 \to K^+K^-/\pi^+\pi^-$ decays, where the FS is a **CP** *even* eigenstate. The decay rate evolution in time is given by a single pure exponential function, controlled by the width for the **CP** even state[41]:

$$\Gamma(D^0(t) \to K^+K^-/\pi^+\pi^-) \propto e^{-\Gamma_+ t}|A(D^0 \to K^+K^-/\pi^+\pi^-)|^2 . \tag{6.127}$$

One can calibrate these transitions with the ones leading to the Cabibbo favoured $K^-\pi^+$ decay. This FS comes in equal measure from **CP** *even* and *odd* states, therefore its time dependence is:

$$\Gamma(D^0(t) \to K^-\pi^+) \propto \frac{1}{2}(e^{-\Gamma_+ t} + e^{-\Gamma_- t})|A(D^0 \to K^-\pi^+)|^2 . \tag{6.128}$$

Their ratio is

$$\frac{\Gamma(D^0 \to K^-\pi^+)}{\Gamma(D^0(t) \to K^+K^-)} \propto \frac{1}{2}\left(1 + e^{(\Gamma_+ - \Gamma_-)t}\right) \simeq 1 + y_{CP} \cdot \frac{t}{\tau_D} \; , \; y_{CP} \equiv \frac{\Gamma_+ - \Gamma_-}{2\bar{\Gamma}} \; .$$

Data from [174] and [8] lead to $y_{CP} \simeq (+6.4 \pm 0.8) \cdot 10^{-3}$. We also know that D^0 transitions are close to **CP** symmetry: $y_D \sim y_{CP}$.

6.7.2 *Indirect* CP *violation in* D^0 *transitions*

We can distinguish three roads for finding indirect **CP** violation in similar patterns as for neutral B mesons, except the scales, (see **Sect. 6.4**).

(1) As shown in Eq. (6.29), *semi*-leptonic D^0 transitions one can probe $|q|_{D^0} \neq |p|_{D^0}$. The Heavy Flavor Averaging Group (HFLAV) performs a global fit to relevant mixing measurements to obtain world average values for fitted parameters, including *semi*-leptonic and *non*-leptonic D^0 transitions. They give[42]: $|q/p|_{D^0} = 0.969^{+0.050}_{-0.045}$.

[40] ΔM_D gets contributions from virtual states; thus it is sensitive to ND that could raise it to percent level, but also below $\mathcal{O}(10^{-3})$. $\Delta\Gamma_D$, being driven by *on*-shell transitions, is hardly sensitive to ND; at the same time, it is very vulnerable to violation of *local* duality: a nearby narrow resonance could wreck a GIM cancellation and raise the value of $\Delta\Gamma_D$ of an order of magnitude. Two possible lessons: (1) HQE has been applied to inclusive weak $\Delta B = 1$ and 2 transitions successfully, at orders $(1\,\mathrm{GeV}/m_b)^2$ and $(1\,\mathrm{GeV}/m_b)^3$. It has also been applied semi-quantitatively to inclusive $\Delta C = 1$ rates at order $(1\,\mathrm{GeV}/m_c)^2$ and $(1\,\mathrm{GeV}/m_c)^3$, but the situation is more 'complex' for $\Delta C = 2$. (2) "Duality" has been somewhat successfully for y_B. Can it be applied qualitatively to y_D, where the threshold is 0.7–1 GeV?

[41] Similar for $D^0 \to K_S\pi^0$ being *odd* eigenstate: $\Gamma(D^0(t) \to K_S\pi^0) \propto e^{-\Gamma_- t}|A(D^0 \to K_S\pi^0)|^2$.

[42] see Reviews of 2020 PDG for details [211].

(2) A *qualitative* analogy to $B \to \pi^+\pi^-$ decays can be found in three D^0 decays where the FS is a **CP** eigenstate: $K_S\pi^0$ and K^+K^- and $\pi^+\pi^-$. The first one is described mostly by $[c\bar{u}] \to [s\bar{d}][u\bar{u}]$ – Cabibbo favored (CF)] – while the other ones $[c\bar{u}] \to [s\bar{u}][u\bar{s}]$ and $[c\bar{u}] \to [d\bar{u}][u\bar{d}]$ – are Cabibbo suppressed (SCS). One expects that two-body FS show indirect **CP** violation on the level of 10^{-3}.

(3) Another promising way to probe **CP** is comparing $\Gamma(\bar{D}^0(t) \to K^-\pi^+)$ vs. $\Gamma(D^0(t) \to K^+\pi^-)$. The latter is driven by $[c\bar{u}] \to [u\bar{s}][d\bar{u}]$ and it is doubly Cabibbo suppressed (DCS) – it should a priori exhibit a higher sensitivity to NS. The FS $K^\pm\pi^\mp$ can be fed by D^0 as well as $\bar{D}^0$ decays via a CF or DCS transition. Thus $D^0(t) \to K^+\pi^-$ and $D^0(t) \to K^-\pi^+$ can both occur even with*out* oscillations.

In the following, we talk about the second and third items.

6.7.3 *Single Cabibbo suppressed transitions*

For $h = K, \pi$ we describe time-dependent rates to first order in x_D and y_D:

$$\frac{\Gamma(D^0(t) \to h^+h^-)}{|A(D^0 \to h^+h^-)|^2} \propto e^{-\Gamma_1 t}\left[1 + y_D\frac{t}{\tau_D}\left(1 - \mathrm{Re}\frac{q}{p}\bar{\rho}(h^+h^-)\right) - x_D\frac{t}{\tau_D}\mathrm{Im}\frac{q}{p}\bar{\rho}(h^+h^-)\right]$$
$$\frac{\Gamma(\bar{D}^0(t) \to h^+h^-)}{|A(\bar{D}^0 \to h^+h^-)|^2} \propto e^{-\Gamma_1 t}\left[1 + y_D\frac{t}{\tau_D}\left(1 - \mathrm{Re}\frac{p}{q}\rho(h^+h^-)\right) - x_D\frac{t}{\tau_D}\mathrm{Im}\frac{p}{q}\rho(h^+h^-)\right].$$

With only indirect **CP** violation one sets $|q/p| = 1 - 2\epsilon_D$ and $(q/p)\bar{\rho}(h^+h^-) = (1 - 2\epsilon_D)e^{i\phi_{h\bar{h}}}$; assuming $|A(D^0 \to h^+h^-| = |A(\bar{D}^0 \to h^+h^-|$ (CKM dynamics inducing a tiny asymmetry), one has:

$$A_{\Gamma_{h\bar{h}}}(t) \equiv \frac{\Gamma_+ - \bar{\Gamma}_+}{\Gamma_+ + \bar{\Gamma}_+} = \frac{\Gamma(\bar{D}^0(t) \to h^+h^-) - \Gamma(D^0(t) \to h^+h^-)}{\Gamma(\bar{D}^0(t) \to h^+h^-) + \Gamma(D^0(t) \to h^+h^-)}$$
$$\simeq x_D\,\frac{t}{\tau_D}\sin\phi_{h\bar{h}} - 2\,y_D\,\frac{t}{\tau_D}\,\epsilon_D\cos\phi_{h\bar{h}}\,. \tag{6.129}$$

One can look at the present averaged data[43]:

$$A_{\mathbf{CP}}(D^0 \to K_S\pi^0)|_{\mathrm{PDG2020}} = (-2.0 \pm 1.7)\cdot 10^{-3} \tag{6.130}$$
$$A_{\mathbf{CP}}(D^0 \to K^+K^-)_{\mathrm{PDG2020}} = (-0.7 \pm 1.1)\cdot 10^{-3} \tag{6.131}$$
$$A_{\mathbf{CP}}(D^0 \to \pi^+\pi^-)_{\mathrm{PDG2020}} = (+1.3 \pm 1.4)\cdot 10^{-3} \tag{6.132}$$
$$A_{\Gamma_{h\bar{h}}}|_{\mathrm{PDG2020}} = (-0.123 \pm 0.526)\cdot 10^{-3}\,; \tag{6.133}$$

no indirect **CP** violation has been found yet – but 'soon' we will find it. However, direct **CP** asymmetry has been established, see **Chapter 7**.

6.7.4 *Doubly Cabibbo suppressed transitions*

The measured ratio of DCS vs. CF decays of two-body FS – $D^0 \to K^+\pi^-$ vs. $\bar{D}^0 \to K^-\pi^+$ – is not surprising: $\sim 3\cdot 10^{-3}$ [210].

[43] In the first equation we talk about a CF transition.

Let us consider $D^0 \to K^+\pi^-$ vs. $D^0 \to K^-\pi^+$ into 'wrong-sign' kaons; i.e., not respecting the selection rule $\Delta C = \Delta S$. The latter decay has two sources, the DCS $c \to d\bar{s}u$ transition and the $D^0 - \bar{D}^0$ oscillation followed by a CF $\bar{c} \to \bar{s}d\bar{u}$ transition. The decay rate evolution as a function of (proper) time yields:

$$\frac{\Gamma(D^0(t) \to K^+\pi^-)}{|A(D^0 \to K^+\pi^-)|^2} \propto \left[1 + \left(\frac{t}{\tau}\right)^2 \left(\frac{x_D^2 + y_D^2}{4}\right) \left|\frac{q}{p}\bar{\rho}(K^+\pi^-)\right|^2 - \left(\frac{t}{\tau}\right) \left[y_D \,\mathrm{Re}\left(\frac{q}{p}\bar{\rho}(K^+\pi^-)\right) + x_D \,\mathrm{Im}\left(\frac{q}{p}\bar{\rho}(K^+\pi^-)\right)\right]\right] \tag{6.134}$$

$$\frac{\Gamma(\bar{D}^0(t) \to K^-\pi^+)}{|A(\bar{D}^0 \to K^-\pi^+)|^2} \propto \left[1 + \left(\frac{t}{\tau}\right)^2 \left(\frac{x_D^2 + y_D^2}{4}\right) \left|\frac{p}{q}\rho(K^-\pi^+)\right|^2 - \left(\frac{t}{\tau}\right) \left[y_D \,\mathrm{Re}\left(\frac{p}{q}\rho(K^-\pi^+)\right) + x_D \,\mathrm{Im}\left(\frac{p}{q}\rho(K^-\pi^+)\right)\right]\right] \tag{6.135}$$

plus $\mathcal{O}[(t/\tau)^3]$. In deriving these equations one can note that $\bar{\rho}(K^+\pi^-)$ and $\rho(K^-\pi^+)$ are enhanced by $1/\tan^2\theta_C$ in the SM; therefore one keeps terms linear and quadratic in x_D or y_D and $\bar{\rho}(K^+\pi^-)$ or $\rho(K^-\pi^+)$. One can predict the time-dependent **CP** violation as:

$$\frac{\Gamma(\bar{D}^0(t) \to K^-\pi^+) - \Gamma(D^0(t) \to K^+\pi^-)}{\Gamma(\bar{D}^0(t) \to K^-\pi^+) + \Gamma(D^0(t) \to K^+\pi^-)} \simeq \left(\frac{t}{\tau}\right) \frac{|\hat{\rho}(K^-\pi^+)|}{\tan\theta_C^2}(2\,\epsilon_D\, y_D' \cos\phi - x_D' \sin\phi) + \left(\frac{t}{\tau}\right)^2 \frac{|\hat{\rho}(K^-\pi^+)|^2}{\tan\theta_C^4}\,\epsilon_D\,(x_D^2 + y_D^2) \tag{6.136}$$

with ϕ being *weak* phase between DCS and CF amplitudes. As said before, the phases of $V_{cd}^* V_{us}$ and $V_{cs}^* V_{ud}$ are basically the same in the SM; therefore we hardly expect **CP** violation there. On the other hand, it is a 'hunting region' for ND, if one has gotten much more data and deals with systematic uncertainties.

PDG2020 provides the relative strong phase of this two-body FS: $\cos\delta_{K^+\pi^-} = 0.97 \pm 0.11$. We realize that FSI has large impact in general, which is not surprising.

LHCb analysis from proton-proton collisions at 7 and 8 TeV led to [212]: $y_D' = (4.8 \pm 1.0) \cdot 10^{-3}$ and $(x_D')^2 = (5.5 \pm 4.9) \cdot 10^{-5}$; it is not yet clear which lesson one can learn about $D^0 - \bar{D}^0$ oscillations from Run-1.

6.8 Neutron-antineutron oscillations

Neutron-antineutron oscillations can be described in somewhat analogy to the cases of neutral mesons and neutrinos, but they 'paint' different 'landscapes'. Measured neutrino masses through oscillations are likely to constrain a Majorana component, as discussed above in **Sect. 4.6**. The Majorana component changes lepton number by *two* units, giving rise to possible neutrino-antineutrino oscillations, and maybe lead to leptogenesis in our Universe, as we will discuss below in **Sect. 15.4.4**.

On the other hand, neutron-antineutron oscillation calls for more 'exotic' extension of the SM: $\Delta(n \Rightarrow \bar{n})$ changes *baryon* number by *two* units. However, the possible violation of baryon number is not upsetting. Both baryon and lepton number conservation are accidental symmetries, and in the SM they do not represent charges of a corresponding gauge field. Moreover, baryon-violating processes are needed to generate the baryon asymmetry in our Universe as it expands and cools, and the present predominance of matter over antimatter in the Universe is widely interpreted as indirect evidence for baryon violation. The SM allows for non-perturbative processes involving $SU(2)$ instantons (sphalerons) which violate baryon number, but conserve the difference $B - L$ [213, 214]. However, since the violation of B[44] and L by sphalerons in the SM is exponentially suppressed at low temperatures, these global symmetries are effectively preserved in the SM at zero temperatures.

No baryon number violating process has been observed so far, and proton stability imposes very tight bounds on it. Does it mean our community should not search for $n - \bar{n}$ oscillations? One can build models of ND, where baryon number violation can proceed only (or mainly) by *two* rather than *one* unit and still achieve baryogenesis in our Universe. An example is 'post-sphaleron baryogenesis' occurring below the electroweak phase transition temperature and requiring baryogenesis to proceed via high dimensional B-violating operators. In general, it seems 'natural' to connect $\Delta B = 2$ and $\Delta L = 2$ while $\Delta(B - L) = 0$ – in the views of theorists.

6.8.1 *Measure neutron-antineutron oscillations*

There are two 'roads' to probe neutron-antineutron oscillations experimentally: the conversion of a neutron in an antineutron using a neutron beam from a reactor or from an accelerator source *or* in a nuclear matter, searching for a nuclear explosion induced by a neutron-antineutron conversion *within* a nucleus. Both types of experiments are very challenging. The first one is complicated by the difficulty to trap a large amount of neutron and by the fact that a free neutron is not stable; one has to track a neutron beam in the 'vacuum', or trap a large amount of neutrons, or use intense beams and search for anti-neutrons in these beams. The second one pays the huge price of having continual loss of coherence due to nucleon collisions within a nucleus, namely searches for the tell-tale sign of antineutrons in their violent annihilation with a neutron or proton. This price is mostly compensated by the fact that the neutrons in the nucleus are stable and that a large numbers of nuclei can be observed at the same time, so the two techniques offer similar limits. So far, we

[44] Here B means baryon numbers, not beauty mesons.

have gotten only a lower bound on the oscillation time:

$$\tau_{\text{free}}(n \to \bar{n}) > 0.86 \cdot 10^8 \text{ s} \tag{6.137}$$

$$\tau_{\text{bound}}(n \to \bar{n})|_{\text{PDG2020}} > 2.7 \cdot 10^8 \text{ s} . \tag{6.138}$$

The limit (6.137) comes from an experimental search performed at the ILL in Grenoble from 1989 to 1991 [215]. We have just indicated that there is a long journey to transform a neutron into an antineutron in different situations; however, the values are quite similar.

The literature comes mostly from the theoretical side [218, 219], but not only. Beyond needing 'exotic' ND, there are important experimental aspects in searches for $n - \bar{n}$ oscillations that one has to consider. We sketch them below:

(a) See Eq. (6.137): One needs intense neutron sources. Intense sources of free neutrons are usually created either through fission in a suitable nuclear reactors or by means of an accelerator, through neutron spallation in high Z targets struck by GeV proton beams.

The neutrons start moving slowly in a 'vacuum'. One needs a detector to find an antineutron at a distance around 1 - 2 meter or possibly more [220].

(b) See Eq. (6.138): Most neutrons are found in nuclei. If a neutron transforms itself into an antineutron in that environment, it would annihilate itself with one of the other nucleons disrupting the nucleus and releasing a large amount of detectable energy. A limit from the decay of ^{16}O is [221]:

$$\tau_{\text{nuclei}} > (1.9 \pm 1.1) \cdot 10^{32} \text{ years} \sim 6 \cdot 10^{39} \text{ s} . \tag{6.139}$$

This bound exceeds the limit in Eq. (6.137) by a 'mere' 31 orders of magnitude; one would view the vacuum searches as truly quixotic. However a more careful consideration shows that the two bounds in Eq. (6.139) and Eq. (6.137) are actually quite equivalent in their sensitivity. This most surprising conclusion is based on some subtle quantum mechanical features. A first orientation had been obtained by an hand-waving argument invoking the 'collapse of the wave function' or the 'quantum Zeno effect' [216, 217].[45]

A neutron bound inside the nuclei will move around with some mean free path. Thus there is an averaged time τ_{between} *between* collisions with other nucleons. The neutron can oscillate only during this brief interval; for the next collision with another nucleon represents a measurement of the baryon number of the neutron in question: its annihilation would reveal it had transmogrified itself into an antineutron; no annihilation mean it had again become validated as a neutron thus setting its oscillation clock back to zero. Thus [222]

$$\tau_{\text{nuclei}} \sim \tau_{\text{vacuum}} \cdot \frac{\tau_{\text{vacuum}}}{\tau_{\text{between}}} \sim 10^{38} \text{ s} \simeq 3 \cdot 10^{31} \text{ years} \tag{6.140}$$

[45] In broad terms, the effect states that if the system undergoes frequent measurements, to check whether it is still in its initial state, it does not evolve. Fans of old history might remember very old examples, namely the 'arrow paradox' – arrows do not move – or Achilles cannot overcome a tortoise; fundamental theorem of calculus overcame that.

using a typical nuclear reaction time of 10^{-23} s for τ_{between}. In more rigorous terms, inside the nuclear medium neutrons and antineutrons are no longer fully degenerate: they experience different potentials due to their differences in the isospin components, magnetic moments etc. State-of-the-art calculations can be expressed as follows: from the observed nuclear stability, Eq. (6.139), one infers [223]:

$$\tau_{\text{vacuum}}(n \to \bar{n}) \simeq 2\sqrt{\tau_{\text{nuclei}}/\Gamma_{\bar{n}}} \geq (2.1 \pm 0.2) \cdot 10^8\ s\ , \tag{6.141}$$

where $\Gamma_{\bar{n}} \sim 100$ MeV is a typical nuclear annihilation width for antineutrons. This bound (that we could call 'second road') slightly better than the one from the direct analysis of "free" neutron beams. Still we find with more recent data [221]:

$$\tau_{\text{'free neutron'}}(n \to \bar{n}) \simeq 2\sqrt{\tau_{\text{nuclei}}/\Gamma_{\bar{n}}} > 2.7 \cdot 10^8\ s\ . \tag{6.142}$$

6.8.2 *Future prospects*

These numbers show that there is still a future for the competition in this 'game'. Maybe it is part of 'luck' to find a nuclear medium with $\Gamma_{\bar{n}} \sim 1$ MeV– or do we need a deeper understanding of nuclear medium? ND mediating neutron-antineutron oscillations require a combination of effective six-quark operators violating baryon number by two units. Previous estimates were mostly based on the 'MIT bag model' estimates of these six-quark matrix elements.[46] This year for the first time we have gotten calculations of the neutron-antineutron matrix elements needed from LQCD with physical quark masses [224]. This LQCD result are 4–8 times larger than from the MIT bag model; present and future experiments have larger sensitivity of ND, namely experiments should observe 16–64 times more neutron-antineutron oscillation events as expected.

We have already seen that at least there are ideas and options for "free" $n \to \bar{n}$ oscillations with a factor of hundred (or even more), see [220] and [225]. It has been suggested to use existing technologies. It is crucial to combine improvements in different regions. The goal is to find $n - \bar{n}$ oscillations, not to produce higher scales of limits. One has to do that in steps: (a) would produce wonderful training place for young physicists and engineers. (b) would allow our community to go beyond what LHC can probe directly.

Again, the possible impact is greatly on the shoulders of our experimental colleagues and the availability of new facilities. The future HyperKamiokande detector, for example, will enhance the sensitivity of SuperKamiokande by more than a factor 20. It was discussed how one can probe free $n - \bar{n}$ oscillations at an European Spallation Source (ESS) in Sweden [220]; the main challenge seemed how to get enough funding (it is not cheap).

6.9 Oscillations of $K^0 - \bar{K}^0$, $D^0 - \bar{D}^0$, $B^0_{(s)} - \bar{B}^0_{(s)}$ and $n - \bar{n}$

1964 was "annus mirabilis" (wonderful year) for fundamental dynamics [12]:

[46] One of us had admitted freely and without shame that he had used the 'MIT bag model' himself.

- The Higgs mechanism for a broken local symmetry in a QFT was first developed.
- The *quark model* and the first elements of *current algebra* were put forward.
- The charm quark was first introduced to establish *quark-lepton symmetry.*
- $SU(6)$ symmetry was proposed.[47]
- The *first storage ring for e^+e^- collisions* was built in Frascati close to Rome.
- The Ω^- baryon was found at Brookhaven National Laboratory.
- **CP** violation was discovered at the same laboratory. Our community understood that a revolution had happened – but not yet where it will lead us.

It was not realized right away the impact of these discoveries and breakthroughs, both on the experimental and theoretical sides. Okun was basically the only person who said that **CP** invariance had to be probed in $K_L \to \pi\pi$ decays, in a (Russian) book in 1963 [1] – *before* the discovery of **CP** violation in 1964. One can learn from the history of fundamental forces that one should not follow the fashions.

As seen before, "mixing" is a general item in QM. There is a special case of "mixing", namely "oscillations", for which time dependent rates are crucial. That can happen only with neutral hadrons, whose flavor eigenstates are not mass eigenstates in general. Oscillations have been established from the data for four mesons, K^0, D^0, B^0 and B^0_s, in different situations:

$$\Delta S = 2: \quad \Delta M_K \simeq \bar{\Gamma}_K \,, \; \Delta\Gamma_K \simeq 2\bar{\Gamma}_K \,, \; \Delta\Gamma_K \sim \Delta M_K \tag{6.143}$$

$$\Delta C = 2: \quad \Delta M_D \ll \Gamma_D \,, \; \Delta\Gamma_D \ll \Gamma_D \,, \; \Delta\Gamma_D \sim \Delta M_D \tag{6.144}$$

$$\Delta B = 2: \quad \Delta M_{B^0} \sim \Gamma_{B^0} \,, \; \Delta\Gamma_{B^0} \ll \Gamma_{B^0} \,, \; \Delta\Gamma_{B^0} \ll \Delta M_{B^0} \tag{6.145}$$

$$\Delta M_{B^0_s} \gg \Gamma_{B^0_s} \,, \; \Delta\Gamma_{B^0_s} \sim \mathcal{O}(\Gamma_{B^0_s}) \,, \; \Delta\Gamma_{B^0_s} \ll \Delta M_{B^0_s} \tag{6.146}$$

We know how to 'paint' these landscapes:

- From $K^0 \to \bar{K}^0$ oscillations we had learnt to need charm quarks with $m_c \sim 1 - 2$ GeV. Still ΔM_K does not give us an accurate test of our understanding its underlying dynamics, as it has been pointed out in Ref. [12] in details. The SM gives short (SD) and long distance (LD) contributions: $\Delta M_K|_{\rm SM} = \Delta M_K|_{\rm SD} + \Delta M_K|_{\rm LD}$ and $\Delta M_K|_{\rm LD} \sim 0.4\,\Delta M_K|_{\rm exp.}$. There has been sizable progress in describing LD, mostly (but not exclusively) due to LQCD; it has lead to a smaller uncertainties in the value of $\Delta M_K|_{\rm LD}$. It has been suggested that one uses the measured value of ΔM_K together with $\Delta M_{B^0_d}$, $\Delta M_{B^0_s}$ and ϵ_K to understand fundamental dynamics. However, our understanding of ΔM_K is not on the same level of accuracy as the other ones.
- Both $\Delta M_{D^0}/\Gamma_{D^0}$ and $\Delta\Gamma_{D^0}/2\Gamma_{D^0}$ are of the same order $O(10^{-3})$, although for different reasons [209]; here it is hard to test our understanding of strong forces with accuracy.
- Our community can 'brag' about our success with a true *pre*diction, that is $m_t > 100$ GeV, based on $\Delta M_{B^0_d}/\Gamma_{B^0_d} \sim 0.7$ and $\Delta M_{B^0_s} \gg \Gamma_{B^0_s}$ and $y_{B^0_s} \sim 0.1 \gg y_B$.

[47] $SU(6)$ was a model – not a theory – to combine $SU(3)_{\rm Fl} \times SU(2)_{\rm spin}$ for spectroscopy.

In order to summarize our lessons about oscillations and *indirect* **CP** asymmetries, we follow a well known Austrian 'statement': "They are the same, only different."

- It was a breakthrough to establish **CP** violation in kaon transitions with $\Delta S = 2$. We learnt that we need three families (at least), as it was pointed in a published paper in 1973 [128], although it was hardly realized outside Japan.
- There was no reason why **CP** asymmetries should be small; actually, they could be close to 100% somewhere. It was predicted that the transitions of neutral beauty hadrons would be large due to the special pattern of the CKM matrix.
- Still sizable **CP** violation can happen in $\Delta C = 2$ transition in $D^0 - \bar{D}^0$ oscillation. It is a challenge to find them, but it is very interesting to learn about the pattern of underlying dynamics.
- **CP** violation in strange, beauty and charm neutral mesons seems to have nothing to do with the huge asymmetry in matter vs. anti-matter in our Universe.
- There is a special case of oscillation that has not been found yet: $n - \bar{n}$ oscillation; however, our community should not give up on that.
- Usually one can measure *indirect* **CP** violation best with two-body FS: $K_L \to \pi^+\pi^-$ (or $\bar{K}^0 \to \pi^+\pi^-$ vs. $K^0 \to \pi^+\pi^-$), $B^0 \to J/\psi K_S$ and $B^0_s \to J/\psi f_0(980)$ (or $B^0_s \to J/\psi\phi$ with more work). So far *indirect* **CP** violation has not been found in the D^0 decays.

Chapter 7

Direct CP violation for flavor hadrons

Indirect **CP** violation can happen only for neutral mesons which undergo oscillations, namely K^0, B^0, B_s^0 and D^0; it depends on the *initial* state of the transitions. *Direct* **CP** asymmetries depend on the FS of flavor hadrons, in different ways and at different levels. The landscapes are quite different, with strange and beauty mesons on one side, and charm mesons on the other one:

- Indirect **CP** violation has been established first in K^0 (or K_L) decays, then in B^0 ones (but not in B_s^0 decays); direct **CP** asymmetries (mostly with smaller values) have been observed later.
- While indirect **CP** violation has not been found yet in $D^0 \to K^+K^-$ or in $D^0 \to \pi^+\pi^-$, direct **CP** asymmetry $\Delta A_{\mathbf{CP}}^{\text{charm}} \equiv A_{\mathbf{CP}}(D^0 \to K^+K^-) - A_{\mathbf{CP}}(D^0 \to \pi^+\pi^-) \neq 0$ has been established in 2019.

First one analyzes two-body FS, and that not only for strange hadrons, but also for beauty and charm hadrons. However, in the latter case, it is crucial to continue analyzing many-body FS. To be realistic, one can hardly go beyond four-body FS. No **CP** violation has been established yet for baryons (except 'our' existence[1]).

7.1 General comments about direct CP asymmetries

Direct **CP** asymmetries can happen with and with*out* oscillations, thus they can be searched for in the decays of K^+, D^+, D_s^+ and B^+ mesons as well as in the decays of Λ, Λ_c^+, Λ_b^0 and Ξ_b baryons.

When one searches for **CP** asymmetries in the weak decays of strange, charm and beauty hadrons, one has to worry about production asymmetries, in particular for pp collisions at LHC.

The production rates of b and $\bar{b}$ hadrons at the LHC are not expected to be strictly equal, as the $\bar{b}$ quark produced in the interaction might combine with the valence quark of the colliding protons, whereas the same is not true for the b quarks. For this reason, one could expects a slight excess in the production of B^+ and B^0

[1]We are referring to **CP** violation in connection with baryogenesis.

over B^- and $\bar{B}^0$ mesons, which has to be compensated by an opposite asymmetry in the other b meson and baryon species.

This phenomenon, commonly referred to as production asymmetry, can mimic **CP** violation and it is thus important to measure it in order to distinguish the physical asymmetries. An obvious way to solve this problem is to calibrate using decays with a FS where one hardly expects **CP** violation, namely Cabibbo favoured (CF) decays as $B \to D\pi/Dl\nu$ or $D \to K\pi/Kl\nu$.[2] This problem is much less so for $\bar{p}p$ and e^+e^- collisions, since these initial states are **CP** symmetric.

As seen at the end of **Chapter 2**, we have:

$$|T(\bar{H} \to \bar{f})|^2 - |T(H \to f)|^2 = 4 \sum_{f \neq a_j} T^{\text{resc}}_{a_j f} \,\text{Im}\, T^*_f T_{a_j} \;;$$

for a **CP** asymmetry becoming an observable, one needs interference between (at least) two coherent amplitudes with both relative *weak* **CP**-odd – $T^*_f \neq T_f$ – and *strong* **CP**-even phases – $T^{\text{resc}}_{a_j f}$.

In the SM the CKM matrix provides weak **CP**-violating phases. The **CP**-conserving phase are supplied by QCD (or QED). A 'good side': there is *no other candidate* for strong forces. Yet it is a true challenge to apply non-perturbative QCD to study FSI.

It is difficult to reliably evaluate the FS phase because of hadron dynamics; one needs help. In the case of resonances, one can use the Breit-Wigner formula; it works well when one analyzes FS with two pseudo-scalars or pseudo-scalar + vector (or axial-vector) in the weak decays of mesons. The situation is more complex for "scalar" resonances. The usual Breit-Wigner parametrization does not describe the impact of broad resonances like $f_0(500)/\sigma$ or $K^*(700)/\kappa$.[3] In these cases subtle tools like dispersion relations are more suitable, see Refs. [104, 105, 226, 227].

So far the weak decays of strange, charm and beauty *baryons* have not established **CP** asymmetries. One can follow the same 'road' followed for mesons, namely to analyze first two-body FS and then three- or four-body FS. Obviously, the analysis is more 'complex' due to the spin-1/2 both in the initial and final states.

Multi-body FS of the weak decays of beauty and charm hadrons are the largest parts of these widths. To be realistic, one shortens the analyses to three- and four-body FS that can manifest their impact on **CP** asymmetries in their distributions. The goal is to understand the underlying dynamics.

7.1.1 *Dalitz plots*

For channels with two-body FS, **CP** asymmetry can manifest itself only in a difference between two partial widths, as already discussed several times. If, however, the final state is more complex, then it contains more dynamical information than expressed by its partial width and **CP** violation can emerge also through asymmetries in final state distributions.

[2] Those are described in the world of quarks: $b \to c\bar{u}d/cl^-\bar{\nu}$ or $c \to su\bar{d}/sl^+\nu$.

[3] The scalar resonance $K^*(700)/\kappa$ had first appeared as a $K^*(800)/\kappa$.

To probe three-body FS, our community has a 'pope', namely two-dimensional Dalitz plots. Developed by Dalitz in 1953 to study three-body kaon decays [228], two-dimensional Dalitz plots illustrate interference between amplitudes.[4] Their plots are flat in the case of purely non-resonant three-body decays. Thus they show visual structures: intermediate states can come from resonances (like well-known $\rho \to \pi\pi$, $K^* \to K\pi$ and $\phi \to K^+K^-$) or thresholds; they show the impact of non-perturbative QCD. Our goals are to find **CP** asymmetries in the weak transitions of beauty and charm hadrons of $P \to h_1 h_2 h_3$ with h_i being light hadrons. In the simplest scenario one compares **CP** conjugate Dalitz plots.

First one measures *averaged* **CP** asymmetries. However, it is quite possible that different regions of a Dalitz plot exhibit **CP** asymmetries of varying signs that largely cancel each other when one integrates over the whole phase space. With more data and more refined analyses in *regional* **CP** asymmetries our community will get more information about the underlying dynamics. One can apply Breit-Wigner formula to model narrow resonances, as said above. However, interference of narrow and broad resonances can*not* be described as being 'inside' and 'outside' the centers of narrow resonances. It must be described in subtle manners; it depends on the situations. One needs help from tool boxes like dispersion relations. Furthermore, one should not focus on the best fitted results of the data; it is crucial to work on the correlations between transitions.

7.1.2 *Four-body FS and T-odd distributions*

For more complex final states, containing four-body FS, other probes than Dalitz plots have to be employed. One possibility is to form T-odd correlations with the momenta, which may give important information about underlying dynamics. One can probe these in non-leptonic transitions in charm and beauty hadrons. It was first pointed out that triple product correlations involving momenta and polarization in four-body FS should be probed in special situations, namely $B \to V_1 V_2 \to h_1 h_2 h_3 h_4$ with V_i describing narrow vector resonances [229]. The weak decays of heavy flavor hadrons with four-body FS can give us more information in general, by comparing $H_Q \to h_1 h_2 h_3 h_4$ vs. $\bar{H}_Q \to \bar{h}_1 \bar{h}_2 \bar{h}_3 \bar{h}_4$ [230].

In the rest frames of H_Q vs. $\bar{H}_Q$ one defines $C_T \equiv \vec{p}_1 \cdot (\vec{p}_2 \times \vec{p}_3)$ and $\bar{C}_T \equiv \vec{\bar{p}}_1 \cdot (\vec{\bar{p}}_2 \times \vec{\bar{p}}_3)$. The latter involves momenta of the conjugate states. Under time reversal T one has $C_T \to -C_T$. First one measures T-odd moments:

$$\langle A_T \rangle \equiv \frac{\Gamma_{H_Q}(C_T > 0) - \Gamma_{H_Q}(C_T < 0)}{\Gamma_{H_Q}(C_T > 0) + \Gamma_{H_Q}(C_T < 0)}$$
$$\langle \bar{A}_T \rangle \equiv \frac{\Gamma_{\bar{H}_Q}(-\bar{C}_T > 0) - \Gamma_{\bar{H}_Q}(-\bar{C}_T < 0)}{\Gamma_{\bar{H}_Q}(-\bar{C}_T > 0) + \Gamma_{\bar{H}_Q}(-\bar{C}_T < 0)} \tag{7.1}$$

[4] At that time it was assumed that the weak transitions also follow **P**, **C** and **T** symmetries.

FSI can produce $\langle A_T \rangle$, $\langle \bar{A}_T \rangle \neq 0$ even with*out* **CP** and **P** violation. However,

$$\langle A_{\mathbf{CP}}^{T-odd} \rangle \equiv \frac{1}{2}(\langle A_T \rangle - \langle \bar{A}_T \rangle) \neq 0$$
$$\langle A_{\mathbf{P}}^{T-odd} \rangle \equiv \frac{1}{2}(\langle A_T \rangle + \langle \bar{A}_T \rangle) \neq 0 \tag{7.2}$$

would establish **CP** and **P** asymmetry. Maybe with more data (and more thinking) we might have some ideas about 'better' values that do not depend only on experimental uncertainties; however, one should not bet on that.

We need one-dimensional observables at least, for instance angles between planes (as opposite to just zero-dimensional ones, as their average values[5]); we have to understand the reasons why different observables are used, and compare their results.

Let us remark that we are referring in general to direct **CP** violation in non-leptonic $\Delta B = 1 = \Delta C$ transitions, that crucially depend on FSI (as we discussed in **Sect. 2.10**[6]).

With infinite data and perfect detectors one can probe local **CP** asymmetries in $H_Q \to h_1 h_2 h_3 h_4$. Realistic goals are to probe *regional* ones. Suggestions are to measure an angle between two planes; actually one can define two planes in three ways: $[h_1h_2][h_3h_4]$ or $[h_1h_3][h_2h_4]$ or $[h_1h_4][h_2h_3]$. In principle it does not matter, if we understand the underlying dynamics. However, the situations are quite different in our real world both on the theoretical and experimental side: we cannot truly control the impact of strong forces, and the goal is to go after ND. We learn by comparing the results from three angles. We give two examples:

- One can measure the angle ϕ between the planes of $[h_1h_2]$ and $[h_3h_4]$ and describe its dependence [12, 137, 231, 232]:

$$\frac{d\Gamma}{d\phi}(H_Q \to h_1 h_2 h_3 h_4) = \Gamma_1 \cos^2\phi + \Gamma_2 \sin^2\phi + \Gamma_3 \cos\phi \sin\phi \tag{7.3}$$
$$\frac{d\Gamma}{d\phi}(\bar{H}_Q \to \bar{h}_1 \bar{h}_2 \bar{h}_3 \bar{h}_4) = \bar{\Gamma}_1 \cos^2\phi + \bar{\Gamma}_2 \sin^2\phi - \bar{\Gamma}_3 \cos\phi \sin\phi \,. \tag{7.4}$$

  Their widths are given by

$$\Gamma(H_Q \to h_1 h_2 h_3 h_4) = \frac{\pi}{2}(\Gamma_1 + \Gamma_2) \;,\; \Gamma(\bar{H}_Q \to \bar{h}_1 \bar{h}_2 \bar{h}_3 \bar{h}_4) = \frac{\pi}{2}(\bar{\Gamma}_1 + \bar{\Gamma}_2) \,. \tag{7.5}$$

  Assuming **CPT** invariance, **CP** symmetry gives $\Gamma_{1,2} = \bar{\Gamma}_{1,2}$. Γ_3 and $\bar{\Gamma}_3$ represent **T** odd correlations; by themselves they do not necessarily indicate **CP** violation, since they can be induced by strong FSI. In other terms, **T** invariance leads to $\Gamma_3 = \bar{\Gamma}_3 \neq 0$ due to FSI (in general); thus their integrated *forward-backward* asymmetries become $\langle A^{FB} \rangle \equiv \Gamma_3/\pi(\Gamma_1 + \Gamma_2) = \bar{\Gamma}_3/\pi(\bar{\Gamma}_1 + \bar{\Gamma}_2)$. In conclusion, direct **CP** and **T** asymmetries are represented by:

$$\langle A_{\mathbf{CP}}(H_Q \to h_1 h_2 h_3 h_4) \rangle = \frac{\Gamma_1 + \Gamma_2 - \bar{\Gamma}_1 - \bar{\Gamma}_2}{\Gamma_1 + \Gamma_2 + \bar{\Gamma}_1 + \bar{\Gamma}_2} \tag{7.6}$$
$$\langle A_{\mathbf{T}}^{FB}(H_Q \to h_1 h_2 h_3 h_4) \rangle = \frac{\Gamma_3 - \bar{\Gamma}_3}{\pi(\Gamma_1 + \Gamma_2 + \bar{\Gamma}_1 + \bar{\Gamma}_2)} \,. \tag{7.7}$$

[5]Patterns in 'averaged Dalitz plots' would be two-dimensional observables, and so on.
[6]There is a special case for **CP** violation in four-body FS of $K_L \to \pi^+\pi^- e^+ e^-$, see **Sect. 6.5.9**.

When one has more data, one could disentangle Γ_1 vs. $\bar{\Gamma}_1$ and Γ_2 vs. $\bar{\Gamma}_2$ by tracking the distribution in ϕ, see Eqs. (7.3,7.4). If there is a *production* asymmetry, it gives global relations $\Gamma_{1,2} = c\bar{\Gamma}_{1,2}$ and $\Gamma_3 = -c\bar{\Gamma}_3$.

- One can discuss four-body FS in more details with unit vectors:

$$\vec{k} = \frac{\vec{p}_1 \times \vec{p}_2}{|\vec{p}_1 \times \vec{p}_2|} \; , \; \vec{l} = \frac{\vec{p}_3 \times \vec{p}_4}{|\vec{p}_3 \times \vec{p}_4|} \; , \; \vec{m} = \frac{\vec{p}_1 + \vec{p}_2}{|\vec{p}_1 + \vec{p}_2|} \tag{7.8}$$

$$\sin\phi = (\vec{k} \times \vec{l}) \cdot \vec{m} \;\; [\mathbf{CP} = -, \mathbf{T} = -] \;\; , \;\; \cos\phi = \vec{k} \cdot \vec{l} \;\; [\mathbf{CP} = +, \mathbf{T} = +] \tag{7.9}$$

$$\begin{aligned} \frac{d}{d\phi}\Gamma(H_Q \to h_1 h_2 h_3 h_4) &= |c_Q|^2 - [b_Q \cos 2\phi + a_Q \sin 2\phi] \\ &= |c_Q|^2 - [b_Q\,(2\cos^2\phi - 1) + 2a_Q \sin\phi \cos\phi] \end{aligned} \tag{7.10}$$

Likewise for **CP** conjugated decays:

$$\begin{aligned} \frac{d}{d\phi}\Gamma(\bar{H}_Q \to \bar{h}_1 \bar{h}_2 \bar{h}_3 \bar{h}_4) &= |\bar{c}_Q|^2 - [\bar{b}_Q \cos 2\phi - \bar{a}_Q \sin 2\phi] \\ &= |\bar{c}_Q|^2 - [\bar{b}_Q\,(2\cos^2\phi - 1) - 2\bar{a}_Q \sin\phi \cos\phi] \, . \end{aligned} \tag{7.11}$$

One can measure asymmetries comparing these widths

$$\Gamma(H_Q \to h_1 h_2 h_3 h_4) = |c_Q|^2 \;\; \text{vs.} \;\; \Gamma(\bar{H}_Q \to \bar{h}_1 \bar{h}_2 \bar{h}_3 \bar{h}_4) = |\bar{c}_Q|^2 \tag{7.12}$$

and normalized moments:

$$\langle A^Q_{\mathbf{CP}} \rangle = \frac{(\int_0^{\pi/2} d\phi - \int_{\pi/2}^{\pi} d\phi)\frac{d\Gamma}{d\phi}}{(\int_0^{\pi/2} d\phi + \int_{\pi/2}^{\pi} d\phi)\frac{d\Gamma}{d\phi}} = \frac{2(a_Q - \bar{a}_Q)}{|c_Q|^2 + |\bar{c}_Q|^2} \; ; \tag{7.13}$$

Let us observe that there is no impact from b_Q and $\bar{b}_Q$ terms.[7]
One can continue by probing semi-regional asymmetries like:

$$A^Q_{\mathbf{CP}}|_e^f = \frac{(\int_e^f d\phi - \int_e^f d\phi)\frac{d\bar{\Gamma}}{d\phi}}{(\int_e^f d\phi + \int_e^f d\phi)\frac{d\bar{\Gamma}}{d\phi}} \tag{7.14}$$

where b_Q and $\bar{b}_Q$ contribute due to $0 \neq e < f \neq \pi/2$. One needs 'judgment' to choose values of $e < f$ that make 'sense', given resonances and thresholds.

These examples are correct on general theoretical grounds. However, some deal better than others with experimental uncertainties, cuts and/or probe better the impact of ND. One cannot focus on only one example, but has to compare the results from current data and simulations of future data; once one has analyzed real data, one might change the 'road'; of course, that is not easy.

[7] It is similar what was discussed in $K_L \to \pi^+\pi^- e^+ e^-$ [185].

7.2 Direct CP asymmetries in $K_L \to \pi\pi$

At the beginning of this **Chapter** we have said it is crucial to probe **CP** asymmetries *beyond* two-body FS. Yet the 'landscape' for direct **CP** asymmetries in $\Delta S = 1$ is very different from $\Delta B = 1 = \Delta C$, already on the qualitative level. Its 'stage' is with two 'actors' to compare and describe one observable, see Eq. (6.81):

$$\text{Re}\left(\frac{\epsilon'}{\epsilon_K}\right) = \frac{1}{6}\,\frac{|\eta_{+-}|^2 - |\eta_{00}|^2}{|\eta_{+-}|^2}\,. \tag{7.15}$$

Data [8] yields $|\eta_{+-}| \simeq 2.23 \cdot 10^{-3}$ and $|\eta_{+-}|^2 \sim |\eta_{00}|^2 + 0.01$, one has:

$$\text{Re}(\epsilon'/\epsilon_K)|_{\text{exp.}} = (1.66 \pm 0.23) \cdot 10^{-3}\,; \tag{7.16}$$

i.e., direct **CP** violation in $\Delta S = 1$ transition alone has been found around the level of $3.7 \cdot 10^{-6}$ (see Eq. (6.85)).

This value has excited different theoretical interpretations and estimates. The 'Buras school' has said for a long time that the SM can hardly go above 10^{-3} for $\text{Re}(\epsilon'/\epsilon_K)$; more precisely, its 'dual approach'[8] to QCD implied in 2018[9] [165]: $\text{Re}(\epsilon'/\epsilon_K)|_{\text{dual}} = (0.5 \pm 0.2) \cdot 10^{-3}$. The 'LQCD culture' had given its first result about the SM prediction in 2015 [233]: $\text{Re}(\epsilon'/\epsilon_K)|_{\text{'LQCD'}} = (0.138 \pm 0.515 \pm 0.443) \cdot 10^{-3}$. 'Marriage' had appeared between the 'Buras school' and LQCD: their central values are within one σ uncertainty. However, another 'actor' had entered the 'stage' based on chiral symmetry, namely the 'Pich school', in 2018 [234]: it said that short- and long-distance forces led to $\text{Re}(\epsilon'/\epsilon_K)|_{\text{'SM'}} = (1.5 \pm 0.7) \cdot 10^{-3}$.

However, the situation has changed in 2019 and 2020.

- The second result of LQCD led to $\text{Re}(\epsilon'/\epsilon_K)|_{\text{'LQCD'}} = (2.17 \pm 0.84) \cdot 10^{-3}$ [235]. Indeed, the RBC and UKQCD Collaborations have made large progress with systematic uncertainties; the extrapolation to the continuum is a true challenge.
- The 'Buras school' has mostly accepted the new LQCD result. First it gave $\text{Re}(\epsilon'/\epsilon_K) = (1.74 \pm 0.61) \cdot 10^{-3}$, but has soon moved to $\text{Re}(\epsilon'/\epsilon_K)|_{\text{'BS2020'}} = (1.39 \pm 0.52) \cdot 10^{-3}$ [236].
- The 'Pich School' has been 'stable': $\text{Re}(\epsilon'/\epsilon_K)|_{\text{'PS2019'}} = (1.4 \pm 0.5) \cdot 10^{-3}$ [237]. We point out that this SM prediction came *before* the 2020 LQCD result.

These three camps are led by true professional experts.[10] We give only two comments: (a) The 'Pich School' said in its 'Abstract' [237]: '$\text{Re}(\epsilon'/\epsilon) = (14 \pm 5) \cdot 10^{-4}$ ... *excellent* agreement with the experimentally measured value.' Of course, it is consisted with it; however, it is also consistent with $\text{Re}(\epsilon'/\epsilon) \sim 9 \cdot 10^{-4}$. (b) This time the 'Buras School' is somewhat conservative in its predictions.

[8]The word of "duality" is used very often in different situations. We use it to describe the connection of the worlds of hadrons vs. quarks in **Sect. 3.10**.

[9]It is difficult to disagree with this talk given at the Epiphany Conference, Cracow, January 2018, for someone who works for a catholic university.

[10]The 'Buras' and the 'Pich' Schools use the measured strong phases δ_0 and δ_2, while LQCD can predict them (in principle).

7.3 About Italo Mannelli and Heinrich Wahl

In this **Section** we talk about Heinrich Wahl and Italo Mannelli; they had impact on our understanding of $\Delta S \neq 0$ processes, and of direct **CP** asymmetry. They have been leaders of very successful experiments at CERN (Geneva), namely NA31[11] (1982–1989) and NA48 (1990s–2007):

- NA31 found the first evidence for direct **CP** violation, with a result that was about three standard deviations from zero: $\mathrm{Re}(\epsilon'/\epsilon_K) = (3.3 \pm 1.1) \cdot 10^{-3}$; and finally $\mathrm{Re}(\epsilon'/\epsilon_K) = (2.0 \pm 0.7) \cdot 10^{-3}$ [238]. The competitor E731 at FNAL had not found a non-zero value.
- Some HEP theorists had predicted $\mathrm{Re}(\epsilon'/\epsilon_K)|_{\mathrm{SM}} < 10^{-3}$, as mentioned in **Sect. 7.2**.
- This ambiguous situation prompted the design of a new generation of detectors, both at CERN (NA48) and at Fermilab (KTeV).
- NA48 established direct **CP** asymmetry: $\mathrm{Re}(\epsilon'/\epsilon_K) = (1.47 \pm 0.22) \cdot 10^{-3}$ [239–241]; KTeV gives [242]: $\mathrm{Re}(\epsilon'/\epsilon_K) = (2.07 \pm 0.28) \cdot 10^{-3}$.

The NA31 and NA48 collaborations have a consistent – actually amazing – record about the data based on detectors and analyses, and their result was fully confirmed by KTeV.

A Summer School in Italy happens year after year in a special place, namely in the Villa Monastero, which is located in Varenna on the shore of Lake Como in Italy.[12] There Italo Mannelli told one of our co-authors long time ago that he expects theorists speaking only in good *faith*, not about *truth* all the time. That was fitting, since lectures were given in a chapel.

The past work of Wahl and Mannelli and of the NA31 and NA48 collaborations can be summarized quickly with these titles:
(1) NA31 Collaboration: "First Evidence for direct **CP** Violation" in 1988 and "A New Measurement of direct **CP** violation in the neutral kaon system" in 1993 [238].
(2) NA48 Collaboration: "A New Measurement of direct **CP** violation in two pion decays of the neutral kaon" 1999 and "A Precision Measurement of direct **CP** violation in the decay of neutral kaons into two pions" in 2002 [239].
(3) NA48 Collaboration again: "Performance of an electromagnetic liquid krypton calorimeter" in 1994 and "Performance of an electromagnetic liquid krypton calorimeter based on a ribbon electrode tower structure" in 1996 [240].

As we had said just above: maybe the 'landscape' of fundamental dynamics had changed with clear sign of New Dynamics? The Panofsky Prize from the APS was given to Italo Mannelli, Heinrich Wahl and Bruce Winstein in 2007 for this

[11] Wahl, who who was spokesperson of NA31 and NA48, has said in his *Physics Report*, V. 403-404, Dec.2004, that the origin of the NA31 experiment was discussed in Jack Steinberger's office in 1981.

[12] The International School of physics Enrico Fermi is one of the cultural activities of the Italian Physical Society, and it was founded in 1953.

reason. In [240] 'hardware' is mentioned since crucial discoveries usually require new experimental tools such as the 'electromagnetic liquid krypton calorimeter' that Mannelli produced; yet the words cannot show his impact.

A 'small' collaboration cannot produce a true new detector or large parts of it. They needed 'recycling' from older detectors. The next step led to the NA62 Collaboration with $\sim$ 333 members (i.e., 'small' now), where Cristina Lazzeroni is the spokesperson from 2019, coming after Augusto Ceccucci. This experiment began in 2006 and will continue for a long time for excellent physical reasons, outlined in **Sect. 6.5.10**. The goal of the NA62 collaboration is to establish 'hyper-rare' decays $K^+ \to \pi^+ \bar{\nu}\nu$ with $\sim$ 80 events, as predicted by the SM. Later the goal is to probe $K_L \to \pi^0 \bar{\nu}\nu$. In the experiment, it has been re-used as much hardware as possible from NA48 ('suitably' re-shaped and re-assembled) in order to 'transfer' an old detector into a new one. Italo himself and his team worked out every details, including the dimensions of each element, how many elements should be built and how should they be built, the precise geometry,[13] how to install the detector and which electronics to use with it.

As said above: Italo expects theorists speaking only in good *faith*. To be more subtle, we suggest to go to the opera 'The Magic Flute' from Mozart and listen to the aria advocating: 'patience when in peril'.

7.4 Direct CP asymmetries in $B_{(s)}$ mesons

7.4.1 *Two-body FS*

Experiments probe **CP** asymmetries first in two-body FS of B mesons including narrow resonances. Although the kinematics is straightforward, there is a rich landscape where finding new lessons on fundamental dynamics.

PDG2020 gives a list of branching ratios for CKM suppressed decays. Mostly it covers the $10^{-6} - 10^{-5}$ region, but with six outliers:

$$\begin{aligned}
\mathrm{BR}(B^0 \to K^+K^-) &= (7.8 \pm 1.5) \cdot 10^{-8} \\
\mathrm{BR}(B^+ \to K^+D_{\mathbf{CP}(+1)}) &= (1.80 \pm 0.07) \cdot 10^{-4} \\
\mathrm{BR}(B^+ \to K^+\eta') &= (7.04 \pm 0.25) \cdot 10^{-5} \\
\mathrm{BR}(B^0_s \to \eta'\eta') &= (3.3 \pm 0.7) \cdot 10^{-5} \\
\mathrm{BR}(B^0_s \to K^+K^-) &= (2.66 \pm 0.22) \cdot 10^{-5} \\
\mathrm{BR}(B^0_s \to \eta'\eta') &= (7.0 \pm 1.0) \cdot 10^{-7}
\end{aligned} \tag{7.17}$$

[13] Italo's trademark is octagonal shape for detectors.

(a) PDG2020 lists three **CP** asymmetries in two-body FS of B^0 decays:

$$A_{\mathbf{CP}}(B^0 \to K^+\pi^-) = -0.083 \pm 0.004 \; , \tag{7.18}$$

$$A_{\mathbf{CP}}(B^0 \to [K^*(892)]^+\pi^-) = -0.27 \pm 0.04 \; , \tag{7.19}$$

$$A_{\mathbf{CP}}(B^0 \to [K^*(892)]^0\eta) = +0.19 \pm 0.05 \; ; \tag{7.20}$$

the second and third ones are parts of Dalitz plots.

(b) Again, PDG2020 lists three **CP** asymmetries in two-body FS of B^+ ones:

$$A_{\mathbf{CP}}(B^+ \to K^+D_{\mathbf{CP}(+1)}) = +0.120 \pm 0.014 \; , \tag{7.21}$$

$$A_{\mathbf{CP}}(B^+ \to K^+\eta) = -0.37 \pm 0.08 \; , \tag{7.22}$$

$$A_{\mathbf{CP}}(B^+ \to K^+\rho^0) = +0.37 \pm 0.10 \; , \tag{7.23}$$

$$A_{\mathbf{CP}}(B^+ \to K^+f_2(1270)) \simeq -0.68 \pm 0.19 \; . \tag{7.24}$$

This time the third and fourth one are parts of Dalitz plots.

(c) So far, we have only one **CP** asymmetry for B^0_s one:

$$A_{\mathbf{CP}}(B^0_s \to \pi^+K^-) = +0.221 \pm 0.015 \; . \tag{7.25}$$

(d) In **Sect. 4.3** we already mentioned that one can use U-spin symmetry to describe the *spectroscopy* of beauty and charm hadrons, and perform a SM test suggested by Lipkin, based on broken U-spin symmetry to compare weak decays of B^0 and B^0_s [124]:

$$\Delta = \frac{A_{\mathbf{CP}}(B^0 \to K^+\pi^-)}{A_{\mathbf{CP}}(B^0_s \to \pi^+K^-)} + \frac{\Gamma(B^0_s \to \pi^+K^-)}{\Gamma(B^0 \to K^+\pi^-)} = 0 \; .$$

Present data [125] lead to:

$$A_{\mathbf{CP}}(B^0 \to K^+\pi^-) = -0.083 \pm 0.004 \;\; , \;\; A_{\mathbf{CP}}(B^0_s \to \pi^+K^-) = 0.221 \pm 0.015$$
$$\Delta = -0.11 \pm 0.04 \pm 0.03 \; .$$

One cannot disagree with the statement: 'no evidence for a deviation from zero of Δ is observed with the present experimental precision'. $\Delta = 0$ is consistent with present data, but a value $\Delta = -0.1$ seems more 'natural'.[14]

We remark again that FSI have important impact on weak amplitudes in general and in particular on **CP** asymmetries; we can*not* focus only on two-body FS in weak decays. As shown in Eqs. (2.58, 2.59, 2.60), intermediate two states strongly re-scatter into FS with two-, three-, four-body etc. states.

We continue with the B^0 and B^0_s decays to FS that are **CP** eigenstates; they produce oscillations including *indirect* **CP** violation.

Direct **CP** asymmetries can be probed in different FS, following Eq. (6.38); we show examples from PDG2020[15]:

$$S_{B^0\to\pi^+\pi^-} = -0.65 \pm 0.04 \;\; , \;\; C_{B^0\to\pi^+\pi^-} = -0.32 \pm 0.04 \tag{7.26}$$

$$S_{B^0\to\phi K^0} = +0.59 \pm 0.14 \;\; , \;\; C_{B^0\to\phi K^0} = +0.01 \pm 0.14 \; . \tag{7.27}$$

[14]As one of the co-authors pointed out several times, the last time being in the 2018 Epiphany Conference in Krakow [243].

[15]They are calibrated by $S_{B^0\to J/\psi(nS)K^0} = +0.701 \pm 0.017$ and $C_{B^0\to J/\psi(nS)K^0} = +0.005 \pm 0.020$.

Direct **CP** violation has been established so far in Eq. (7.26). The results from PDG2020 for two-body FS of B_s^0 give

$$S_{B_s^0\to K^+K^-} = 0.30 \pm 0.13 \ , \ C_{B_s^0\to K^+K^-} = 0.14 \pm 0.11 \ . \qquad (7.28)$$

The LHCb collaboration has shown results from 7 and 8 TeV collisions [125]:

$$S_{B_s^0\to K^+K^-} = +0.18 \pm 0.06 \pm 0.02 \ , C_{B_s^0\to K^+K^-} = +0.20 \pm 0.06 \pm 0.02 \ ,$$
$$S_{B^0\to \pi^+\pi^-} = -0.63 \pm 0.05 \pm 0.01 \ , C_{B^0\to \pi^+\pi^-} = -0.34 \pm 0.06 \pm 0.01 \ ; \qquad (7.29)$$

they found evidence for **CP** violation in $B_s^0 \to K^+K^-$ decay for the first time [125].

7.4.2 *Three-body FS*

Charmless decays of B mesons to three hadrons are dominated by quasi-two-body processes involving intermediate resonant states. The branching ratios for CKM suppressed B^+ decays given by PDG2020 are:

$$\mathrm{BR}(B^+ \to K^+\pi^-\pi^+) = (5.10 \pm 0.29) \cdot 10^{-5}$$
$$\mathrm{BR}(B^+ \to K^+K^-K^+) = (3.40 \pm 0.14) \cdot 10^{-5} \ . \qquad (7.30)$$

They are similar, which is not surprising, since kaons are light on the scale of M_B. The large samples of charmless B decays collected by the LHCb experiment allow direct **CP** violation to be measured in regions of phase space. LHCb data based on *pp* collisions of run-1 have first showed averaged **CP** asymmetries [244]:

$$A_{\mathbf{CP}}(B^+ \to K^+\pi^+\pi^-) = +0.025 \pm 0.004|_{\rm stat} \pm 0.004|_{\rm syst} \pm 0.007|_{\psi K^\pm}$$
$$A_{\mathbf{CP}}(B^+ \to K^+K^+K^-) = -0.036 \pm 0.004|_{\rm stat} \pm 0.002|_{\rm syst} \pm 0.007|_{\psi K^\pm} \qquad (7.31)$$

where the third uncertainty is due to the **CP** asymmetry of the $B^\pm \to \psi K^\pm$ reference mode. Averaged **CP** asymmetries based on more refined LHCb analyses are shown by PDG2020

$$\langle A_{\mathbf{CP}}(B^+ \to K^+\pi^+\pi^-)\rangle = +0.027 \pm 0.008$$
$$\langle A_{\mathbf{CP}}(B^+ \to K^+K^+K^-)\rangle = -0.033 \pm 0.008 \ . \qquad (7.32)$$

Based on our experience with the impact of penguin diagrams on the measured $B^0 \to K^+\pi^-$, the sizes of these averaged asymmetries are not really surprising; however, that does not mean that we could have really predicted them. On the other hand, it is interesting that they come with the opposite sign, as we explain. **CPT** invariance give the total widths of states and anti-states; in this case:

$$\Gamma(B^+) \equiv \sum_{f^+} \Gamma(B^+ \to f^+) = \Gamma(B^-) \equiv \sum_{f^-} \Gamma(B^- \to f^-) \ . \qquad (7.33)$$

It means: when one finds a **CP** asymmetry in one width, one has to find a sum of FS with the opposite sign. One example is Eq. (7.32); is it just luck?

LHCb data have also shown 'regional' **CP** asymmetries, namely asymmetries observed in localized regions of phase space [244]; it was surprising that the regional asymmetries are much larger:

$$A_{\mathbf{CP}}(B^+ \to K^+\pi^+\pi^-)|_{\text{'region'}} = +0.678 \pm 0.078|_{\text{stat}} \pm 0.032|_{\text{syst}} \pm 0.007|_{\psi K^\pm}$$
$$A_{\mathbf{CP}}(B^+ \to K^+K^+K^-)|_{\text{'region'}} = -0.226 \pm 0.020|_{\text{stat}} \pm 0.004|_{\text{syst}} \pm 0.007|_{\psi K^\pm}$$

'Regional' **CP** asymmetries are defined by the LHCb collaboration for run-1: positive asymmetry at low $m_{\pi^+\pi^-}$ just below m_{ρ^0}; negative asymmetry both at low and high $m_{K^+K^-}$ values. Again, one should note the opposite signs above. It is not surprising that 'regional' asymmetries are very different from averaged ones. Even when one uses states only from the SM – $SU(3)_C \times SU(2)_L \times U(1)$ – one expects that; it shows the *impact of re-scattering* due to $SU(3)_C$ (actually $SU(3)_C \times$ QED) in general. Our community needs more data, but that is not enough. The best fitted analyses do not always give us the best understanding of the underlying fundamental dynamics. There are questions in need of more precise answers:

- What is the best way to define regional asymmetries and probe them on the experimental and theoretical sides?
- The phase space of three body FS allows a rich tapestry of resonant structures to emerge. Can it show the impact of broad resonances like $f_0(500)/\sigma$ and $K^*(700)/\kappa$?

Data of even more CKM suppressed B^+ decays are also available

$$\text{BR}(B^+ \to \pi^+\pi^-\pi^+) = (1.52 \pm 0.14) \cdot 10^{-5}$$
$$\text{BR}(B^+ \to \pi^+K^-K^+) = (0.52 \pm 0.04) \cdot 10^{-5} \,. \tag{7.34}$$

PDG2020 show larger averaged **CP** asymmetries with the opposite signs:

$$\langle A_{\mathbf{CP}}(B^+ \to \pi^+\pi^+\pi^-)\rangle = +0.057 \pm 0.013$$
$$\langle A_{\mathbf{CP}}(B^+ \to \pi^+K^+K^-)\rangle = -0.122 \pm 0.021 \,.$$

Looking just at the SM penguin diagrams it is surprising, since $b \Rightarrow d$ penguin diagrams are more suppressed than $b \Rightarrow s$ ones. Again data are available for **CP** asymmetries which focus on small regions in the Dalitz plots [244].

$$A_{\mathbf{CP}}(B^\pm \to \pi^\pm\pi^+\pi^-)|_{\text{'region'}} = +0.584 \pm 0.082|_{\text{stat}} \pm 0.027|_{\text{syst}} \pm 0.007|_{\psi K^\pm}$$
$$A_{\mathbf{CP}}(B^\pm \to \pi^\pm K^+K^-)|_{\text{'region'}} = -0.648 \pm 0.070|_{\text{stat}} \pm 0.013|_{\text{syst}} \pm 0.007|_{\psi K^\pm}$$

The LHCb experiment has the total data of run-2 at 13 TeV collisions. However, our community has not gotten yet the analyses of these data including regional asymmetries. We expect that these new analyses will enhance our understanding of fundamental dynamics – with the help of theorists. 'Soon' the Belle II Collaboration will measure also other FS of neutral beauty mesons like $B^0 \to K^+\pi^-\pi^0/K^0\pi^+\pi^-/K^0K^+K^-/\pi^+\pi^-\pi^0/K^+K^-\pi^0$ and $B_s^0 \to K^+K^-\pi^0/\pi^+\pi^-\pi^0/\bar{K}^0\pi^+\pi^-/\bar{K}^0K^+K^-$. Measuring those transitions will allow us to refine our analyses, and give us better lessons in different directions and levels.

7.4.3 *Future lessons from three- and four-body FS of beauty mesons*

The analysis of three-body FS is based on well-tested and experimentally solid tools: Dalitz plots. We need more data, of course, but that is not enough, because new theoretical tools are needed as well to interpret future data. Particularly:

- Diagrams may help us about the directions, in particular about penguin diagrams. The situations are very different for $\Delta S = 1$ vs. $\Delta B = 1$; the first one leads to two- and three-body FS, while the second one mostly produces many-body ones.
- First one analyzes the data using insensitive techniques [245, 246], compares them and discuss the results – but that is not the end of our 'travel'.
- Dispersion relations are not just another model. Their analyses are based on chiral symmetry, namely non-perturbative QCD – with some 'judgment'.
- One can use **CP** asymmetries as hunting regions for ND, since their impact can be seen just in its amplitudes.
- The impact of broad resonances such as $f_0(500)/\sigma$ and $K^*(700)/\kappa$ will have to be understood thoroughly.

Four-body FS will have to be probed with the best possible and realistic tools. We focus on FS with only charged mesons, which LHCb can experimentally probe with good efficienly and relatively clean experimental signature.

There are two classes of suppressed transitions in neutral B mesons: (a) $B^0 \to K^+\pi^-\pi^+\pi^-/\ K^+K^-K^+\pi^-$ and $B^0_s \to K^+K^-\pi^+\pi^-/\ K^+K^-K^+K^-/\ \pi^+\pi^-\pi^+\pi^-$; (b) $B^0 \to \pi^+\pi^-\pi^+\pi^-/\ K^+K^-\pi^+\pi^-/\ K^+K^-K^+K^-$ and $B^0_s \to K^-\pi^+\pi^-\pi^+/\ K^-K^+K^-\pi^+$. So far we have only limits. Of course, our community cannot give up. Future LHCb data, in particular, may bring interesting results.

FSI produces non-zero **T**-odd distributions with*out* **CP** violation (see **Sect. 7.1.2**); yet one has to compare **T**-odd observables from $B_{(s)}$ and $\bar{B}_{(s)}$ to probe **CP** asymmetries. As a first step one measures moments. It would not be surprising to find very small values; one has to go beyond moments. To be realistic, one can measure the angles between two planes. Furthermore one can use different definitions of the planes. We give examples about $B^0 \to K^+\pi_1^-\pi^+\pi_2^-$ and $B^0 \to K_1^+K^-K_2^+\pi^-$ beyond moments. Coming back to Eqs. (7.3 – 7.7) we have two 'roads' somewhat including **CPT** tools, where the data come from the rest frame of B^0. First 'road': one defines one plane of $\vec{p}(K^+) \times \vec{p}(\pi_1^-)$ with the other from $\vec{p}(\pi^+) \times \vec{p}(\pi_2^-)$ and measure the angle between them. Next one can measure the angle between the planes of $\vec{p}(K_1^+) \times \vec{p}(\pi^-)$ and $\vec{p}(K_2^+) \times \vec{p}(\pi^-)$. Second 'road': one defines the plane of $\vec{p}(K^+) \times \vec{p}(\pi^+)$ with the other one from $\vec{p}(\pi_1^-) \times \vec{p}(\pi_2^-)$ and measure the angle between them. How one can differentiate between π_1^- vs. π_2^- and K_1^+ vs. K_2^+? One has to be realistic: the obvious way is to use cuts in energy; however, one could use more subtle 'roads', as discussed above in Eqs. (7.10 – 7.14).

For the future the Belle II collaboration can measure or show the impact of full **CPT** invariance in decays such as $B^+ \to K^+\pi^-\pi^+\pi^0[\eta,\eta']/K^+K^-K^+\pi^0[\eta,\eta']$ and $B^+ \to \pi^+\pi^-\pi^+\pi^0[\eta,\eta']/\ \pi^+K^+K^-\pi^0[\eta,\eta']$.

7.5 Direct CP asymmetries in $D_{(s)}$ mesons

The existence of DCS channels has a subtle, yet significant impact on charged CF D-meson decays producing neutral kaons. In the SM one expects zero **CP** asymmetry in the charged CF decays $D^+ = [c\bar{d}] \to s\bar{d}u\bar{d} \Rightarrow \bar{K}^0\pi^+$ vs. $D^- = [\bar{c}d] \to \bar{s}d\bar{u}d \Rightarrow K^0\pi^-$, where the final states have different strangeness. The observed $K_S\pi^+$ final state is a coherent sum of amplitudes for the CF $D^+ \to \bar{K}^0\pi^+$ decay and the DCS $D^+ \to K^0\pi^+$ one. In 1995 it was predicted that the SM gives non-zero asymmetry there due to indirect **CP** violation in $\Delta S = 2$ oscillations [247]:

$$\text{SM}: \frac{\Gamma(D^+ \to K_S\pi^+) - \Gamma(D^- \to K_S\pi^-)}{\Gamma(D^+ \to K_S\pi^+) + \Gamma(D^- \to K_S\pi^-)} \simeq -2\,\mathrm{Re}\,\epsilon_K = -(3.32 \pm 0.06)\cdot 10^{-3}. \tag{7.35}$$

Thus in the absence of direct **CP** violation in CF and DCS decays (as expected within the SM), one still finds **CP** asymmetry in $D^+ \to K_S\pi^+$ decay within the SM because of indirect **CP** violation in the kaon system. On the other hand, if BSM physics generates additional weak phases other than the CKM one, interference between CF and DCS decays could generate direct **CP** asymmetry due to the D-meson decay. Recent data give [8]:

$$\text{2020 data}: \frac{\Gamma(D^+ \to K_S\pi^+) - \Gamma(D^- \to K_S\pi^-)}{\Gamma(D^+ \to K_S\pi^+) + \Gamma(D^- \to K_S\pi^-)} = -(4.1 \pm 0.9)\cdot 10^{-3}\,. \tag{7.36}$$

Can the impact of ND in $\Delta C \neq 0$ hide here? Maybe, but we would not bet on that.

Direct **CP** violation in D decays can be probed more effectively in SCS decays $D^0 \to K^+K^-$ and $D^0 \to \pi^+\pi^-$. One can throw out the approximately universal contribution of indirect **CP** violation (and also production asymmetry) in the difference (data from PDG2018): $\Delta A_{\mathbf{CP}} = A_{\mathbf{CP}}(D^0 \to K^+K^-) - A_{\mathbf{CP}}(D^0 \to \pi^+\pi^-) = (-12 \pm 13)\cdot 10^{-4}$. In March 2019 we entered a *novel era*: direct **CP** violation in D^0 decays has been established at pp collisions of 13 TeV (including results from run-1) by the LHCb collaboration [174]

$$A_{\mathbf{CP}}(D^0 \to K^+K^-) - A_{\mathbf{CP}}(D^0 \to \pi^+\pi^-) = -(15.4 \pm 2.9)\cdot 10^{-4}\,; \tag{7.37}$$

this pioneering achievement has opened a new gate to $\Delta C \neq 0$! Furthermore, this is the beginning of a new 'travel'.

The decays of $D_{(s)}$ mesons are mostly described by two-, three- and four-body non-leptonic FS. Two-body FS are only small parts of them and it is crucial to probe non-leptonic three- and four-bodies. In the case of charm we cannot apply with confidence heavy-quark expansion tools, and we rely on approximate symmetries and phenomenological models, where the knowledge of branching fractions and resonant structures in the case of multi-body decays are key inputs. Several

Dalitz analysys of three body decays are currently available, such as, e.g. Ref. [248]. Not surprisingly, these analysis show a rich spectrum of intermediate hadronic states.

7.5.1 D^0 *and* D^+

Direct **CP** asymmetry has been established in $D^0 \to h^+h^-$, see Eq. (7.37). The LHCb collaboration will analyze the data from run-2 looking for indirect **CP** violation in $D^0 \to K^+K^-/\pi^+\pi^-$ – see Eq. (6.129) – or probe three- and four-body FS, that are larger parts of D^0 and D^+ decays. Relevant data is also expected from BES-III and Belle II experiments. The field of amplitude analyses for multi-body decays remains very active. Here we list are some examples from [8]:

$$\begin{aligned} A_{\mathbf{CP}}(D^0 \to \pi^+\pi^-\pi^0) &= +(3 \pm 4) \cdot 10^{-3} \\ A_{\mathbf{T}}(D^0 \to K^+K^-\pi^+\pi^-) &= +(2.9 \pm 2.2) \cdot 10^{-3} \\ A_{\mathbf{CP}}(D^+ \to K^+K^-\pi^+) &= +(3.7 \pm 2.9) \cdot 10^{-3} \\ A_{\mathbf{T}}(D^+ \to K_S K^\pm\pi^+\pi^-) &= -(12 \pm 11) \cdot 10^{-3} \end{aligned} \tag{7.38}$$

- One has to go *beyond* averaged **CP** violation for three- and four-body FS. It is a true challenge to understand the underlying dynamics both on the experimental and theoretical sides.
- As in the case of $\Delta B \neq 0$ decays, multibody decays of D meson allow the **CP** asymmetries to be probed across the phase space of the decay, and to analyze 'regional' **CP** asymmetries. In principle the latter ones may be larger than global **CP** asymmetries.
- Long-range hadronic effects such as re-scattering limit the accuracy of SM predictions [249]. Re-scattering connects different *two-body* FS – like $\bar{K}K \leftrightarrow \pi\pi$ – but even more with *two-body* $\to$ *many-body* FS.
  Amplitude analyses go towards an increasing sophistication; however, they depends on non-perturbative QCD, for which we have little control in general. In special situations tools such as chiral symmetry, dispersion relations, U- and V-spin symmetries and their violations (see **Sect. 4.3**) may be useful.

On the theoretical side, DCS decays like $D^+ \to K^+\pi^+\pi^-$, $D^+ \to K^+K^+K^-$ are easier to deal with: penguin diagrams cannot contribute and the SM gives basically zero **CP** asymmetries. The challenges for experimenters are obvious. Exclusive branching ratios are $\mathcal{O}(10^{-4})$ for D^0 and D^+. Our community has to produce more data and to deal with large backgrounds even from the SM. We expect impressive experimental achievements from LHC run-3, Belle II and BES-III.

7.5.2 D_s^+

The 'landscape' of measured **CP** asymmetries is even slimmer for D_s^+ than for D^+. From PDG2020 we have:

$$\begin{aligned} A_{\mathbf{CP}}(D_s^+ \to K^+ K_S) &= +(0.09 \pm 0.26) \\ A_{\mathbf{CP}}(D_s^+ \to K^+ K^- \pi^+) &= -(0.5 \pm 0.9) \\ A_{\mathbf{T}}(D_s^+ \to K_S K^+ \pi^+ \pi^-) &= -(1.4 \pm 0.8) \end{aligned} \tag{7.39}$$

These asymmetries come from Cabibbo *favored* decays and there is little chance to find **CP** violation there.

Data have reached the level for probing SCS transitions:

$$\mathrm{BR}(D_s^+ \to K^+ \pi^- \pi^+) = (6.5 \pm 0.4) \cdot 10^{-3} \tag{7.40}$$

$$\mathrm{BR}(D_s^+ \to K^+ K^- K^+) = (2.16 \pm 0.20) \cdot 10^{-4} \; ; \tag{7.41}$$

and DCS transitions:

$$\mathrm{BR}(D_s^+ \to K^+ K^+ \pi^-) = (1.28 \pm 0.04) \cdot 10^{-4} \; . \tag{7.42}$$

The branching ratio in Eq. (7.41) latter is smaller than the one in Eq. (7.40) because of different kinematics and phase space.

We expect to see the results for asymmetries of charged three-body FS that have been measured by the LHCb collaboration, based on pp collisions at $\sqrt{s} = 13$ TeV. These results might be published soon; we expect they will follow the 'roads' of the Dalitz plots for weak transitions of $B^\pm$ as discussed in **Sect. 7.4.2**, with regional asymmetries larger than averaged ones:

- The SM predicts averaged **CP** asymmetries of $\mathcal{O}(10^{-3})$ for SCS transitions. Can regional asymmetries be of $\mathcal{O}(10^{-2})$ (or more)?
- The SM predicts basically zero **CP** asymmetries for DCS transitions. Thus it is a hunting region for ND – if one has enough data to measure them with non-zero values and deal with the large backgrounds.

7.6 Novel challenges from Λ_b^0 decays?

Beauty baryons are now being observed in significant numbers and have therefore started to offer complementary means to probe **CP** violation. As baryons do not oscillate, only direct **CP** violation in decay is expected. Since the same classes of quark-level transitions as in the meson case are possible, one expects asymmetries of similar magnitude and that the best chance to find **CP** asymmetries is in charmless decays mediated by $b \to u\bar{u}s$ and $b \to u\bar{u}d$ transitions.

The lightest b baryon is the Λ_b^0 (udb) state. Searches for CP violation have been performed in $\Lambda_b^0 \to p\pi^-$ and $\Lambda_b^0 \to pK^-$ decays. Both decays depend on production asymmetries in the same way. PDG2020 data give us:

$$A_{\mathbf{CP}}(\Lambda_b^0 \to p\pi^-) = -0.025 \pm 0.029 \tag{7.43}$$

$$A_{\mathbf{CP}}(\Lambda_b^0 \to pK^-) = -0.025 \pm 0.022 \; ; \tag{7.44}$$

no **CP** asymmetry has been found in these two-body FS, but, with the current uncertainties, these results would also be consistent with a few % effects, as seen in charmless B meson decays.

Contributions to $\Lambda_b^0 \to p\pi^-(K^-)$ come from both color-allowed $b \to u\bar{u}d$ tree-level diagram and $b \Rightarrow d$ ($b \Rightarrow s$) QCD penguin diagrams. The branching ratios are:

$$\mathrm{BR}(\Lambda_b^0 \to p\pi^-) = (0.45 \pm 0.08) \cdot 10^{-5} \tag{7.45}$$

$$\mathrm{BR}(\Lambda_b^0 \to pK^-) = (0.54 \pm 0.10) \cdot 10^{-5} \,. \tag{7.46}$$

Basically their values are the same.

The rich kinematic distributions of multibody baryon decays provide us with other opportunities to search for **CP** violation. PDG2020 shows branching ratios for 4-body FS:

$$\begin{aligned} \mathrm{BR}(\Lambda_b^0 \to p\pi^-\pi^+\pi^-) &= (2.11 \pm 0.23) \cdot 10^{-5} \\ \mathrm{BR}(\Lambda_b^0 \to p\pi^-K^+K^-) &= (0.41 \pm 0.06) \cdot 10^{-5} \,; \qquad (7.47) \\ \mathrm{BR}(\Lambda_b^0 \to pK^-\pi^+\pi^-) &= (5.1 \pm 0.5) \cdot 10^{-5} \\ \mathrm{BR}(\Lambda_b^0 \to pK^-K^+K^-) &= (1.27 \pm 0.14) \cdot 10^{-5} \,. \qquad (7.48) \end{aligned}$$

It is easy to 'paint' diagrams for those, but we have to go beyond that [250].

In **Sect. 6.5.9**, we have described how **T**-odd moments of $K_L \to \pi^+\pi^-e^+e^-$ had been very successfully applied to establish **CP** violation. In **Sect. 7.1.2** we have continued to use **T**-odd moments in several ways to probe **T** and **CP** asymmetries in non-leptonic decays. One can build **T**-odd moments also in b-baryon decays, combining the momentum of three final state particles in the mother C.M. frame. In pp collisions one gets different numbers of Λ_b^0 vs. $\bar{\Lambda}_b^0$ due to *production* asymmetries. They do not affect **CP** asymmetries based on **T**-odd moments.

CP violation has not yet been observed in b-baryon decays, although LHCb data had shown 'evidence' for **CP** asymmetry in $\Lambda_b^0 \to p\pi^-\pi^+\pi^-$ based on run-1 of 3 fb^{-1} [251].[16] By probing **T**-odd observables, the collaboration has measured the angle between two planes: one is formed by the momenta of p and π^-_{fast}, while the other one by the momenta of π^+ and π^-_{slow}.[17]

Maybe if **CP** asymmetry will be found in a decay of a baryon for the first time (except 'our existence'...), it will be in the decay of a beauty baryon. It is self evident that many-body FS are *not* just a background for the information our community gets from two-body FS, but it has the potential to give significative information on the (not yet well known) underlying dynamics.

In a later (2018) paper, the LHCb collaboration has searched for **CP** asymmetries in $\Lambda_b^0 \to p\pi^-\pi^+\pi^-$, $\Lambda_b^0 \to pK^-K^+K^-$ and $\Xi_b^0 \to pK^-K^-\pi^+$ with data collected from pp collisions at 7 and 8 TeV [254]: *no* significant deviation from **CP** (or **P**) symmetry was found either integrated over all phases spaces or within specific regions. Our community was eagerly waiting for the analyses of weak decays of

[16] A summary of this search can be read in the CERNCOURIER of March 2017 [252].
[17] The general approach is presented in Ref. [253].

beauty hadrons based in pp collisions at $\sqrt{s} = 13$ TeV. Since the impact of regional asymmetries in the Dalitz plots of $B^\pm$ decays is sizable, one naturally wonders if the 'landscapes' of beauty baryons are different. At present, Ref. [8] lists only averaged **CP** asymmetry for $\Lambda_b^0 \to p\pi^-\pi^+\pi^-$ decays:

$$\Delta A_{\mathbf{CP}} = 0.011 \pm 0.026 \tag{7.49}$$

The LHCb collaboration can combine run-1 and run-2 data with 9 fb^{-1}; so far it has analyzed data of $\Lambda_b^0 \to p\pi^-\pi^+\pi^-$ from 6.6 fb^{-1}; i.e., 3 fb^{-1} from 7 and 8 TeV and 3.6 fb^{-1} from 13 TeV collisions [255]. In Eq. (7.1) and Eq. (7.2) we described the hadron decays $H_Q \to h_1 h_2 h_3 h_4$ and $\bar{H}_Q \to \bar{h}_1 \bar{h}_2 \bar{h}_3 \bar{h}_4$ in general. Now we have a special case, namely $\Lambda_b^0 \to p\pi^-\pi^+\pi^-$ and $\bar{\Lambda}_b^0 \to \bar{p}\pi^+\pi^-\pi^+$ leading to

$$\begin{aligned} C_T &\equiv \vec{p}_p \cdot (\vec{p}_{\pi^-_{\text{fast}}} \times \vec{p}_{\pi^+}) \Longrightarrow \langle A_T \rangle \neq 0 \\ \bar{C}_T &\equiv \vec{p}_{\bar{p}} \cdot (\vec{p}_{\pi^+_{\text{fast}}} \times \vec{p}_{\pi^-}) \Longrightarrow \langle \bar{A}_T \rangle \neq 0 \end{aligned} \tag{7.50}$$

with*out* **CP** and **P** violation. However, finding

$$\begin{aligned} \langle A_{\mathbf{CP}}^{T-odd} \rangle &= \frac{1}{2}(\langle A_T \rangle - \langle \bar{A}_T \rangle) \neq 0 \\ \langle A_{\mathbf{P}}^{T-odd} \rangle &= \frac{1}{2}(\langle A_T \rangle + \langle \bar{A}_T \rangle) \neq 0 \end{aligned} \tag{7.51}$$

implies **CP** asymmetry and **P** asymmetry, respectively.[18] Indeed:

- **P** asymmetry with 5.5 standard deviation has been established in a novel landscape [255]. It is quite an achievement after hard work. However, there is not yet a clear lesson we have learnt on fundamental dynamics.
- This result give **CP** asymmetry with 2.9 standard deviation. Still we are on the right road – or not? We can hope to get analyses from the remaining 3.0 fb^{-1} from run-2 – or do we have to wait for run-3 data?

That are excellent reasons for a upgrade, where LHCb could collect $\mathcal{L} = 50$ fb^{-1} and be able to probe **CP** asymmetries also in the decays of Ξ_b^0 and Ω_b^-, as well as *regional* asymmetries.

7.7 CP asymmetries for strange baryons Λ and Σ

We know that direct **CP** asymmetry in $\Delta S = 1$ decays is tiny, as observed from comparisons of $K_L \to \pi^+\pi^-$ vs. $K_L \to \pi^0\pi^0$ meson decays. A novel way to probe **CP** asymmetries involves hyperon decays:

[18]One of us (IB) has attended a CERN seminar in person; it was excellent to understand the experimental side [256].

- The narrow resonance J/ψ can produce pairs of strange baryons: $J/\psi \to \bar{\Lambda}\Lambda \to [\bar{p}\pi^+][p\pi^-]$ and $J/\psi \to \bar{\Sigma}^-\Sigma^+ \to [\bar{p}\pi^0][p\pi^0]$ [257].[19] From PDG2020 we have:

$$\mathrm{BR}(J/\psi \to \bar{\Lambda}\Lambda) = (1.89 \pm 0.09) \cdot 10^{-3} \tag{7.52}$$

$$\mathrm{BR}(J/\psi \to \bar{\Sigma}^-\Sigma^+) = (1.50 \pm 0.24) \cdot 10^{-3} \, . \tag{7.53}$$

- Re-scattering gives sizable impacts, as already observed for Λ and Σ^+:

$$\mathrm{BR}(\Lambda \to p\pi^-) = 0.639 \pm 0.005 \, , \, \mathrm{BR}(\Lambda \to n\pi^0) = 0.358 \pm 0.005$$
$$\mathrm{BR}(\Sigma^+ \to p\pi^0) = 0.5157 \pm 0.0030 \, , \, \mathrm{BR}(\Sigma^+ \to n\pi^+) = 0.4831 \pm 0.0030 \, .$$

- *T-odd* moments for $\Lambda \to p\pi^-$ and $\bar{\Lambda} \to \bar{p}\pi^+$ have been measured, see PDG2020:

$$\alpha_- \equiv \langle \vec{s}_\Lambda \cdot (\vec{s}_p \times \vec{p}_p) \rangle = +0.732 \pm 0.014 \, , \tag{7.54}$$

$$\alpha_+ \equiv \langle \vec{s}_{\bar{\Lambda}} \cdot (\vec{s}_{\bar{p}} \times \vec{p}_{\bar{p}}) \rangle = -0.758 \pm 0.012 \, . \tag{7.55}$$

  Assuming **CPT** invariance, one probes **CP** asymmetry:

$$\langle A_{\mathbf{CP}}(\Lambda \to p\pi^-) \rangle \equiv \frac{\alpha_- + \alpha_+}{\alpha_- - \alpha_+} \, . \tag{7.56}$$

- The asymmetries of interest for **CP** violation study in $\Lambda/\bar{\Lambda}$ decays can be related to the spins of $\Lambda/\bar{\Lambda}/p/\bar{p}$, $(\vec{s}_{\Lambda,\bar{\Lambda},p,\bar{p}})$ and to polarization observables by means of the Jacob-Wick helicity formalism [258], devised almost 60 years ago. In these decays, the J/ψ rest frame is along the Λ out-going direction, and the solid angle $\Omega_0(\theta, \phi)$ is between the incoming e^+ and the out-going Λ. For $\Lambda \to p\pi^-$ the solid angle of the 'daughter' particle $\Omega_1(\theta_1, \phi_1)$ is referred to the Λ rest frame (although as out-going direction); likewise for $\bar{\Lambda} \to \bar{p}\pi^+$. We describe the angular distribution for this process following Ref. [259]:

$$\frac{d\Gamma}{d\Omega} \propto (1 - \alpha_{J/\psi}) \sin^2\theta \cdot \left[1 + \alpha_-\alpha_+ \left(\cos\theta_1 \cos\bar{\theta}_1 + \sin\theta_1 \sin\bar{\theta}_1 \cos(\phi_1 + \bar{\phi}_1)\right)\right]$$
$$- (1 + \alpha_{J/\psi})(1 + \cos^2\theta) \left(\alpha_-\alpha_+ \cos\theta_1 \cos\bar{\theta}_1 - 1\right) \, , \tag{7.57}$$

  where $d\Omega \equiv d\Omega_0 d\Omega_1 d\bar{\Omega}_1$ and

  - $\alpha_{J/\psi}$ is the angular distribution parameter for Λ;
  - $\alpha_-[\alpha_+]$ is the $\Lambda[\bar{\Lambda}]$ decay parameter;
  - data depend only on the product of $\alpha_-\alpha_+$ (see Eq. (7.57)).

By fitting Eq. (7.57) to the data, one can determine $\alpha_{J/\psi}$ and $\alpha_-\alpha_+$, and make a replacement by defining:

$$\alpha_-\alpha_+ \equiv \frac{A_{\mathbf{CP}} - 1}{A_{\mathbf{CP}} + 1} \alpha_-^2 \, , \tag{7.58}$$

where $A_{\mathbf{CP}}$ describes a **CP** asymmetry observable.

The BESIII collaboration is able to collect about 10^{10} J/ψ. With this sample, **CP** asymmetries will be probed at the level of 0.06%, at least at statistical level.

[19] One can measure $\Lambda \to p\pi^-\gamma$ vs. $\bar{\Lambda} \to \bar{p}\pi^+\gamma$ that is independent of asymmetries production.

In the future the LHCb collaboration could enter this competition of $J/\psi(3100) \to \bar{\Lambda}\Lambda$ with a much larger data-set from 13–14 TeV pp collisions, and probe **CP** asymmetry below 10^{-4}. It is a real challenge to find it in the huge backgrounds[20] and the possibility to maintain systematic errors at the level of the statistical sensitivity will have to be demonstrated. The goal, however, justifies the effort: measuring regional asymmetries would also give us novel lessons on non-perturbative QCD at least – or maybe on ND.

7.8 CP asymmetries for charm baryon Λ_c^+

The Cabibbo favored decays of $\Lambda_c^+ \to pK^-\pi^+$ are the best measured among charmed baryon decays:

$$\mathrm{BR}(\Lambda_c^+ \to pK^-\pi^+) = (6.28 \pm 0.32) \cdot 10^{-2}. \tag{7.59}$$

We anticipate that with new data this uncertaintly will be reduced relatively soon. One can use this decay to calibrate the rates of SCS $\Lambda_c^+ \to p\pi^+\pi^-$ and $\Lambda_c^+ \to pK^+K^-$ and DCS $\Lambda_c^+ \to p\pi^-K^+$ *including Dalitz plots.* Charm baryons might turn out the 'Poor Princesses' for establishing **CP** asymmetries there and even the impact of ND [260].

PDG2020 gives for SCS decays

$$\begin{aligned} \mathrm{BR}(\Lambda_c^+ \to p\pi^+\pi^-) &= (4.61 \pm 0.28) \cdot 10^{-3} \\ \mathrm{BR}(\Lambda_c^+ \to pK^+K^-) &= (1.06 \pm 0.06) \cdot 10^{-3} \end{aligned} \tag{7.60}$$

For the DCS one it gives

$$\mathrm{BR}(\Lambda_c^+ \to pK^+\pi^-) = (0.111 \pm 0.018) \cdot 10^{-3} \tag{7.61}$$

These values will be updated from the run-2 of the LHCb experiment 'soon' and later by Belle II.

The measurements of these branching fractions represent the first step towards probing direct **CP** asymmetries.

7.9 Short summary of direct CP asymmetries

- Most of the FS of charm and beauty (but not strange) hadrons are described by many-body FS; their importance is not to be underestimated.
- No direct **CP** violation has been found yet for strange, charm and beauty baryons.
- In the data of run-1 the LHCb collaboration had found 'evidence' – i.e., 3.3 standard deviation – for **CP** asymmetry in $\Lambda_b^0 \to p\pi^-\pi^+\pi^-$. Subsequently the same collaboration has added parts of run-2, obtaining a standard deviation of 2.9– so? 'We' should not give up!

[20] G. Punzi had suggested that for the run-3/4 with a dedicated trigger.

7.9.1 $\Delta S = 1$

Direct **CP** asymmetry has been established a long time ago in $K_L \to \pi\pi$:

$$\mathrm{Re}(\epsilon'/\epsilon_K)|_{\mathrm{PDG2020}} = (1.66 \pm 0.23) \cdot 10^{-3} \, . \tag{7.62}$$

The situation came to a turn in 2019/2020, when SM calculations could predict this value. Two 'actors' said: yes! Yet the third one said: not sure. See the list of Refs. [235–237]:

$$\mathrm{Re}(\epsilon'/\epsilon_K)|_{\text{`LQCD2020'}} = (2.17 \pm 0.84) \cdot 10^{-3} \, ; \tag{7.63}$$

$$\mathrm{Re}(\epsilon'/\epsilon_K)|_{\text{`PS2019'}} = (1.4 \pm 0.5) \cdot 10^{-3} \, ; \tag{7.64}$$

$$\mathrm{Re}(\epsilon'/\epsilon_K)|_{\text{`BS2020'}} = (1.39 \pm 0.52) \cdot 10^{-3} \, . \tag{7.65}$$

These values from the SM are consistent with the PDG2020 data – but also with possible contributions from the ND.

7.9.2 $\Delta B = 1$

PDG2020 has shown 'averaged' **CP** asymmetries:

$$\begin{aligned}
\langle A_{\mathbf{CP}}(B^+ \to K^+\pi^+\pi^-)\rangle &= +0.027 \pm 0.008 \\
\langle A_{\mathbf{CP}}(B^+ \to K^+K^+K^-)\rangle &= -0.033 \pm 0.008 \\
\langle A_{\mathbf{CP}}(B^+ \to \pi^+\pi^+\pi^-)\rangle &= +0.057 \pm 0.013 \\
\langle A_{\mathbf{CP}}(B^+ \to \pi^+K^+K^-)\rangle &= -0.122 \pm 0.021 \, .
\end{aligned}$$

LHCb data have shown large 'regional' **CP** asymmetries from run-1 [244]:

$$\begin{aligned}
A_{\mathbf{CP}}(B^+ \to K^+\pi^+\pi^-)|_{\text{`region'}} &= +0.678 \pm 0.078|_{\mathrm{stat}} \pm 0.032|_{\mathrm{syst}} \pm 0.007|_{\psi K^\pm} \\
A_{\mathbf{CP}}(B^+ \to K^+K^+K^-)|_{\text{`region'}} &= -0.226 \pm 0.020|_{\mathrm{stat}} \pm 0.004|_{\mathrm{syst}} \pm 0.007|_{\psi K^\pm} \\
A_{\mathbf{CP}}(B^\pm \to \pi^\pm\pi^+\pi^-)|_{\text{`region'}} &= +0.584 \pm 0.082|_{\mathrm{stat}} \pm 0.027|_{\mathrm{syst}} \pm 0.007|_{\psi K^\pm} \\
A_{\mathbf{CP}}(B^\pm \to \pi^\pm K^+K^-)|_{\text{`region'}} &= -0.648 \pm 0.070|_{\mathrm{stat}} \pm 0.013|_{\mathrm{syst}} \pm 0.007|_{\psi K^\pm} \, .
\end{aligned}$$

7.9.3 $\Delta C = 1$

For the first time **CP** asymmetry has been established by the LHCb collaboration [174]:

$$A_{\mathbf{CP}}(D^0 \to K^+K^-) - A_{\mathbf{CP}}(D^0 \to \pi^+\pi^-) = -(15.4 \pm 2.9) \cdot 10^{-4} \, . \tag{7.66}$$

One expects to find **CP** asymmetries in other SCS processes, including the ones with three-body FS.

Chapter 8

CP asymmetries in the transitions of top quarks

"Thus Spoke Zarathustra"

Friedrich Nietzsche
Richard Strauss

The top quark, unsuccessfully searched for by means of electron-positron colliders (TRISTAN in Japan in the 80s and LEP at CERN in the early 90s) and by means of the $p\bar{p}$ collider SPS at CERN, was finally discovered at the Fermilab Tevatron by the CDF and D0 collaborations in 1995. It was a remarkable achievement at that time, because of the relatively poor signal and the large top quark mass. Though unsuccessful because of the very large top quark mass, the efforts made at electron-positron colliders, TRISTAN and LEP in particular, played two key roles: the first one was to develop the technology and the analysis tools, paving the way to the asymmetric colliders, which have been the key for the great successes of BaBar and Belle in the first decade of this new millennium; the second one was the precise prediction of the top quark mass made by means of the careful calculation of forward-background production asymmetries of fermion-anti fermion pairs from Z^0 decay at LEP.

The existence of top quarks in the production of a pair $\bar{t}...t$ in $\bar{p}p$ collisions at $\sqrt{s} = 1.96$ TeV was established at the Tevatron. One can call it a 'legacy' measurement including the first $\bar{t}...t$[1] forward-backward asymmetry [261].

The situation has changed since then after the very successful run-1 at the LHC pp collisions with $\sqrt{s} = 7$ and 8 TeV and recently in the run-2 with $\sqrt{s} = 13$ TeV at CERN: the ATLAS and CMS collaborations have measured pairs of $\bar{t}...t$ quarks (and even two pairs in the same collisions); likewise in associations with 'the' Higgs field and the gauge $W^\pm/Z^0$ bosons. Furthermore, they have analyzed single t and $\bar{t}$ quarks and added hard photons, $W^\pm$ and Z^0 bosons. Many more important results will be coming in the next decade to improve our understanding of fundamental dynamics and maybe find ND. There are suggestions for top quark factories from e^+e^- collisions to show an impact of ND in the future.

[1]We 'paint' the FS with a pair of top and anti-top quarks as $\bar{t}...t$, where ... always includes $q\bar{q}$.

From indirect searches, we had realized long ago that top quarks had to be very heavy. To produce them directly, one needs large energies in collisions. However, that is not enough: one has to think where and how to search. Artists showed a metaphor a long time ago, see **Fig. 8.1**: the Greek Goddess Athena needs 'power', but also 'thinking'.

Fig. 8.1 "Pensive" Greek Goddess Athena, Relief (around 460 BC) [Acropolis Museum, Athens].

The *ultra*heavy top quark is not just very heavy compared with beauty quark: top quarks decay semi-weakly[2] due to $t \to bW^+$ *before* they can produce top mesons and baryons [51]. In decay analyses we are less hampered by our failure to accurately evaluate hadronic matrix elements, shaped by non-perturbative dynamics; perturbative calculations of strong interaction effects have a major impact. Does it mean that top quarks are 'free' and thus leave their "confinement"? Not at all, since the local "color" gauge symmetry $SU(3)_C$ = QCD is unbroken! Thus the "color" of the top (or anti-top) quark has to be transferred to the "color" of the another anti-quark (or quark) in the FS: i.e. amplitudes of (anti-)top quarks have to find another consistent source of non-zero "color" to describe the transitions of initial to final states of hadrons with zero "color" states. A 'team' of a top (or anti-top) quark plus gluons can*not* produce zero "color" states; one needs the 'team' of a top quark plus anti-quark [+ gluons] to do that (or anti-top quark plus quark [+gluons]) as QCD tells us. Thus the impact of strong forces is not only perturbative even for top quarks.

When one talks about **CP** asymmetries here, the situation appears quite 'complex'. Due to the huge phase space that is available in top decays, there can be only little interference between different sub-processes. **CP** sensitive observables as discussed above for other hadrons do *not* work for top quarks; strong forces are no

[2]We call it 'semi-weakly', since the charged boson $W^\pm$ is weakly coupled to quarks, while the transitions of top quarks produce $W^\pm$ *on* shell.

longer sufficiently efficient in 'cooling' different transitions into a 'coherent' state (unless a 'miracle' happens).

Top quarks represent one piece in the flavor mystery: why are top quarks so unusual 'massive' compared with other quarks – never mind leptons. Top quarks have a mass comparable to the electroweak scale $\sim$ 250 GeV – not the one that is expected for quark flavor scales. The coupling of the top quark with 'the' Higgs boson is large, namely $\sim \mathcal{O}(1)$; maybe is it a novel gateway for ND to a different landscape?

- The top quark mass is *not* observable by itself, because they decay before they can produce hadrons. Since the mass of the top quark is a fundamental parameter of the SM, its determination is of paramount importance.
- 'Simulations' are used by experimenters to plan and construct elements of a detector and to prepare tools for analyses of future data. Usually experimental papers use the *pole* mass for top quarks [262].
  The 2020 PDG review of "60. Top Quark" [263] mostly uses its 'pole' mass: $m_t^{\rm Pole} = 172.4 \pm 0.7$ GeV from cross-section measurements and its width $\Gamma_t \sim 1.4$ GeV. One of us had pointed out [243] that by using the 'pole' mass one ignores the impact of non-perturbative QCD – in particular, when one wants to go for accuracy.[3] Using the pole mass means that the SM expectations are based only on the impact of perturbative QCD, not on non-perturbative one. A better way is to measure the impact of top quarks by comparing theory predictions with experimental data. In this case the measured mass has to be identified with the top mass parameter employed in the calculations.
- One cannot ignore renormalons in IR dynamics, as we discussed in **Sect. 3.8**. We signal a short, but interesting paper with the title 'Renormalons and the Top Quark Mass Measurement' [264] where progress with *perturbative* expansion is discussed.

8.1 Transitions of $e^+e^- \to \bar{t}$ ... t

First we consider very high energy e^+e^- scattering into a $\bar{t}$ ... t state. Top decays through the EW interaction into a W boson and (usually) a bottom quark. Decays into strange or down quarks are suppressed by the small CKM elements. Thus the experimental signature of a decaying top quark is a jet containing a bottom quark and the W decay products. The W boson decays into all of its possible final states: charged lepton-neutrino pairs or into jets. The simpler situations (in the view of

[3] In several cases it has been assumed that the measured top-quark mass corresponds to the pole mass and errors of the order of few hundreds MeV have been added to account for possible deviations from this identification.

theorists) are the dilepton transitions, where both W bosons decay leptonically:

$$e^+e^- \to \gamma^*/Z^* \to [(\bar{b})_{\rm jet}W^-]_{\bar{t}} \ ... \ [(b)_{\rm jet}W^+]_t \to [l^-\bar{\nu}]_{\rm W^-} \ [l^+\nu]_{\rm W^+} \ h_{\bar{b},1} \, h_{b,2} \ X$$

with (anti-)beauty hadrons h_b and $h_{\bar{b}}$ in the FS. X can include even numbers of beauty hadrons; they carry "color" zero, while their charge can be $0, \pm 1$. These processes could lead to forward-backward asymmetries (or parity ones) and charge conjugation asymmetries in the FS.

As of today, no existing accelerator has sufficient energy to study this reaction. There are many projects for future accelerators among which the Future Circular Collider in the first electron-positron implementation (FCC-ee) [265], the Compact Linear Collider (CLIC) [266] and the ILC [267]; if built, they will have enough center of mass energy to produce a rich sample of $\bar{t}$ ... t final states. Also a Muon Collider [268] may probe the same physics.

8.1.1 *Very special case:* $\sigma(e^+e^- \to$ "$\bar{t}_L t_L$") *vs.* $\sigma(e^+e^- \to$ "$\bar{t}_R t_R$")

As said above, top quarks are not 'free'; we remark that using the notation "$\bar{t}t$", rather than $\bar{t}t$. Yet there are cases where the lack of hadronisation of the top-quark becomes an asset for **CP** violation studies. For a pair of "$\bar{t}t$" quarks the threshold is about 346 GeV in e^+e^- collisions. The couplings of the photon and the Z boson to top quarks conserve chirality. Chirality and helicity coincide only for fast particles; close to threshold there is a significant component when t and $\bar{t}$ both have either helicity $+1$ or -1: $e^+e^- \to \bar{t}_R t_R$ or $e^+e^- \to \bar{t}_L t_L$. With t_L and $\bar{t}_R$ being conjugate to each others – i.e., $t_R \overset{\mathbf{CP}}{\Longrightarrow} \bar{t}_L$ – a difference in "$\bar{t}_R t_R$" vs. "$\bar{t}_L t_L$" production

$$\sigma(e^+e^- \to \text{“}\bar{t}_R t_R\text{”}) \neq \sigma(e^+e^- \to \text{“}\bar{t}_L t_L\text{”}) \tag{8.1}$$

implies **CP** violation. The usual conditions have to be satisfied: (i) two amplitudes have to contribute *coherently* to the production process that involve a **CP** asymmetry in the underlying dynamics, and (ii) an absorptive component has to be generated. One can employ **P** violating decays

$$t \to b_L + W^+ \quad , \quad \bar{t} \to \bar{b}_R + W^- \tag{8.2}$$

to analyze the polarization of the decaying top quark and thus probe the production process where the asymmetry resides. It was suggested in 1992 by Ref. [269] that one could find **CP** asymmetry as 'large' as $\mathcal{O}(10^{-3})$ because of interference of tree diagrams and one-loop diagrams with a neutral Higgs field.[4] Note that in transitions close to a threshold:

$$\text{“}\bar{t}_L t_L\text{”} \to \text{energetic } W^- + \text{slow } W^+ + (\bar{b})_{\rm jet} + (b)_{\rm jet} \tag{8.3}$$

$$\text{“}\bar{t}_R t_R\text{”} \to \text{slow } W^- + \text{energetic } W^+ + (\bar{b})_{\rm jet} + (b)_{\rm jet} \, . \tag{8.4}$$

Thus a difference in the production of "$\bar{t}_L t_L$" and "$\bar{t}_R t_R$" leads to a charge asymmetry in the energy distributions of W's or their decay leptons, which is a **CP** violation observable.

[4] To be more specific: top-antitop vertexes and one-loop vertexes where a neutral Higgs field is exchanged between the top states.

'The' Higgs boson has been established in 2012 by the ATLAS and CMS collaborations with mass about 125 GeV. Unlike other quarks, it is natural for top quarks to possess large Yukawa couplings to Higgs fields in general. Higgs exchanges thus can generate significant absorption and **CP** violation as well.

CP violation could be probed at a possible future top quark factory based on e^+e^- collisions with *novel* Higgs fields with masses *larger* than the $\bar{t}t$ threshold.

8.1.2 CP *asymmetries in* $\sigma(e^+e^- \to$ "$\bar{t}...t$")

Now we briefly discuss a general 'landscape' with a pair of top quarks without thresholds. We consider a simple situation:

$$e^+e^- \to [(\bar{b})_{\rm jet}W^-]_{\bar{t}} \,...\, [(q)_{\rm jet}W^+]_t \;\; \text{vs.} \;\; e^+e^- \to [(\bar{q})_{\rm jet}W^-]_{\bar{t}} \,...\, [(b)_{\rm jet}W^+]_t \quad (8.5)$$

with $q = s, d$ with*out* a single b. $q/\bar{q}$ jets do not have a single beauty jet; they can include two beauty jets and likewise beauty jets can include additional two beauty jets.

By comparing the rates and features of $e^+e^- \to [(\bar{b})_{\rm jet}W^-]_{\bar{t}} \,...\, [(q)_{\rm jet}W^+]_t$ vs. $e^+e^- \to [(\bar{q})_{\rm jet}W^-]_{\bar{t}} \,...\, [(b)_{\rm jet}W^+]_t$ one can probe **CP** asymmetries, in principle. The SM produces such events, but very rarely. Thus a few events could show the impact of ND with*out* a real background – in the world of theorists. It is important to analyze the productions of top quarks and their decays together with their possible impact on ND. Even if new possibilities open up, the challenge to measure e^+e^- collisions with $\sqrt{s} > 350$ GeV goes down on the shoulders of our experimenters.

8.2 The present era of $pp \to$ "$\bar{t}...t$" $+ X$ at LHC

The existence of top quarks has been established *directly* in 1995 by the CDF and D0 collaborations at Fermilab.

The cross-section for the process $pp \to \bar{t}...t + X$ has been predicted by the SM to be about ~ 830 pb at $\sqrt{s} = 13$ TeV, which is consistent with the data from the ATLAS (~ 830 pb) and CMS experiments (~ 834 pb) [263].

The fast accumulation of high-energy data at the LHC has contributed to a renovated interest in 'painting' the top quark dynamics.

Recent specialized workshops can help getting an overview:

- The 10th International Workshop on Top Quark Physics in 2017 reviewed an amazing progress on the experimental side, including refined analyses with large data sets.[5] The speaker of the Summary Talk used a nice title for his introduction: 'the Age of Plenty' [270]. It refers to the present age, where measurements that used to be statistics-limited are reaching a state where accumulation of

[5]The beam luminosity has been so far calibrated by mean of Z^0 boson production at LHC. It has been suggested to use instead the production of $\bar{t}t$ in the future [270].

more data has very little direct benefit on the precision or the sensitivity of a search because most measurement are already limited by systematic effects.

- Likewise in the 'Top2018: Experimental Summary' [271] at the 11th International Workshop on Top Quark Physics. The author pointed out that in the future ATLAS and CMS will produce smaller statistical uncertainties, but not systematic ones. On the other hand, the LHCb detector has smaller systematic uncertainties. The inclusive production cross section is smaller in the very forward direction useful for LHCb acceptance (σ_{ttX}=126 $\pm 19_{(stat)} \pm 16_{(syst)} \pm 5_{(lumi)}$ fb), but it will get much more data in the future and the perspectives are very interesting [272].
- In the (slides of the) summaries of the 12th International Workshop on Top Quark Physics in 2019 [273, 274], it was underlined that the special characteristics of the top quark (heavy mass, large couplings with the Higgs field $y_t \sim 1$, lack of hadronization) make it an excellent probe of physics beyond the SM. One possibility in pp collisions at CERN is to study gluons collisions leading to H^0 and $\bar{t}H^0t$ FS; one aims at determining all properties of the Higgs boson including possible deviations from the SM predictions. Run-2 had produced $\sim$ 275 millions of top quarks in 139 fb^{-1} at 13 TeV. Top physics has been at the core of the Tevatron and the LHC physics program and will continue to be for the High luminosity/High Energy LHC upgrades, as well as for all future colliders currently under discussion.

8.2.1 *Production and decays of a pair of top quarks*

At hadron colliders, top quarks are dominantly produced in pairs via the strong interaction. The leading order QCD production processes are the $q\bar{q}$ annihilation $q\bar{q} \to \bar{t}t$ and the gluon-gluon fusion $gg \to \bar{t}t$. The former process is dominant to the $p\bar{p}$ collider Tevatron, where approximately 85% (at 1.96 TeV) of the production is from $q\bar{q}$ annihilation. The tree diagram for $q\bar{q}$ annihilation $q\bar{q} \to \bar{t}t$ is mediated by a virtual gluon. This proportion is approximately reversed for the pp collider LHC, where gluon-gluon fusion is the dominant process. The tree diagram for gluon-gluon fusion $gg \to \bar{t}t$ has an initial state of two gluons that couple to an intermediate gluon, producing the $t\bar{t}$ pair. At LHC the ATLAS and CMS collaborations have measured the inclusive $\bar{t}t$ cross-section in the region $\sqrt{s} = 7, 8, 13$ TeV.

It is an achievement by itself, due to the huge background. It is consistent with the SM predictions; on the other hand, the experimental and the theoretical uncertainties are still sizable.

With a mass well above the Wb threshold, and the coupling to the b quark very close to unity, the decay width of the top quark is expected to be dominated by the two-body channel $t \to Wb$. Thus one measures the transitions of $pp \to [(\bar{b})_{\rm jet}W^-]_{\bar{t}}...[(b)_{\rm jet}W^+]_t$ for the leading pair-production process as much as one can, in order to learn new lessons on QCD (and even to find the impact of ND).

The enhancement at forward rapidity of $t\bar{t}$ production via $q\bar{q}$ and qg scattering, relative to gg fusion, can result in larger charge asymmetries, which may be sensitive to physics beyond the SM. The rapidity y of a particle is defined as $y = 1/2\log(E + p_z)/(E - p_z)$, where E is the energy of the particle, and p_z is the momentum of the particle along the proton beam axis (z component). The rapidity y quantifies the direction the top quark or antiquark is moving. The case $y = 0$ indicates the quark is moving perpendicular to the beam line, positive [negative] y in the forward [backward] direction; i.e. the more positive [negative] the *rapidity*, the smaller the angle between the momentum of the quark and the forward [backward] direction. In this sense the value of y is directly related to the definition of forward and backward.

Current strategies to search for new physics in top quark events include searches for new resonant states through decay processes involving the top quarks. For example, our community might find a resonance Y with $M_Y \sim 400$ GeV; its decay to "$\bar{t}...t$" would be enhanced close to its threshold due to non-perturbative QCD, in particular as a boosted resonance in the forward and backward region. Even so, it would be unlikely to find **CP** asymmetry in these situations, (although possible).

A recent paper by the ATLAS collaboration has discussed the 'color flow' in $\bar{t}...t$ events at $\sqrt{s} = 13$ TeV [275]. It is interesting to analyze observables sensitive to the colour flow, which are rarely seen in our literature. The diagram of Fig. 2 in Ref. [275] follows the colors of the (anti-)quarks. It 'paints' the landscape, and sketches the pp collisions leading first to intermediate states (a pair of $t...\bar{t}$ quarks), then to $(b)_{\rm jet}W^+...(\bar{b})_{\rm jet}W^-$ and finally to $(b)_{\rm jet}[l^+\nu]_{W^+}...(\bar{b})_{\rm jet}[q\bar{q}']_{W^-}$. The measured distributions are compared to several theoretical predictions obtained from MC simulation. It is a good idea; yet we are not convinced that it works in the real world. The results agrees poorly with the SM predictions.

8.3 Measuring CP asymmetries with on-shell Higgs dynamics

The Higgs boson H^0 was first observed with the loop induced $H^0 \to \gamma\gamma$ decay by the ATLAS and CMS experiments and then established in 2018 with the direct channel $H^0 \to$ "$\bar{b}\,b$" jets: $\mathrm{BR}(H^0 \to \gamma\gamma) \simeq 0.0023$ and $\mathrm{BR}(H^0 \to$ "$\bar{b}b$"$) \simeq 0.58$.

The SM H^0 is 100% scalar, but it is not excluded by data a possible non-zero pseudo-scalar contribution. The ratios between scalar and pseudoscalar couplings might differ from channel to channel in the presence of **CP** violation, making important to analyze the greatest number of channels. In view of its large Yukawa coupling, it is of particular interest connecting "$\bar{t}\,t$" with a neutral zero-spin boson H^0.

A thoughtful way to parametrize the top-Higgs interaction is [276]

$$\Delta\mathcal{L} = -\frac{m_t}{v}\,K\,\bar{t}\,(\cos\alpha + i\gamma_5\sin\alpha)\,t\,H^0\,, \tag{8.6}$$

where K is a real number and α a **CP**-phase. The **CP**-*even* Higgs 0^+ boson is described by $(K,\alpha) = (1,0)$. Yet $\alpha \neq 0$ (and $K \neq 1$) might show the impact of ND.[6] The associated Higgs with $t\bar{t}$ pair production is a good hunting region, although not an easy one.

8.3.1 *Probing* $e^+e^- \to \bar{t}...H^0...t$

In 1996 it had been suggested to probe **CP** asymmetry in the center-of-momentum frame (or c.m. rest frame) of the beams [277]:

$$e^+(-p_{\text{beam}})e^-(p_{\text{beam}}) \to \gamma^*, Z^* \to \bar{t}(p_{\bar{t}})...H^0(p_{H^0})...t(p_t) \; . \tag{8.7}$$

One can measure $O_- \equiv \langle \vec{p}_{\text{beam}} \cdot (\vec{p}_t \times \vec{p}_{\bar{t}}) \rangle$, which is odd under **CP** $O_- \overset{\mathbf{CP}}{\Longrightarrow} -O_-$; hence

$$O_- \neq 0 \implies \mathbf{CP} \text{ asymmetry !} \tag{8.8}$$

Reconstructing the t and $\bar{t}$ momenta is not an easy task. Instead one could measure the momenta of the b and $\bar{b}$ initiated jets emerging from the t and $\bar{t}$ decays:

$$O_-^{(b,\bar{b})} \equiv \langle \vec{p}_{\text{beam}} \cdot (\vec{p}_b \times \vec{p}_{\bar{b}}) \rangle \; . \tag{8.9}$$

Finding $O_-^{(b,\bar{b})} \neq 0$ would manifest **CP** violation. From the theorists' viewpoint it is a very good idea – if we have e^+e^- collisions at very high energy: $E_{\text{CM}} \sim 500$ GeV, or more.

8.3.2 *Probing* $pp \to H^0 + X$ *and* $pp \to \bar{t}\, H^0\, t + X$

Now we come back to the present and future situations at pp collisions at CERN. At pp collisions at 13 TeV one expects (59 ± 5) pb [278].

The next step is to produce tree-level amplitudes:

$$pp \to gg + X \to \bar{t}...H^0...t + Y \; . \tag{8.10}$$

One can 'paint' the situation with the items of the SM:

$$gg \to \bar{t}...H^0...t \; . \tag{8.11}$$

The results from the ATLAS Collaboration [279] at $\sqrt{s} = 13$ TeV led to

$$\sigma(pp \to \bar{t}...H^0...t + X) \sim 0.5 \text{ pb} \; . \tag{8.12}$$

[6] A value $\alpha = \pi/2$ would lead to a *pure* **CP**-*odd* Higgs boson!

8.4 Single top quark in pp collisions at LHC

With the experiments at the CERN LHC our community has entered a new era: to produce significant amounts of top quarks due to electroweak dynamics. Indeed, top quarks are produced mainly in pairs through the strong interaction, but can also be produced individually via a charged-current electroweak interaction involving a $\bar{b}tW$ vertex. Single-top quark production was observed in 2009 by the CDF and D0 collaborations at the Tevatron. It allows probing of BSM and it also becomes a background to the precision $\bar{t}t$ physics and to other SM processes.

We discuss the two simple examples of a top quark produced together with $W^{\pm}$ or H^0:

- Single top quark together with $W^{\pm}$
  The associated production of top quark and W boson, also known as the tW-channel, is the second most dominant process of single-top quark production, after the t channel, at the LHC. Evidence for the tW production was found at centre-of-mass energy $\sqrt{s} = 7$ TeV with 2.05 fb^{-1} of data at ATLAS experiment and its observation was found at $\sqrt{s} = 8$ TeV with 12.2 fb^{-1} of data at CMS experiment.
  One can measure single top quark decays with $W^{\pm}$ and compare $pp \to (b)_{\rm jet} W^- “t” X$ with $pp \to (\bar{b})_{\rm jet} W^+ “\bar{t}” \bar{X}$ processes [280]. The leading diagram at the parton level is a tree diagram with a gluon and a b quark in the initial state, an intermediate b state, and a final state with the top quark and the W meson; hence by comparing $bg \to W^- t$ vs. $\bar{b}g \to W^+ \bar{t}$ one could probe **CP** asymmetry.
- Single top quark with the Higgs field.
  One can measure a single top quark with H^0, namely $pp \to t...\bar{q}\ H^0 + X$ and $pp \to \bar{t}...q\ H^0 + X$. The cross section is dominated by the so-called t-channel W exchange process, where the W boson emitted from a quark or anti-quark in a proton scatters with a b or $\bar{b}$ quark in the other proton to produce a pair of H^0 and t or $\bar{t}$. With LHC detectors (and their teams), we can probe the existence of ND and **CP** asymmetries in the production of top quarks together with Higgs bosons and their transitions.

8.5 Present summary of top quark dynamics

We give a short summary on the experimental status of the top quark dynamics investigated by the ATLAS and CMS collaborations (see [273] and references therein).

- Productions of the W or H^0 boson, together with an inclusive hadronic state X, a $\bar{t}t$ pair or a single top: in pp collisions at $\sqrt{s} = 13$ TeV we have $\sigma(pp \to \bar{t}...t + X) \sim 830$ pb, $\sigma(pp \to \bar{t}/t\ W + X) \sim 63$ pb, $\sigma(pp \to H^0 + X) \sim 49$ pb and $\sigma(pp \to \bar{t}...H^0...t + X) \sim 0.5$ pb. It is amazing what these teams have done already.

- Productions of $\bar{t}t$ pairs: (a) ATLAS and CMS have probed the combination of *charge asymmetries* using their data at $\sqrt{s} = 7$ and 8 TeV [281]. (b) ATLAS has measured the *highly boosted* top quarks to all-hadronic final states at $\sqrt{s} =$ 13 TeV [282].
- Productions of single top quarks: (a) CMS has measured its top quark mass using the data with single top quarks at $\sqrt{s} = 8$ TeV [283]. (b) ATLAS has measured differential cross-sections of single top quark in association with a W boson at $\sqrt{s} = 13$ TeV with ATLAS [284].

These are rich hunting regions to understand the dynamics and even to find the existence of ND. One needs more analyses, more thinking about which directions to follow and what are the connections between different directions. The situations will change with the more refined analyses of run-2 of LHC and future run-3. The goal is 'accuracy'.

8.6 CP asymmetries from top quarks (with H^0 and $W^\pm$ bosons)

The **CP** landscape is very different for the dynamics of the ultra-heavy top quarks with respect to other heavy quarks: they decay (semi-)weakly before they can produce top hadrons [51]. However, one cannot ignore the non-perturbative impact of QCD, since the local color symmetry is *not* broken and the top quark must in a way or another transfer its color. The strategy for finding ND here has therefore to be very different:

- We have found a neutral boson with spin zero and its mass 125 GeV in the region that is somewhat acceptable in our understanding of cosmology. In the SM the Higgs field is scalar, while the data tell us it is mostly a scalar boson. It is a qualified statement.
  It is not heavy enough to affect the production of a pair of $\bar{t}...t$ of a virtual exchange. The good sign is: it could open a new 'road' for ND like for non-minimal Higgs sector – including indirect sign of SUSY.
- The best effective tool is to combine production and decay amplitudes. Even with much more data we need more refined tools for analyzing them to deal with their large backgrounds.
- The SM does not give us a realistic background for **CP** asymmetries in the transitions of top quarks. On the other hand, the data give us backgrounds that are not connected with top quark dynamics.
- We follow the old rules in fundamental physics: one goes after miracles; however, experience tells as one cannot find two miracles at the same time. In this case we discuss FS with top quarks in connections with H^0 or $W^\pm$ bosons.

Experiments at LHC are searching for **CP** violating effects in top dynamics both in top pair productions (QCD) and in single top production (EW). Although no clear SM effect is expected, **CP** violation may be generated by ND through the existence of anomalous top quark couplings or in the presence of extended scalar sectors.

The CMS collaboration has performed a careful search of **CP** violating effects in top quark pair production, seeing no effect [285]. CMS searched for asymmetries in T-odd, triple-product correlation observables, using a sample of pp collisions at $\sqrt{s} = 8$ TeV corresponding to an integrated luminosity of 19.7 fb^{-1}. The T-odd observables are measured using four-momentum vectors associated with $t\bar{t}$ production and decay using an event sample with a least a muon or an electron and four jets in the final state. The measured asymmetries exhibit no evidence for **CP**-violating effects, in agreement with the SM.

A complementary approach was followed by ATLAS collaboration [286], which has published an analysis of the search of **CP** violating asymmetries in $t\bar{t}$ pair decays exploiting the correlation between lepton charge from W decays and b-flavor. In particular, the search for charge asymmetries is based on the charge of the lepton from the top-quark decay and the charge of the soft muon from the semileptonic decay of a b-hadron. Four **CP** asymmetries (one mixing and three direct) are measured and found to be compatible with zero and consistent with the SM.

Taking into account that both analysis are based on a relatively small statistical sample and that much more data will be collected in next runs and High Luminosity LHC program, our community may hope that new interesting results may come in the future.

8.7 Strategy for going after New Dynamics

So far we have mostly focused on the impact of the SM, although we have given comments beyond that. In the following chapters the focus will shift on physics beyond the SM (BSM): we know that the SM is not wrong – it is incomplete. Hence it seems appropriate to derail from the argument of this chapter, namely top quark physics, and give a brief outline of the ND horizons that we are going to face:

- As we will discuss in **Chapter 9** with details, neutrino oscillations have been found. The existence of BSM physics has been established in lepton dynamics with real data. We have found not only a new road towards fundamental dynamics: also a novel one – maybe – about one of our crucial puzzles, the huge asymmetry in matter vs. anti-matter in our Universe.

- Our community knows there is only one candidate among local QFTs to describe strong forces, the unbroken non-abelian $SU(3)$ = QCD. It was seen that **CP** is 'naturally' conserved by strong dynamics. However, later it was realized that in principle strong dynamics can produce **CP** (and also **P**) *violation* adding a special feature: a dimension-4 operator $\tilde{G} \cdot G$ which has no flavor structure, as will be discussed in **Chapter 10**. Furthermore, this $\tilde{G} \cdot G$ operator could produce measurable impact on neutron electric dipole moments see **Sect. 10.4**. In any case, we have not seen any sign of such a feature of QCD – in particular when investigating EDMs of the neutron, where we have only tiny limits. It is seen as 'un-natural', at least in the views of many theorists.
- The search for a solution to the strong **CP** problem outlined before has lead to the concept of "axions", that we will discuss in **Chapter 11**. The idea came from theorists; on the other hand, it stimulated experimenters to come up with new ideas and tools. In our view it is a good example of combining the works of theorists and experimenters. Furthermore, "axions" are good candidates for Dark Matter. However, "Axions" (or similar ones) have not been found yet, but our community has to continue searching! At the very least, we learn that fundamental dynamics is truly 'complex', and it may be another example where "thinking" is an excellent tool, but not on the short time schedule.
  Even if we do not find axions, we have still to probe EDMs of leptons – see **Sect. 12.2** – and baryons – see **Chapter 13**.
- No member of any SUSY has been found (yet) directly. We can point out only – maybe – that indirect impact of SUSY helps to solve puzzles on the predictions from the SM and present data for some rare transitions. We will come back to SUSY in **Chapter 14**.
- As said in the **Prologue**, our Universe consists of Dark Matter $\sim 26.5\%$, compared with known matter $\sim 4.5\%$, and assuming the rest is Dark Energy (or Vacuum Energy). Except that for gravity forces, so far we have no signs of DM in the transitions of known matter. It is a crucial challenge to establish its existence and hopefully its features, at least indirectly. We will discuss that in **Sect. 15.1**.
- The tale tells us that the Danish physicist Niels Bohr said: "Predictions are very difficult, in particular for the future." We will discuss that as a summary in **Chapter 16**, as best as we can.

There are two crucial points: (a) Real data are 'the' referees – in the end. (b) We need true collaboration between theorists and experimentalists. This also means we should *not* follow the fashion, both on theory and/or experiments. To be honest: now the load is mostly on experimentalists' shoulders and their analyses.[7]

[7] At least one co-author claims he did not follow fashion: there are deep reasons to follow the landscape of super-symmetry. It is not a 'road' – it is a concept! There are many, many classes of SUSY; we cannot see why our Universe should prefer a version somewhat close to the minimal one.

Chapter 9

CP violation in the transitions of leptons

The SM had been constructed under the assumption that the three known neutrinos are massless; therefore they cannot undergo flavor oscillations. This assumption was very well supported by experiments in the early days, in particular by the measurement of neutrino fixed helicity made by Goldhaber, Grodzhins and Sunyar [146] and by the failed attempts to measure the neutrino mass, particularly by studying the ^3H β spectrum near the end-point (for a review of the early attempts see [287]; for a recent review see [288]). However, it was quite clear from the beginning that no fundamental symmetry prevented neutrino masses[1] and speculations about flavor oscillations [148] in the neutrino sector date back to the late 60's.[2] Now we know that neutrinos do oscillate among the three known flavors and likewise for anti-neutrinos – i.e., $\Delta L = 0$. *No* evidence for neutrino - antineutrino oscillations nor for neutrino-less double beta decay has been found yet – i.e., $\Delta L = 2$. We will come back to this in **Sect. 9.5.1**.

Except leptonic **CP** violation and $\Delta L = 2$ transitions, oscillation parameters have been measured with at most 4% uncertainties. With three families of leptons (without a sterile neutrino) one gets three mixing angles θ_{12}, θ_{13}, θ_{23} and a phase $\delta^{\rm lept}_{\bf CP}$ from the PMNS matrix; besides, data prove the existence of three mass squared differences Δm^2_{21}, Δm^2_{32} and Δm^2_{13}, as seen in **Sect. 4.6**. PDG2020 lists the values[3]:

$$\sin^2(\theta_{12}) = 0.307 \pm 0.013 \tag{9.1}$$
$$\sin^2(\theta_{13}) = 0.0218 \pm 0.0007 \tag{9.2}$$
$$\sin^2(\theta_{23}) = 0.545 \pm 0.021 \;\; \text{(Normal order)} \tag{9.3}$$
$$\sin^2(\theta_{23}) = 0.547 \pm 0.021 \;\; \text{(Inverted order)} \tag{9.4}$$
$$\delta^{\rm lept}_{\bf CP} = (1.36 \pm 0.17)\,\pi \text{ rad}\,. \tag{9.5}$$

While there is a sign for a sizable **CP** violation, the evidence is not conclusive and no high precision measurement of the **CP** phase is available yet. Furthermore,

[1] Prescribing zero mass for neutrinos was, ultimately, again 'par ordre du mufti'.

[2] It was suggested to probe neutrino - antineutrino oscillations at the very beginning.

[3] In the literature it is custom to call 'Hierarchy' the neutrino mass ordering. We do not follow this fashion in this book.

we have:

$$\Delta m_{21}^2 = +(7.53 \pm 0.18) \cdot 10^{-5}\,(\text{eV})^2 \tag{9.6}$$
$$\Delta m_{32}^2 = +(2.453 \pm 0.034) \cdot 10^{-3}\,(\text{eV})^2 \quad (\text{Normal order}) \tag{9.7}$$
$$\Delta m_{32}^2 = -(2.546\ ^{+0.034}_{-0.040}) \cdot 10^{-3}\,(\text{eV})^2 \quad (\text{Inverted order})\,. \tag{9.8}$$

where $\Delta m_{ij}^2 \equiv m_i^2 - m_j^2$. Data are consistent with mixing angles and mass splittings being the same for neutrinos and anti-neutrinos, in agreement with **CPT** conserving oscillation mechanism.

The search for **CP** asymmetry in the lepton sector has begun, and its discovery is expected in the next decade, at least for the **CP** violating sector controlled by a PMNS phase, see **Sects. 4.6.2, 4.6.3**. Neutrino oscillations are the key avenue to **CP** violation in the lepton sector, thus in this **Chapter** we focus mostly on this topic, leaving comments on charged leptons, particularly τ decays, in **Sect. 9.7**. Most theorists see a good case for Majorana neutrinos, which leads to continue probing neutrino-less double beta decays, discussed in **Sect. 9.5.1** despite the experimental difficulties.

In **Sect. 4.6** we have 'painted' the general landscape of neutrino oscillations. Here we summarize the main lessons[4]:

- The three fields ν_1, ν_2 and ν_3 have a special pattern in their mass spitting:
$$m_1 < m_2 \ll m_3 \quad \text{or} \quad \Delta m_{32}^2 \gg \Delta m_{21}^2 > 0 \quad \text{'Normal Order'} = \text{NO}$$
$$m_3 \ll m_1 < m_2 \quad \text{or} \quad \Delta m_{21}^2 \gg -\Delta m_{31}^2 > 0 \quad \text{'Inverted Order'} = \text{IO}\,.$$
Likewise $\theta_{13} \ll \theta_{12} < \theta_{23}$.
- The source of **CP** violation is $\delta_{\mathbf{CP}}^{\text{lept}} \neq 0, \pi$. Yet to measure a non-zero value, one also needs $\theta_{12} \neq 0, \pi/2$, $\theta_{23} \neq 0, \pi/2$ and $\theta_{13} \neq 0, \pi/2, 3\pi/2$. These conditions are all true, so the possibility that **CP** violating effects in neutrino oscillations are large is more than a hope. The goal is to observe:
$$\Delta_{\alpha\beta}^{\mathbf{CP}} = P(\nu_\alpha(t) \to \nu_\beta) - P(\bar{\nu}_\alpha(t) \to \bar{\nu}_\beta) = |\langle \nu_\beta | \nu_\alpha(t)\rangle|^2 - |\langle \bar{\nu}_\beta | \bar{\nu}_\alpha(t)\rangle|^2 \neq 0$$
regardless of their Dirac or Majorana mass term.
- If neutrinos are Majorana particles, their mixing matrix is given by:
$$\mathbf{U}_{\text{PMNS}} = \mathbf{V}_{\text{PMNS}}(\theta_{12}, \theta_{13}, \theta_{23}, \delta_{\mathbf{CP}}^{\text{lept}})[e^{i\phi}]$$
as we have seen in Eq. (4.76) with some details. It means that one gets two more phases that can be measurable, at least in principle – as a price: the leptonic quantum number is no more a symmetry of Nature.

9.1 The tools for neutrino oscillations

Neutrino oscillations have been discovered and extensively studied by natural and man-made neutrino sources. Both proved to be useful and complementary from

[4] Again, we assume conditions implied by **CPT** invariance, such as $m_i = \bar{m}_i$ and $\mathbf{V}_{\alpha i}^{l*} = \bar{\mathbf{V}}_{\alpha i}^{l}$.

many points of view, and this is likely to remain so in next generation experiments. Some natural sources are interesting by themselves. In that case neutrinos are a powerful probe to understand them, but this goes out of the scope of this book. We will focus on neutrino sources and experimental techniques which have proven to be useful for the discovery of neutrino oscillations and which are likely to find or even establish **CP** asymmetries in the forthcoming years.

Natural neutrino sources important for neutrino oscillations are:

- **The Sun:** A large flux of electron neutrinos with variable energy between zero and about 15 MeV is emitted from the Sun's core and reaches the Earth as an incoherent mixture of all neutrino flavors.[5] This flux is now well understood (though some aspects are still under scrutiny and motivate further precision measurements). Solar neutrinos have been at first detected by means of radiochemical techniques with ^{37}Cl and ^{71}Ga, then by means of water and heavy water Cherenkov detectors (threshold 3 MeV minimum, only ^{8}B neutrinos can be seen), and more recently by means of ultra-pure liquid scintillators (threshold around 40 keV, all neutrino components can be detected, including those produced by CNO cycle [289]).
  In case of solar neutrinos the size of the source (the Sun's core, about 0.1 solar radius) and its high temperature ($\approx$ keV, much higher than neutrino mass splittings) wash out coherence, delivering to Earth an incoherent mixture of electron neutrinos, muon neutrinos and a small component of tau neutrinos. For energies above a few MeV electron neutrinos undergo matter enhanced oscillations along their trip through the Sun (see **Sect. 9.1.1**), while at low energy the oscillations are essentially those in vacuum.
- **Atmospheric neutrinos:** Cosmic radiation, mostly protons, interact in the top atmosphere producing an hadronic shower of pions and muons, and therefore electron and muon neutrinos, with a production ratio at the source roughly 1:2. These atmospheric neutrinos reach the detector from any point of the Earth with variable energies between sub-GeV up to 100s GeV. The Earth is still transparent at these energies,[6] making a perfectly isotropic flux at the detector location in absence of oscillations. Atmospheric neutrinos are mostly detected with Water Cherenkov detectors, either liquid or solid. The separation of muon and electron neutrinos is possible, and also statistical separation of tau neutrinos was recently obtained. The typical baseline is offered by the Earth diameter, which gives a luckily large value of E/L with the observed value of $|\Delta m^2_{32}| \sim 10^{-3}$ (eV)^2.

[5]Coherence is completely lost both because of the high core temperature ($\simeq$ keV) compared to neutrino mass splittings and of the large Sun's core size compared to vacuum oscillation lengths.

[6]This statement is not trivial. At higher energies, such as those detected by the IceCube detector at the South Pole, the absorption of the Earth for up-going neutrinos cannot be neglected.

Man-made sources for neutrino oscillations are:

- **Accelerators:** Muon neutrino beams can be made with accelerators sending intense proton beams on targets and selecting pions of right momenta. Long decay pipes for π^+ produce muon neutrino beams (anti-neutrinos by reversing pion charge). Accelerators are used today mostly with Long Base Line distances from beam target to detectors (295 km in Japan, 730 km from CERN in Switzerland (concluded), 730 km from Fermilab to Minnesota and 1300 km from Fermilab to South Dakota for the future DUNE experiment).
  Accelerator beams are neither pure nor very stable (small contamination of wrong muon sign and electron type are unavoidable from various pion and kaon decay channels) and therefore a Near Detector a few hundred meters down the proton target is often used to monitor the beam and determine its energy spectrum and flavor composition.
  Increasing the length of the oscillation baseline improves the sensitivity to smaller mass splittings. Beam energies can be as low as a few GeV, allowing good sensitivity to $E/L \gtrsim 10^{-3}$ (eV)^2 with distances 100–1000 km. However, for very long baselines matter effects become important. This is a blessing, because matter effects allows to measure the sign of Δm^2_{13} (the neutrino ordering), but is also a problem because they mimic **CP** violating effects (there is one Earth and no anti-Earth to do the experiment through). They must be carefully evaluated, especially for the longer baselines.
- **Nuclear reactors:** They produce abundant fluxes (about 10^{21} ν/s/GW, depending on reactor fuel and age) of very pure electron anti-neutrinos with variable energy between 0 and $\approx$ 10 MeV. Typical detection of reactor anti-neutrinos is done with Reines-Cowan inverse beta decay in liquid scintillators. The baseline can be very short (order of 10–20 m) for Near Detectors designed to calibrate the flux precisely (the reactor flux is *not* precisely known, its uncertainty of the order of 5% in some energy ranges) or for the search of sterile neutrinos in some mass ranges. Longer reactor baselines have been used for the measurement of θ_{13} (about 1 km) or for the measurement of θ_{12} and Δm^2_{21} (about 200 km).
- **Radioactive sources:** Powerful sources of electron neutrinos have been used by the GALLEX [290] and SAGE [291] experiments[7] started in 1991 for calibration.[8] The sources were made of either ^{35}Cr or ^{37}Ar, which both deliver a mono-chromatic electron neutrino beam of high intensity.

9.1.1 *The Mikheyev-Smirnov-Wolfenstein (MSW) effect*

Neutrinos and anti-neutrinos traveling through matter behave differently in principle with respect to vacuum oscillations, in particular through the Sun's core, but also through the Earth. The formalism of neutrino oscillations in matter was

[7]There are two examples when physicists want to give 'literature': GALLEX = [GALL]ium [EX]periment and SAGE = Ru[S]sian-[A]merian [G]allium [E]xperiment.

[8]These have been used for probing the impact of sterile neutrinos.

worked out in 1978 and 1986 [292]. There is a nice contribution by A. Smirnov at the International Conference on History of the Neutrino, Sept. 5–7, 2018, Paris, France [293]: he describes the connection of neutrinos with *matter* in the period of 1978–1986 in three parts: (a) 1978–1984: Wolfenstein's papers and follow-ups; (b) 1984–1985: the Mikheyev-Smirnov mechanism; (c) 1985–1986: further developments. It is not 'old history': it gives a deeper understanding of the underlying dynamics.

By ignoring matter effects, one follows **Sect. 4.6.3** and finds the solar neutrino flux by integrating the source's position inside the Sun's core; it gives:

$$P(\bar{\nu}_e \to \bar{\nu}_e; \text{solar}) = 1 - 2|\mathbf{U}_{e2}|^2|\mathbf{U}_{e1}|^2 \tag{9.9}$$

Neutrinos oscillate differently while traveling through matter. Consider a ν beam so low in energy that muons cannot be produced through scattering off nuclei and electrons.[9] While both ν_e and ν_μ undergo neutral current scattering, only ν_e can induce charged current interactions off nuclei as well as the electrons:

$$\begin{array}{c} \nu_e + A \overset{\text{NC}}{\longrightarrow} \nu_e + A \;\; , \;\; \nu_e + A \overset{\text{CC}}{\longrightarrow} e + A' \;, \; \nu_e + e \overset{\text{CC}}{\longrightarrow} e + \nu_e \\ \nu_\mu + A \overset{\text{NC}}{\longrightarrow} \nu_\mu + A \;\; , \;\; \nu_\mu + A \overset{\text{CC}}{\not\longrightarrow} \mu + A' \;, \; \nu_\mu + e \overset{\text{CC}}{\not\longrightarrow} \mu + \nu_e \end{array} \tag{9.10}$$

The extra force experienced by ν_e traveling through matter can be approximated by an effective interaction term

$$\mathcal{H}_{\text{eff}} = \frac{G_F}{\sqrt{2}}\, \bar{\nu}_e \gamma_\alpha (1-\gamma_5)\nu_e\, \bar{e}\gamma^\alpha(1-\gamma_5)e \,. \tag{9.11}$$

When the electron is nearly at rest and the neutrino relativistic, only the $\alpha = 0$ component is relevant and $\bar{e}\gamma^0(1-\gamma_5)e = n_e$ denotes the electron density. It is convenient to define $\xi = p/\delta m^2 \mathcal{H}_{\text{eff}}$.

The equation of motion is modified due to 'matter'[10]:

$$i\frac{\partial}{\partial t}\begin{pmatrix} \nu_e \\ \psi_- \\ \psi_+ \end{pmatrix} = \left[\left(p + \frac{\bar{m}^2}{2p}\right)\mathbf{1} + \frac{\Delta m^2}{2p}\begin{pmatrix} 0\,0\,0 \\ 0\,0\,0 \\ 0\,0\,1 \end{pmatrix} + \frac{\delta m^2}{2p}\begin{pmatrix} \xi\,0\,0 \\ 0\,\xi\,0 \\ 0\,0\,0 \end{pmatrix} \right.$$
$$\left. + \frac{\delta m^2}{2p}\begin{pmatrix} \xi - c & s & 0 \\ s & -\xi + c & 0 \\ 0 & 0 & 0 \end{pmatrix}\right]\begin{pmatrix} \nu_e \\ \psi_- \\ \psi_+ \end{pmatrix} \tag{9.12}$$

---
[9]This condition is always met for solar and reactor neutrinos.
[10]For a derivation see **Sect. 16.4** of Ref. [12].

with $\psi_\pm = \frac{1}{\sqrt{2}}(\nu_\mu \pm \nu_\tau)$. The eigenstates of the Hamiltonian are given as follows:

$$|\nu_1^M\rangle = \cos\theta_M|\nu_e\rangle - \sin\theta_M\frac{1}{\sqrt{2}}(|\nu_\mu\rangle - |\nu_\tau\rangle) \quad , \quad m_1^2 = \bar{m}^2 + \delta m^2[\xi - \sqrt{(\xi - c)^2 + s^2}]$$
$$|\nu_2^M\rangle = \sin\theta_M|\nu_e\rangle + \cos\theta_M\frac{1}{\sqrt{2}}(|\nu_\mu\rangle - |\nu_\tau\rangle) \quad , \quad m_2^2 = \bar{m}^2 + \delta m^2[\xi + \sqrt{(\xi - c)^2 + s^2}]$$
$$|\nu_3^M\rangle = \frac{1}{\sqrt{2}}(|\nu_\mu\rangle + |\nu_\tau\rangle) \quad , \quad m_3^2 = \bar{m}^2 + \Delta m^2 \tag{9.13}$$

$$\sin\theta_M = \sqrt{\frac{(\xi - c) + \sqrt{(\xi - c)^2 + s^2}}{2\sqrt{(\xi - c)^2 + s^2}}}$$
$$\cos\theta_M = \sqrt{\frac{-(\xi - c) + \sqrt{(\xi - c)^2 + s^2}}{2\sqrt{(\xi - c)^2 + s^2}}} \,. \tag{9.14}$$

For $\xi \gg 1$ one gets $\sin\theta_M = 1$, and ν_e is also an eigenstate. After its birth it travels freely through the Sun. When ν_e exits its core, ξ decreases, and the eigenvalue changes: $\bar{m}^2 + 2\delta m^2\xi \Longrightarrow \bar{m}^2 + \delta m^2$. The mass eigenstate changes adiabatically from ν_e to ν_2 as it exits the Sun and reaches the Earth. The probability to find again a ν_e while traveling from the Sun to Earth is $\sin^2(\theta_{12})$, the value of $\sin^2(\theta_M)$ for $\xi = 0$. In realistic scenarios ξ is neither large nor small. The equation of motion in Eq. (9.12) has to be integrated numerically to get the mixing angle θ_M; one always gets $\sin^2\theta_{12} < \sin^2\theta_M < 1$ and the limit $\sin^2\theta_M < 0.5$. This resolves an otherwise puzzling feature of the solar neutrino flux, namely the fact that the survival probability of ν_e falls below 0.5. This result is inconsistent with neutrino oscillations in *vacuum* for an effective two-neutrino scenario.

9.2 First hints of neutrino oscillations

The first indirect evidence of neutrino oscillations has been offered by the observed *deficit* of electron neutrinos produced in the Sun by nuclear fusion and detected on earth by various experimental techniques.

This deficit was first observed by the Homestake experiment [150], which was located deep underground in a gold mine in Lead, South Dakota. The deficit, albeit with varying and puzzling factors, was confirmed by many other following experiments, chiefly GALLEX [295] at the Laboratori Nazionali del Gran Sasso (Italy), SAGE [291] in Russia, and KamioKANDE in Japan [296].

Back in the early 60's John Bahcall [297] had started to calculate the rate at which a detector on Earth should capture neutrinos emitted by the 'natural' nuclear fusion *inside* our Sun.

The Homestake experiment was to measure the flux of neutrinos from our Sun's core and prove the validity of the fusion model developed over the previous three decades. The Homestake detector started data taking in 1970, and the final results were published in 1994 by R. Davis [294].[11] Electron neutrinos were detected

[11]Besides the wonderful final result, R. Davis's dedication and perseverance over such a long time span truly deserved him the Nobel prize he got in 2002.

by the radio-chemical reaction $\nu_e +^{37}$ Cl $\rightarrow e^- +^{37}$ Ar, where the Ar atoms were accumulated over a period of a couple of months in a tank of approximately 380.000 liters of perchloroethylene. The Ar atoms were then extracted through careful gas flow, and counted by their subsequent e^-- capture radioactive decay by means of suitable low background counters.[12] The radio-chemical technique is powerful but limited, because it does not measure the energy of the incoming neutrinos nor their direction. It simply counts the integrated neutrino flux above a kinematic threshold, 0.814 MeV for ^{37}Cl neutrino capture. Homestake could therefore measure only neutrinos emitted by the ^{8}B decay in the Sun, a highly energetic but small component of the whole solar neutrino flux, which depends critically on the exact temperature in the Sun's core. The experiment was already seen as a remarkable achievement, if it could find a non-zero value while not worrying about a factor of two or three. The uncertainties both on the experimental side (chiefly, the efficiency in collecting and counting the ^{37}Ar atoms) and on the theoretical side (the precise determination of the expected flux produced by the complex nuclear fusion chain operating in the Sun) did not allow to draw firm conclusions from the observed discrepancy, although the results gave a hint that a real problem existed.

Now we know that Homestake was a true pioneering experiment. It was the first to observe a *deficit* in the flux of neutrinos from our Sun, which since then has consistently remained the same, namely a factor of about 1/3 for ^{8}B neutrinos. We now know that electron neutrinos are transformed into other neutrinos on the flight from the Sun to the detector and the observed deficit is fully explained by the fact that only electron neutrinos can contribute to the radiochemical reaction (solar neutrino energies are well below muon mass).

Two other radiochemical detectors – GALLEX [290] and SAGE [291] – had measured $\nu_e +^{71}$ Ga $\rightarrow e^- +^{71}$ Ge from 1990–2000[13]: they also yielded *deficits* in the flux of neutrinos from our Sun with a factor of about 1/2 this time. GALLEX and SAGE gave a big boost to the interpretation of the neutrino deficit in terms of oscillations, although they were not able to completely prove it. The Standard Solar Model gained credibility also thanks to an impressive set of elio-seismological measurements, which confirmed most basic features of the theory. Everything was quite under control, except the lack of neutrinos, which was named 'Solar Neutrino Problem', see **Sect. 9.1.1**.

The importance of Gallex and SAGE cannot be under-estimated for this field. Being able to measure solar neutrinos at low energy threshold where the flux is dominated by *pp* neutrinos, they provided a much more solid evidence that the problem was relying on neutrino physics and not on our poor understanding of solar core temperature.

---

[12]The detector was a tank containing 615 tons of dry cleaning fluid. Could theorists come up with this idea? That is chemistry in the real world.

[13]J. Ellis coined the *bon mot* of the 'Alsace-Lorraine reaction': Gallium $\rightarrow$ Germanium $\rightarrow$ Gallium ...

These experiments could not prove the existence of neutrino oscillations, but they have shown that solar neutrinos were consistently measured with a deficit. Their 'road' was followed by other experiments with different techniques and different neutrino sources: it finally yielded to the discovery of atmospheric and solar neutrino oscillations with matter enhanced oscillations.

9.3 'First and Second Era' for neutrino oscillations

Three generations of crucial experiments were, are and will be based on water as detector medium, in which neutrino detection is made by means of Cherenkov radiation produced by prompt electrons/positrons, either emitted by inverse β decay on protons or by elastic scattering on electrons:

(1) KamiokaNDE[14] had 3,000 tons of pure water and produced very important results on solar, supernova (SN1987A), and atmospheric neutrinos. Its main goal was to find the decays of protons, but it finally got beautiful results on solar, atmospheric and super-novae neutrinos.

(2) Super-KamiokaNDE (or SK) has still today 50,000 tons of pure water and has produced even more important results on solar and atmospheric neutrinos. It is also used as the far detector of the T2K experiment, a long base line neutrino oscillation experiment made by sending a muon neutrino beam produced with the J-PARK accelerator and directed to SK. SK goals are much broader, including precise measurements of oscillations of solar, atmospheric and accelrator neutrinos and search for proton decay. Super-KamiokaNDE has provided the first clear evidence of neutrino oscillations by means of atmospheric neutrinos. It is still running, now with Gd loading in the water in order to enhance neutron detection efficiency.

(3) There is a future project called 'Hyper-KamiokaNDE' made with 260,000 tons of pure water and starting $\sim$ 2027. It has rich program from **CP** asymmetries in the leptonic section, neutrino mixing parameters, neutrinos from astronomical origin with accuracy – and proton decays.

Heavy water (D_2O) as a detection medium was used at the Sudbury Neutrino Observatory (SNO), that was sited 2 km underground in a mine near Sudbury in Canada [298], the current location of SNOLab. This experiment was initiated in 1984 to provide a definitive answer to the Solar Neutrino Problem and stopped data-taking, after three experimental phases, in 2006.[15] The main results from SNO for solar neutrinos had shown that electron neutrinos from ^{8}B in the solar core changed their flavor in the transit to Earth, proving beyond doubts that neutrino oscillations explain the solar neutrino problem.

Summarizing, the experiments that have given clear and unambiguous evidence of neutrino oscillations started in the end of the previous century and were

[14] KamiokaNDE = [Kamioka] [N]uclear '[D]ecay [E]xperiment.

[15] As a historical note, SNO was not the first heavy-water solar-neutrino experiment. In 1965, Tom Jenkins, along with other members of what was then the Case Institute of Technology, began the construction of a 2 tonne small heavy-water Cherenkov detector in the Morton salt mine in Ohio. This experiment finished in 1968 and served to place an upper limit on the solar ^{8}B flux.

the Super-KamiokaNDE experiment in Japan (still running in 2020) and SNO in Canada (completed in 2006). The use of water and heavy water allowed the clear identification of neutrino oscillations both in solar neutrinos (SNO) and in atmospheric neutrinos (SK) by means of excellent detectors, clever tricks, and a little luck.[16] The two main teams in Japan and Canada came from different areas of fundamental physics. The story of these experiments is very well described by the 2016 Nobel winners McDonald and Kajita in their lectures [152, 153].

9.3.1 *Discovery of atmospheric neutrino oscillations*

Starting from the late 1980's, the KamiokaNDE-II [299] and the Irvine-Michigan-Brookhaven detector (IMB) experiments [300][17] started to observe a clear deficit in atmospheric muon neutrinos, expected to be roughly double the number of electron neutrinos, but actually observed to be about 1/2. This deficit was called the 'atmospheric neutrino anomaly'.

Both experiments were based on Water Cherenkov technology, which allowed the determination of energy, incoming direction and also the electron or muon type of the incoming neutrinos. Long muon tracks produced by charged current muon neutrino interactions could be disentangled from electromagnetic showers produced by electron neutrinos. Neutral current interactions made the selection blurred, but it was still statistically possible to separate the two samples effectively.

The Super-KamiokaNDE detector (SK) started to operate in 1996 as a greatly improved version of the KamiokaNDE-II experiment, with much more pure water and better PMT (photomultiplier tube) coverage. It was able not only to improve the statistics, but also study the complete angular distribution of the incoming neutrinos both from above and from below and for both muon and electron type neutrinos. It has determined $\sin^2\theta_{23} = 0.588^{+0.031}_{-0.064}$ and $\Delta m^2_{32} = (2.50^{+0.13}_{-0.20}) \cdot 10^{-3}(\text{eV})^2$ [301]. Furthermore, SK has shown the change of muon neutrinos to tau neutrinos during their propagation with significance level of 4.6 sigma [302].

The SK result was a game changer. Neutrino oscillation physics was real and the community started to take it seriously,[18] particularly with the approvals of accelerators based on long base line experiments in Japan (K2K and later T2K), Europe (CERN to Gran Sasso beam with OPERA and Icarus experiments) and USA with the NOvA experiment from Fermilab to Minnesota. This long base line experiments have thoroughly confirmed SK result and contributed to the precise determination of oscillation parameters. They are also providing hints of **CP** violation in the lepton sector.

[16]The fact that the atmospheric neutrino oscillation length at GeV scale is comparable with the Earth diameter was crucial to obtain the result.

[17]Both were built with the purpose of searching for proton decays predicted by most GUT theories.

[18]Until the 1998 SK discovery and actually even a few years later, most of 'our' community was skeptical about the interpretation of (atmospheric and solar) deficits as neutrino oscillations – the SNO result finally convinced 'all'.

The IceCube Collaboration has measured atmospheric neutrino oscillations in a unusual location, the Amundsen-Scott South Pole Station in the continent of Antarctica. Neutrinos are detected in IceCube by observing the Cherenkov light produced in ice by charged particles created when neutrinos interact. The technique is similar to that of SK, but the detector is much bigger (GTon scale) and has a much higher energy threshold because of the sparser PMT coverage with respect to SK. Its primary purpose is the detection of extra-terrestrial high energy neutrinos, a goal that was achieved in 2013 [303].

This experiment has measured disappearance of muon neutrinos from the full sky from 5.6 to 56 GeV. Assuming normal neutrino mass order it has gotten [304]: $\Delta m^2_{32} \simeq (2.31 \pm 0.13) \cdot 10^{-3}(\text{eV})^2$ and $\sin^2\theta_{23} \simeq 0.51 \pm 0.09$.

9.3.2 *Discovery of solar neutrino oscillations*

'Our' Sun produces electron neutrinos in its core mainly[19] through nuclear reactions initiated by the fusion of two protons and continued by a chain that involve the production and decay of ^{7}Be and ^{8}B and, through various branches, always terminates with the production of ^{4}He, neutrinos and energy as:

$$p + p \to d + e^+ + \nu_e \qquad [pp] \tag{9.15}$$

$$p + p + e^- \to d + \nu_e \qquad [pep] \tag{9.16}$$

$$d + p \to {}^3He + \gamma$$

$$^3He + {}^3He \to {}^4He + p + p$$

$$^3He + {}^4He \to {}^7Be + \gamma$$

$$^7Be + e^- \to {}^7Li + \nu_e \qquad [^7Be] \tag{9.17}$$

$$^7Be + p \to {}^8B + \gamma$$

$$^7Li + p \to {}^8Be \to {}^4He + {}^4He$$

$$^8B \to {}^8Be + e^+ + \nu_e \qquad [^8B] \tag{9.18}$$

The labels (in [ ]) identify each neutrino component of the *pp* chain. Although largely subdominant ($\sim 2 \cdot 10^{-4}$ in flux) neutrinos emitted by the β^+ decay of ^{8}B – see Eq. (9.18) – are the easier to detect, being the highest in energy: the end point of the β^+ spectrum reaches about 15 MeV. Particularly, only ^{8}B solar neutrinos can be detected in water via Cherenkov emission of scattered electrons or, crucial for the SNO experiment, via inverse β decay on deuteron with the emission of a prompt electron and two protons.

The SNO experiment has been the key one to establish solar neutrino oscillations. With the aid of a crucial reservoir of heavy water made available by the Canadian government and by means of a very well conceived set of experimental

[19] Another mechanism, known as CNO cycle, is dominant in heavier stars but contributes only at the level of about 1% in the Sun. It has recently been detected by Borexino, but it played no role so far on the study of neutrino oscillations and we do not discuss it therefore here.

phases, SNO was able to measure precisely both the flux of electron neutrinos and, at the same time, the total flux of all neutrino species.

This was done by means of three reactions, which depends on neutrino flavor type differently. The first is the elastic scattering on electrons:

$$\nu_x + e^- \to \nu_x + e^- \tag{9.19}$$

which is possible for neutrinos ν_x of any flavour, but with different cross sections. This reaction was precisely measured by Super-KamiokaNDE as well. The second one is the inverse β decay on deuteron:

$$\nu_e + d \to p + p + e^- \tag{9.20}$$

which is possible for electron neutrinos only, since the incoming neutrinos do not have enough energy to produce a muon or tau in the final state.
And, finally, the inelastic scattering on deuteron via neutral currents:

$$\nu_x + d \to n + p + \nu_x \tag{9.21}$$

which is possible with the very same cross section for all neutrino flavors.

By measuring all three reactions carefully SNO has established, in one shot, that neutrinos do oscillate from electron flavor to either muon or tau flavors during their path from the Sun's core to the Earth, and that the total flux is indeed consistent with the Standard Solar Model expectations, when the effect of matter enhanced oscillations (MSW effect [293]) is taken into account for the ^{8}B neutrinos.

The final result for the flux of electron neutrinos ϕ, obtained analyzing data from 1999 to 2006, has been [151,152]:

$$\phi^{\rm SNO} = (5.54 \pm 0.33|_{\rm stat} \pm 0.36|_{\rm syst}) \cdot 10^6\,{\rm cm}^{-2}\,{\rm s}^{-1}\,, \tag{9.22}$$

in excellent agreement with the standard solar model prediction taking into account MSW oscillations in solar matter.

SNO also identified unambiguously the oscillation parameters relevant for solar neutrinos, pointing to a large mixing angle θ_{12}. The current value from PDG2020 is: $\sin^2\theta_{12} = 0.307{\pm}0.013$ with a tiny mass splitting: $\Delta m^2_{21} = (7.53{\pm}0.18){\cdot}10^{-5}\,({\rm eV})^2$.

9.3.3 *Accelerators*

The possibility to build neutrino beams by means of carefully designed pion beams has been exploited since the 1960s, after the discovery of the muon neutrino by Lederman, Schwarz and Steinberger, and muon neutrino oscillation experiments have been attempted during the 1990s in several labs around the world, particularly at CERN with the Nomad and Chorus experiments.

After SK discoveries in 1998, it became clear that direct study of neutrino oscillations of muon neutrinos would have been possible by observing oscillations over a distance of the order of 1000 km with typical muon neutrino energies of about 10 GeV. Three laboratories around the world started the Long BaseLine (*LBL*) program, namely KEK in Japan, CERN in Europe and Fermilab in the USA. Several experiments exploited the disappearance of muon neutrinos and anti-neutrinos (MINOS experiment from Fermilab to Minnesota and later NOvA along the same beam line), appearance of tau neutrinos (OPERA experiment, with a beam from CERN to Gran Sasso), and appearance of electron neutrinos (K2K and later T2K from KEK to SK detector). We give below a brief account of the main results of these three Long Baseline *LBL* experiments.

- The T2K experiment in Japan sends an intense beam of ν_μ from J-PARC to Super-Kamiokande over a distance of 295 km. Its program started in 2010; in 2014 it also started taking data from a $\bar{\nu}_\mu$ beam. They tallied how many muon neutrinos morphed into their electron counterparts en route. The T2K collaboration has shown analyses of neutrino and antineutrino oscillations both in *appearance* and *disappearance* channels. Its 2017 results are based on $22.5 \cdot 10^{20}$ protons on target (POT). Its results can be summarized with the values (in the "Normal [Inverted] Order"):

$$\Delta m^2_{32} = (2.54 \pm 0.08)\;[(-2.51 \pm 0.08)] \cdot 10^{-3}\,(\text{eV})^2 \tag{9.23}$$

$$\sin^2\theta_{23} = 0.55^{+0.05}_{-0.09}\;[0.55^{+0.05}_{-0.08}]\,. \tag{9.24}$$

It has detected 89 ν_e and 7 $\bar{\nu}_e$. These counts should have been closer to 68 and 9, respectively, if **CP** symmetry had been unbroken. The discrepancy indicates **CP** *violation* within a 95% confidence interval [305]. As said in **Sect. 9.3.1**, that is a strong hint for $\delta^{\text{lept}}_{\mathbf{CP}} \neq 0$.[20]
In 2020 the T2K Collaboration has given $14.9\,[16.4] \cdot 10^{20}$ POT for ν $[\bar{\nu}]$.[21] By assuming normal neutrino mass ordering they find:

$$\Delta m^2_{32} = (2.47^{+0.08}_{-0.09})\;[(2.50^{+0.18}_{-0.13})] \tag{9.25}$$

$$\sin^2\theta_{23} = 0.51^{+0.06}_{-0.07}\;[0.43^{+0.21}_{-0.05}]\,, \tag{9.26}$$

for neutrinos [anti-neutrinos]; no significant difference between results for ν_μ and $\bar{\nu}_\mu$ has been observed [306].
- OPERA was a long baseline neutrino oscillation experiment in *appearing* mode designed to search for $\nu_\mu \to \nu_\tau$ oscillations. The experiment was made between CERN in Geneva and the Laboratori Nazionali del Gran Sasso (LNGS) in Italy at a distance of about 730 km. Its full data set was collected between 2008 and 2012 and it was based on $1.8 \cdot 10^{20}$ POT. It was a hybrid emulsion/electronic apparatus, in which the main target was made of bricks of pure lead and emulsion foils. The interaction vertex of ν_τ candidates were at first roughly located by means of standard tracking techniques that identified the right brick. Then, offline, the brick was opened and the primary interaction vertex located by careful scanning of the emulsion foils. The OPERA experiment has not found any candidate for $\nu_\mu \to \nu_e$[22]: $\sin^2(2\,\theta_{13}) < 0.43$ (90% C.L.) [307].

[20] We would say more than a hint, but we are obviously biased.
[21] The T2K Collaboration has measured $31.3 \cdot 10^{20}$ POT; in the future it expects $\sim 50 \cdot 10^{20}$ POT.
[22] We ignore $\nu_\mu \to \nu_{\text{sterile}}$ here.

On the other hand, it has found ten candidates for $\nu_\mu \to \nu_\tau$. Assuming $\sin^2(2\,\theta_{23}) = 1$, one expects $\Delta m^2_{32} \sim 2.5 \cdot 10^{-3}$ (eV)^2. OPERA results give the value [308]: $\Delta m^2_{32} = (2.7 \pm 0.7) \cdot 10^{-3}$ (eV)^2. They are fully consistent with what expected from SK, KamLAND and solar neutrinos results.

- The MINOS and NOνA detectors are part of the long baseline neutrino oscillation program made with ν_μ and $\bar{\nu}_\mu$ beams from Fermilab. The "Near Detector" is $\sim$ 1 km from the source of neutrinos and the two "Far Detectors" have been built first at a distance of about 730 km in Minnesota (MINOS) and later at a distance of about 810 km (NOνA). The NOνA detector is slightly off the neutrino axis, a trick that reduces the flux but makes the neutrino energy spectrum much narrower, improving therefore the sensitivity to oscillations.
MINOS has taken data from 2005 and has been completed in 2016. It observed neutrino disappearance so its main contribution has been the measurement of Δm_{32} and θ_{23}. Its 2011 results with $7.25 \cdot 10^{20}$ POT report: $\Delta m^2_{32} = (2.32^{+0.12}_{-0.08}) \cdot 10^{-3}$ (eV)^2 and $\sin^2(2\theta_{23}) > 0.90$ with 90% C.L. An initial claim of **CPT** violation through a possible difference of disappearance of neutrinos and antineutrinos has finally been excluded [309].
Its final measurement of Δm^2_{32} and $\sin^2\theta_{23}$ is based on $23.76 \cdot 10^{20}$ POT with ν_μ and $\bar{\nu}_\mu$ beams: $|\Delta m^2_{32}| \simeq (2.40 \pm 0.09)\,[(2.45 \pm 0.0.08)] \cdot 10^{-3}\,(\text{eV})^2$ and $\sin^2\theta_{23} = 0.43^{+0.20}_{-0.04}\,[0.42^{+0.07}_{-0.03}]$ for Normal [Inverted] Order [310].
The NOνA detector is made of 14 tonnes (344000 cells) of extruded, highly reflective plastic PVC cells filled with liquid scintillator. Each cell in the far detector measures 3.9 cm wide, 6.0 cm deep and 15.5 meters long. This fine segmentation allows to separate electron showers initiated by electron neutrino charged current interactions from muons. The off-axis location of the detector maximizes the number of detected 2 GeV neutrinos, the energy at which oscillations are expected to be maximum with the distance of 810 km.
Based on $8.85 \cdot 10^{20}$ POT the NOνA Collaboration has measured appearance of $\nu_\mu \to \nu_e$ and disappearance of $\nu_\mu \to \nu_\mu$ with data from 2014 – 2017. To be more precise: joining data from $\nu_\mu \to \nu_e$ and $\nu_\mu \to \nu_\mu$ it has given information on the parameters $|\Delta m^2_{32}|$, θ_{23}, $\delta^{\text{lept}}_{\mathbf{CP}}$ and the mass order: $\Delta m^2_{32} \sim (2.35 - 2.52) \cdot 10^{-3}$ (eV)^2, $\sin^2\theta_{23} \sim (0.43 - 0.51)$ or $(0.52 - 0.60)$ and $\delta^{\text{lept}}_{\mathbf{CP}} \sim (0 - 0.12\,\pi)$ or $(0.91\,\pi - 2)$[23] [311].
Using $12.33 \cdot 10^{20}$ POT the NOνA experiment has recorded 27 $\bar{\nu}_\mu \to \bar{\nu}_e$ candidates with a background of 10.3 and 102 $\bar{\nu}_\mu \to \bar{\nu}_\mu$ [312]. This new antineutrino data is combined with NOνA neutrino data [311] leading to $|\Delta m^2_{32}| = (2.48^{+0.11}_{-0.06}) \cdot 10^{-3}(\text{eV})^2$ and $\sin^2\theta = 0.56^{+0.04}_{-0.03}$ in the Normal Order.

There are two special cases of *LBL* experiments made with a natural source:

- The IceCube experiment introduced in **Sect. 9.3.1** has been named "IceCube Neutrino Observatory" for good reasons. It has both broad and amazing

[23]The Fig. 11 in Ref. [311] shows the plot of $\Delta m^2_{32}(10^{-3})$ (eV)^2 vs. $\sin^2\theta_{23}$; it is an example on how to compare the results from other experiments.

experimental goals; we list only three: astrophysical neutrinos from our galaxy, cosmic neutrinos from sources outside the Milky Way and the connection with Dark Matter.
The IceCube detector is composed of thousands of optical sensors sunk deep beneath the Antarctic ice at the South Pole of our Earth. It observes the Northern Hemisphere sky, where the Earth serves as a filter to help weed out a background of muons created when cosmic rays crash into the Earth's atmosphere at the energies where the Earth is 'transparent'. At the highest energies (PeV scale) neutrinos are effectively stopped by the Earth. These neutrinos must therefore be detected from above, controlling the possible muon background. Its analyses have confirmed the existence of astrophysical neutrinos from our galaxy as well as cosmic neutrinos from sources *outside* our Milky Way:

$$P(\nu_\mu \to \nu_\mu) \simeq 1 - 4|U_{\mu 3}|^2(1 - |U_{\mu 3}|^2)\sin\left(\frac{\Delta m^2_{32}}{4}\frac{D}{E_\nu}\right) , \tag{9.27}$$

where $U_{\mu 3} = \sin\theta_{23}\cos\theta_{23}$ is one of the PMNS matrix; θ_{12}, Δm^2_{21} and $\delta^{\rm lept}_{\bf CP}$ have no impact on the *present* data.

- In 2018 the IceCube Collaboration has published its measurement of detection and reconstruction of neutrinos produced by the interaction of cosmic rays in Earth's atmosphere at energies as low as $\sim$ 5 GeV. This energy threshold permits to measure the muon neutrino disappearance over a range of baseline up to the diameters of our Earth and thus to probe the ratio of D/E_ν. Assuming normal neutrino mass ordering, this analysis used neutrinos with reconstructed energies from 5.6 to 56 GeV, measuring [304]:

$$\Delta m^2_{32} \simeq (2.31 \pm 0.13) \cdot 10^{-3}\ {\rm eV}^2 \tag{9.28}$$
$$\sin^2\theta_{23} \simeq 0.51 \pm 0.09 . \tag{9.29}$$

- These two values are consistent with those from the T2K, OPERA, MINOS and NOνA experiments. On the other hand, they came from very different situations. Look at the IceCube detector: interactions of cosmic rays in the atmosphere provide a large flux of neutrinos traveling distances from $D \sim$ 20 km (vertically *down*-going) to $D \sim 1.3 \cdot 10^4$ km (vertically *up*-going) to a detector near the Earth's surface.

- The Borexino experiment provided additional important information about solar neutrino oscillations in matter and about solar nuclear reactions dynamics through the complete measurement of the solar neutrino spectrum, both from *pp* chain and from the sub-dominant (in the Sun) CNO cycle. The challenge was to build a detector capable to determine independently and by direct counting each solar neutrino component, see Eqs. (9.15, 9.16, 9.17, 9.18) of the *pp* fusion chain and also those from the CNO cycle.

This result was achieved by Borexino by developing the most ultra-pure detector ever built. Thanks to this extreme purity (several orders of magnitude better than competing detectors), the low energy threshold (about 40 keV, well below *pp* neutrinos energy), and the demonstrated capability to identify each solar neutrino component independently by spectral analysis [313, 314], including those of CNO cycle [289], Borexino has completed the job. It has confirmed the oscillation pattern of solar neutrinos and directly also the MSW effect – see **Sect. 9.1.1** – by proving convincingly that the survival probability of electron neutrinos originating from the Sun's *core* has a clear energy dependence not accounted for by pure vacuum oscillations. The Borexino Collaboration has shown "Comprehensive measurement of *pp*-chain solar neutrinos" [313] and "Simultaneous precision spectroscopy of *pp*, ^{7}Be, and *pep* solar neutrinos with Borexino Phase-II" [314]; **Fig. 9.1** provides a clear summary of Borexino's work.

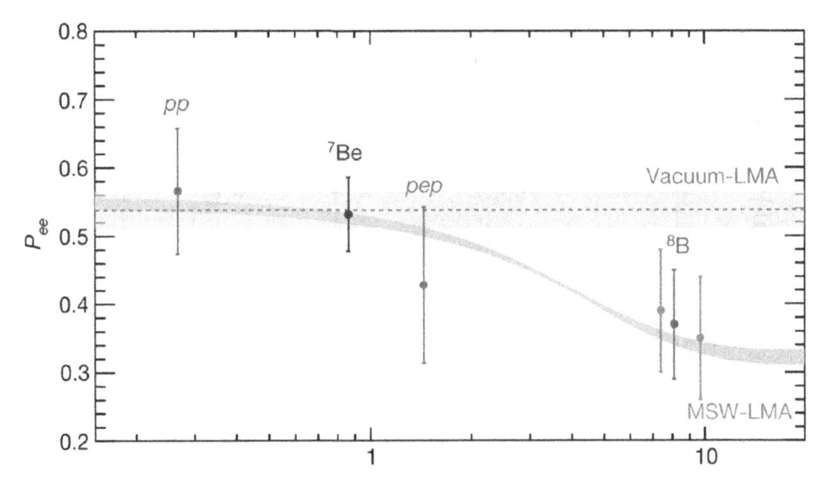

Fig. 9.1 The electron survival probability of all solar neutrinos of the pp chain are measured by the Borexino experiment [313]. Data show a clear energy dependence, explained by the fact that vacuum oscillations dominate at low energy, while a MSW effect is crucial at for ^{8}B neutrino energies.

9.3.4 *Nuclear reactors*

Other 'actors' have entered the 'play' on neutrino, namely data from nuclear reactors; their main result is the determination $\sin^2 2\theta_{13}$. We list four experiments that measure the disappearance of electron neutrinos $\bar{\nu}_e$ from reactors: the first one has a detector that is not close to the reactor, while the second, third and fourth ones are somewhat close to their reactors. In the latter cases they also need a second detector located at a distance of about 1 km; the 'Near Detector' technique proved to be crucial, since the exact fluxes from nuclear reactors are far from being completely understood.

- The oscillation pattern identified by SNO has been confirmed by KamLAND and Borexino in two different directions. The KamLAND experiment was a liquid scintillator detector located at about 200 km on average from Japanese nuclear reactors and was therefore able to probe directly the oscillation pattern identified by previous solar neutrino experiments and particularly by SK and SNO. KamLAND gave a clear visual picture of oscillations and provided a precise measurement of the mass splitting. **Fig. 9.2** provides a very clear oscillation pattern as a function of L/E.

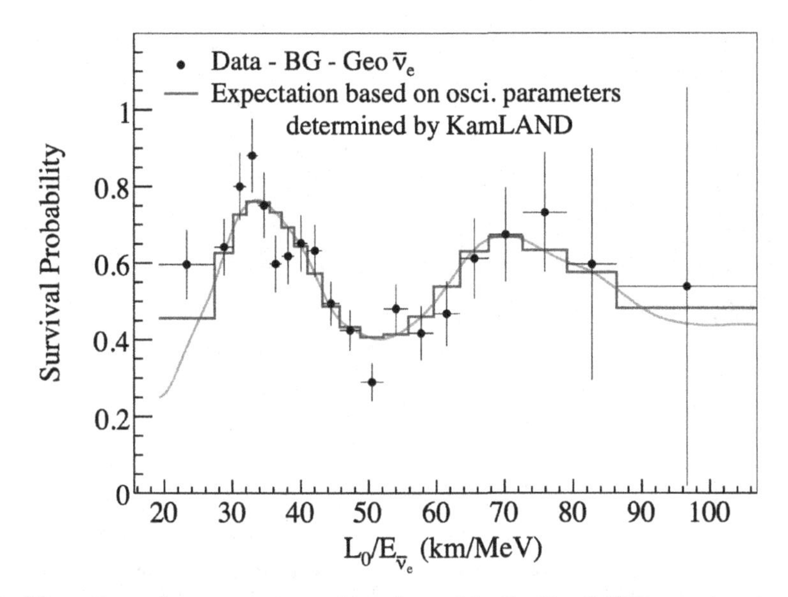

Fig. 9.2 The pattern of reactor anti-neutrinos observed by the KamLAND experiment, see [154]. The first and second oscillation peaks are beautifully visible.

- The Daya Bay collaboration had eight $\bar{\nu}_e$ detectors within 2 km of six reactors in China. Based on 1958 days of operation in 2018 its analyses led to: $\sin^2 2\theta_{13} = 0.0856 \pm 0.0029$ and $\Delta m^2_{32} = (2.471^{+0.068}_{-0.070})\,[-(2.575^{+0.068}_{-0.070})] \cdot 10^{-3}\,(\mathrm{eV})^2$ with normal [inverted] Order [315]. The JUNO experiment is an "update" of the Daya Bay experiment, that has moved by 53 km; its current plan is to start getting data in 2023.
- The RENO (**R**eactor **E**xperiment for **N**eutrino **O**scillation) collaboration had measured flux of disappearing $\bar{\nu}_e$ from six nuclear reactors in Korea with the

distance of 294 m and 1,383 m with two identical detectors. The present result is $\sin^2 2\theta_{13} = 0.0896 \pm 0.048|_{\text{stat}} \pm 0.048|_{\text{sys}}$ [316].

- The Double Chooz experiment close to three reactors (in the northeast of France) has found: $\sin^2 2\theta_{13} = 0.105 \pm 0.014$ in 2018 [317].

The measurements of θ_{13} (and θ_{12} and Δm^2_{31}) has been complemented by the results of other experiments like T2K.[24] In 2014 the T2K collaboration has provided electron neutrino appearance with 7.3σ experimental uncertainty and an independent measurement of θ_{13} [318]. A very recent article by the T2K collaboration had suggested that $\pi < \delta^{\text{lept}}_{\mathbf{CP}} < 2\pi$ comes from neutrino oscillations [319]. A current estimate from the 2018 Review of Particle Physics indicates a possible maximum value $\delta^{\text{lept}}_{\mathbf{CP}} \sim \frac{3}{2}\pi$ [320].

The observables $\sin^2 2\theta_{13}$, Δm^2_{21}, Δm^2_{31} and Δm^2_{32} have been already measured accurately, and future generation experiments such as JUNO, DUNE and Hyper-Kamiokande will augment their precision and provide a pathway to the discovery of **CP** violation in the lepton sector.

9.4 Global analysis and perspectives for forthcoming experiments

With accurate measurement of θ_{13} and the improved results of all solar, atmospheric and long baseline experiments, we have entered the era of global three flavors analysis. These comprehensive fits to all available data improve our knowledge of mixing angles and mass splittings and offer interesting hints on the remaining yet unknown parameters.

An example of such comprehensive analysis is given by Ref. [321], where Normal Order is preferred at about 2σ level with respect to Inverse Order. Atmospheric neutrino data, particularly from the IceCube Collaboration (as referred to in **Sects. 9.3.1, 9.3.3**) are an important source of information, but also cosmological data play a relevant role.

The **CP** violating Dirac phase is still unknown, but more and more data points to the direction that is not zero, and even suggest that it could be quite large. There is a very important 2018 result of the T2K Collaboration [305]: with a total exposure of $14.7\,[7.6] \cdot 10^{20}$ POT in neutrino [antineutrino] mode, the T2K experiment found 89 [7] ν_e [$\bar{\nu}_e$] candidates. The expectation in absence of any **CP** violation ($\delta^{\text{lept}}_{\mathbf{CP}} = 0, \pi$) and normal mass ordering is 67.5 [9.0] events, respectively.

As of today it is quite possible that the combined T2K and Noνa results in the forthcoming decade might yield to firm evidence of $\delta^{\text{lept}}_{\mathbf{CP}} \neq 0, \pi$ and provide a first measurement of its value. Future long base line experiments such as DUNE and Hyper-Kamiokande, featuring higher intensity beams and much larger and better detectors, will most likely discover **CP** violation in the lepton sector (if not done earlier) and provide a quite precise measurement of $\delta^{\text{lept}}_{\mathbf{CP}}$.

[24]The T2K analyses were interrupted abruptly by the 2011 Earthquake and Tsunami.

9.5 The 'seesaw' mechanism

Neutrinos are the only electrically neutral standard model fermions that have been established, and their pattern of small masses and tiny splittings is very different from that among quarks. 'Seesaw' models propose a way to explain the smallness of neutrino masses introducing heavy, right-handed neutrinos–the higher the mass of the right-handed neutrino, the lower the mass of the left-handed neutrinos.

Let us assume the lepton sector to consist of three families of left-handed neutrinos and N_R families of $SU(2)_L$ singlet right-handed neutrinos. Define $\nu \equiv (\nu_e, \nu_\mu, \nu_\tau)^T$ as the neutrino states whose left-handed components enter the weak Lagrangian (T stands for transpose) and $\nu_R \equiv (\nu'_1, \nu'_2, \cdots, \nu'_{N_R})^T$ as the right-handed neutrinos.

The more general Dirac-Majorana Lagrangian takes the form

$$\mathcal{L}_{D-M} = -\overline{\nu_R}\,\mathcal{M}_D\,\nu_L - \frac{1}{2}\,\overline{(\nu_R)^C}\,M_{MR}\,\nu_R + h.c. \tag{9.30}$$

The first term is the usual Dirac mass term, where $\mathcal{M}_D$ is a $N_R \times 3$ Dirac mass matrix. The second term is an additional Majorana mass term for right-handed fermion, allowed by the Lorentz symmetry, and M_{MR} is the corresponding $N_R \times N_R$ Majorana mass matrix. Let us observe that we have omitted the Majorana mass term $\mathcal{L}_{ML}$ already seen in Eq. (4.72), which is equivalent to set to zero the 3×3 mass matrix m_{ML}. This follows assuming that one obtains all mass terms from an $SU(2)$-invariant theory. Since left-handed neutrinos are in the same $SU(2)$ doublet as the charged leptons, without $SU(2)$ breaking any term that exists for neutrinos must also exist for charged leptons. For the latter, however, a Majorana mass term equivalent to $\mathcal{L}_{ML}$ is forbidden because they are charged. The same reasoning does *not* hold for the $SU(2)$ singlet right-handed neutrinos.

By defining a vector n_L as

$$n_L = \begin{pmatrix} \nu_L \\ (\nu_R)^C \end{pmatrix} \tag{9.31}$$

and the mass matrix M_{DM} as

$$M_{DM} = \begin{pmatrix} 0 & m_D^T \\ m_D & M_{MR} \end{pmatrix}, \tag{9.32}$$

the Dirac-Majorana[25] Lagrangian can be expressed in the compact form

$$\mathcal{L}_{DM} = -\frac{1}{2}\,\overline{(n_L)^C}\,M_{DM}\,n_L + h.c. \tag{9.33}$$

The core of the 'seesaw' mechanism is the relative weight of the mass matrix elements involved. The Dirac mass m_D is expected to be generated through the Higgs mechanism, as a consequence of the symmetry breaking, as the other fermion mass

[25]This is another example that physicists are short of words: DM can also indicate Dark Matter, a 'hot' and deep item we have talked about. On the other hand, comparing Dirac vs. Majorana neutrinos is deep, but 'hot' mostly only for theorists!

terms in the SM. Hence, m_D is expected to be proportional to the symmetry breaking scale, which is of the order of 10^2 GeV, as the other fermion masses. On the other hand, the chiral neutrino field ν_R is a SM singlet and the Majorana mass term for the right-handed neutrino fields is invariant under the gauge symmetries of the SM. Therefore, the elements of m_{MR} can have arbitrary large values. If M_{MR} is generated by the Higgs mechanism at a high energy scale of ND beyond the SM, the elements of M_{MR} are expected to be of the order of such high energy scale which could be as high as the grand unification scale of about 10^{15} GeV.

To summarize: by calling m_d and m_R the mass scale for the matrix elements of m_D and m_{MR}, respectively, we assume the hierarchy $m_d \ll m_R$. It can be proven that M_{DM} is symmetric: $M^T_{DM} = M_{DM}$; therefore it can be diagonalized by a unitary matrix. In the 'seesaw' assumptions, one finds[26] that the eigenvalues of the mass matrix become

$$m_1 \simeq \mathcal{O}\left(m_d^2/m_R\right) \ , \ m_2 \simeq \mathcal{O}\left(m_R\right) . \tag{9.34}$$

The three eigenvectors ν_1^i are practically very light and left-handed Majorana fermions, that can be identified with the ordinary neutrinos. The eigenvectors ν_2^j are basically right-handed and very heavy Majorana neutrinos; the heavier the ν_2 neutrino, the lighter the ν_1 one. This mechanism described above is the original formulation, known nowadays as minimal or type-I 'seesaw'. There are several variants, as e.g. type-II and III, where a new scalar triplet or triplet neutrinos are added to the Lagrangian in place of adding singlet right-handed neutrinos.

An additional attractive feature of this minimal scenario: the generation and smallness of neutrino masses can be related through the leptogenesis mechanism to the generation of the baryon asymmetry of the Universe. In the Dirac-Majorana Lagrangian we can add terms of interaction of right-handed neutrinos with the SM leptons through Yukawa interactions aside the mass terms. Such Yukawa couplings, in general, are not **CP** conserving and induce **CP** asymmetries in heavy neutrinos' decays. These may be sufficiently slow to occur out of thermal equilibrium around the energy scale where the Majorana masses emerge. The resulting lepton asymmetry may be transferred into a baryon number through sphaleron mediated processes. In this scenario, the neutrino masses and mixing and the baryon asymmetry have the same origin - the neutrino Yukawa couplings and the existence of (at least two) heavy Majorana neutrinos [214, 322].

9.5.1 *Can one measure Majorana phases?*

As shown above that neutrino oscillation experiments are *not* sensitive to the diagonal phase $[e^{i\alpha}]$ that appears in Eq. (4.76). Is there any way to get this phase? The answer is yes – in principle. Consider double β decays ($\beta\beta$ decays) of a nucleus with the number Z of protons and A as the sum of neutron and protons[27]:

$$_ZX^A \rightarrow_{Z+2} X^A + 2e^- + 2\bar{\nu}_e \ . \tag{9.35}$$

[26] For details one can see, e.g.: **Sects.** 16.6 and 16.7 (pages 353-355) in Ref. [12] (hint!).
[27] One can 'paint' the process at nucleon level as $nn \to pe^-\bar{\nu}_e pe^-\bar{\nu}_e$.

This decay can be seen only if the ordinary β decay ${}_Z X^A \to_{Z+1} X^A + e^- + \bar{\nu}_e$ is forbidden energetically or highly suppressed by angular momentum transition. While the decay of this process was computed by Goeppert-Mayer in 1935 [323], it was not until 1987 that the first observation was made [324] – quite achievements on the theoretical and experimental sides!

The next step is to probe neutrino-*less* double β decays[28]:

$$_Z X^A \to_{Z+2} X^A + 2e^- \ . \tag{9.36}$$

The rate for processes depicted in Eq. (9.35) and Eq. (9.36) are

$$(T^{2\nu}_{1/2})^{-1} \simeq G_{2\nu}(Q_{\beta\beta}, Z)|M_{2\nu}|^2 \ , \tag{9.37}$$

$$(T^{0\nu}_{1/2})^{-1} \simeq G_{0\nu}(Q_{\beta\beta}, Z)|M_{0\nu}|^2 \, \langle m_{\beta\beta}\rangle^2 \tag{9.38}$$

The derivation uses models for nuclear physics and can be found in Ref. [159]. The phase factors $G_{2\nu}$ depends on the Q-value of the decay, while $|M_{2\nu}|$ is a 'nuclear matrix element'. The amplitude is second order in the weak interaction. The double β decay conserves lepton number and does not discriminate between Majorana and Dirac neutrinos. The situation is quite different for the Eqs. (9.36,9.38): $(T^{0\nu}_{1/2})^{-1}$ would be driven by exchange of light Majorana neutrinos. $G_{0\nu}$ is the phase space factor for the emission of two electrons, $|M_{0\nu}|^2$ is another nuclear matrix element and an 'effective' Majorana mass is defined as

$$\langle m_{\beta\beta}\rangle = |\sum_{\nu^m_k} m_k(\mathbf{U_{PMNS}})^2_{1k}| \ .$$

It is a combination of the neutrino mass eigenstates and the neutrino mixing matrix terms. It strongly depends upon the Majorana phases and the absolute neutrino mass scale. A value $(T^{0\nu}_{1/2})^{-1} \neq 0$ would show that lepton number is broken with $\Delta L = 2$.

Through the discovery of neutrino-less double β decay it is possible, in principle, to obtain a non-zero value of $\langle m_{\beta\beta}\rangle$ – and it would be a pioneering achievement. It would prove the Majorana nature of the neutrinos and show that lepton number is violated. However, a direct extraction of the two Majorana phases from the measurement of $\langle m_{\beta\beta}\rangle$ is far from obvious, particularly because of the large uncertainty still existing in the calculation of nuclear matrix elements. A not impossible option is to extract the phases from a large number of independent observations of neutrino-less double beta decays from different nuclei, but there is certainly a long way to go for that. It should be underlined, however, that this is the only known way to probe Majorana phases directly. A review of existing neutrino-less double β decay experiments goes beyond the scope of this book. We can just hope that the searches in progress with many different nuclei (chiefly ^{136}Xe, ^{76}Ge, ^{130}Te and ^{100}Mo) may offer a big discovery in the next decade.

[28] One can 'paint' the process at nucleon level as $nn \to pe^-pe^-$.

9.6 Neutrino masses

Oscillation experiments tell us about the *differences* of neutrino masses. What do we know about the absolute values of their masses? Since one cannot measure the energies vs. momenta of neutrinos directly, one has to deduce their masses from very carefully analyzing the kinematics of reactions where they appear. The highest sensitivity has been achieved for the electron neutrino by studying the endpoint region in tritium beta decay where the neutrino emerges almost at rest. PDG2020 tells us: $m_\nu < 1.1$ eV at 90% CL.

The goal of the "KATRIN" experiment[29] is to reach $m_{\nu_e} < 0.2$ eV at 90% CL. Of course, the real goal is not to produce a smaller limit – it is to find non-zero values. Ref. [325] talks about a 'discovery potential': a neutrino mass of 0.35 [0.30] eV would be discovered with 5 [3]σ experimental uncertainty. It is a true challenge, and actually disfavored by cosmology – but we all hope!

We know that neutrinos are the lightest fundamental states with *non*-zero masses that are established. We also know we need huge detectors to probe their impact in the real world in different 'dimensions' – see **Fig. 9.3**.

Fig. 9.3 Moving the KATRIN experiment (the spectrometer) through the German town of Eggenstein-Leopoldshafen to the Karlsruhe Institute of Technology (2006) [Forschungszentrum Karlsruhe KIT Katrin. Media gallery].

Other options exist for the measurement of the neutrino absolute scale by means of cosmological observations. Next generation galactic surveys will measure red-shifts and lensing for a huge number of galaxies in large portions of the sky and to very far distances, improving sensitivity to neutrino mass through their effect on structure formation, and particularly on the two-point galaxy correlation function. Good examples are the ESA Euclid mission[30] and the Large Synoptic Survey Telescope (LSST).[31] If the total sum of neutrino masses $\Sigma_\nu = m_1 + m_2 + m_3$ is

[29]KATRIN = KA[rlsruhe] TRI[tium] N[eutrino].
[30]https://www.euclid-ec.org.
[31]https://www.lsst.org.

larger that 0.1 eV, Euclid will determine the neutrino mass scale independently of the cosmological model assumed. For Σ_ν below that value, the sensitivity reaches 0.03 eV in the context of a minimal extension of the Lambda-CDM model. Similar results can be obtained by LSST and actually even better ones might be obtained by means of global fits to all available data, including cosmic microwave background (CMB) data. It is quite possible that cosmology will give us the neutrino mass scale before direct kinematic experiments during next decade.

9.7 CP violation for charged leptons

It is crucial to probe fundamental dynamics also of charged leptons, where no **CP** violation has been observed yet. Most of the methods employed effectively in the quark sector do not work here, for instance no particle-antiparticle oscillations can occur for charged leptons due to the conservation of electric charge. The best chance to find **CP** violation is to probe the EDM of electrons, in the first place, and next the EDMs for muons and taus (see **Sect. 12.2**). Yet there are other avenues to this goal that certainly are at least as difficult, namely to probe **CP** symmetry in muon and tau decays.

Muon decays proceed basically through a single channel $\mu \to e\bar{\nu}_e\nu_\mu$ and the only practical way to proceed is to search for a **T**-odd correlation involving the decay of polarized muons into electrons whose spin is measured. The **T**-odd moment $\vec{\sigma}_e \cdot (\vec{p}_e \times \vec{\sigma}_\mu)$ – the component of the decay electron polarization transverse to the electron momentum and the muon polarization– probes **T** invariance. No such effect arises in the SM on a measurable level, and none has been seen for the energy averaged transverse polarization, measured at PSI by the ETH Zurich-Cracow-PSI Collaboration [326].

Tau decays provide a better stage to search for manifestations of **CP** breaking in the leptonic sector because its large mass and large number of possible final states. Several features enhance the chance to find that phenomenon: • there are many more channels than in muon decays making the constraints imposed by **CPT** symmetry much less restrictive; • hadrons can appear in the final state, allowing more types of T odd correlations to be constructed • due to the higher mass of the charged lepton, there is a better chance for ND, as multi-Higgs or leptoquarks models, to create an observable impact; • having polarized tau leptons provides a powerful handle on **CP** asymmetries and control over systematics. It allows us to construct new types of **CP** and **T** odd correlations.

A usual 'road' is to measure **CP** asymmetries in Cabibbo suppressed transitions driven by $\tau^- \to \nu\, s ... \bar{u}$ vs. $\tau^+ \to \bar{\nu}\, u ... \bar{s}$. One can compare $\tau^- \to \nu[K\pi]^-/\nu K^-\eta/\nu[K\pi\pi]^-$ with the **CP** conjugate channels $\tau^+ \to \bar{\nu}[K\pi]^+/\bar{\nu}K^+\eta/\bar{\nu}[K\pi\pi]^+$. There is one case already [327–329]:

$$A_{\mathbf{CP}}(\tau^+ \to \bar{\nu}K_S\pi^+)|_{\mathrm{SM}} = +(0.36 \pm 0.01)\% \tag{9.39}$$

$$A_{\mathbf{CP}}(\tau^+ \to \bar{\nu}K_S\pi^+[+\pi^0\,'\mathrm{s}])|_{\mathrm{BaBar2012}} = -(0.36 \pm 0.23 \pm 0.11)\%\,, \tag{9.40}$$

where the asymmetry is in the measured decay widths and the SM's prediction is due to indirect **CP** violation in $\bar{K}^0 - K^0$ oscillations. There is a 2.8σ difference between the BaBar data and the SM – however, 'experience' tells 'us' that one has to go *above* 3σ uncertainties to draw any conclusion.

Chapter 10

Strong CP violation

Old problems (like old soldiers)
never die – they just fade away

The only known candidate among local QFTs capable to describe the strong forces is QCD, based on the unbroken non-abelian gauge group $SU(3)_C$. **CP** is 'naturally' conserved in the Lagrangian of QCD, given in Eq. (3.1): a very attractive feature. QCD is in agreement with experimental observations, in particular "confinement" and "asymptotic freedom", and it is very well tested in high energy collisions.

P and **T** transformations act on "color" electric and magnetic fields $\vec{E}_a$ and $\vec{B}_a$ as:

$$\vec{E}_a \stackrel{\mathbf{P}}{\Longrightarrow} -\vec{E}_a \qquad \vec{E}_a \stackrel{\mathbf{T}}{\Longrightarrow} +\vec{E}_a$$
$$\vec{B}_a \stackrel{\mathbf{P}}{\Longrightarrow} +\vec{B}_a \qquad \vec{B}_a \stackrel{\mathbf{T}}{\Longrightarrow} -\vec{B}_a$$
$$G \cdot G \equiv G_{\mu\nu;\,a}\tilde{G}^{\mu\nu;\,a} \propto \sum_a |\vec{E}_a|^2 + \sum_a |\vec{B}_a|^2 \stackrel{\mathbf{P,T}}{\Longrightarrow} \sum_a |\vec{E}_a|^2 + \sum_a |\vec{B}_a|^2$$

Alas – there is a 'fly-in-the-ointment': there exists a gauge-invariant four-dimensional operator ignored above:

$$G \cdot \tilde{G} \equiv G_{\mu\nu;\,a}\tilde{G}^{\mu\nu}_a$$

with

$$\tilde{G}_{\mu\nu;\,a} \equiv \frac{i}{2}\epsilon_{\mu\nu\alpha\beta}G^{\alpha\beta}_a \,.$$

This term violates parity and time reversal, while charge conjugation is conserved:

$$G \cdot \tilde{G} \propto \sum_a \vec{E}_a \cdot \vec{B}_a \stackrel{\mathbf{P,T}}{\Longrightarrow} -\sum_a \vec{E}_a \cdot \vec{B}_a \,. \tag{10.1}$$

In QFT we must include all 4-dimensional gauge invariant operators; there is no reason to exclude $G \cdot \tilde{G}$. Thus one has to look at an effective QFT:

$$\mathcal{L}_{\text{eff}} = \mathcal{L}_{\text{QCD}} + \theta\,\frac{\alpha_S}{8\pi} G \cdot \tilde{G} \,, \tag{10.2}$$

where θ is a parameter. Even if this term, which breaks **P** and **T** invariance in QCD, was left out from the original Lagrangian, it would be induced in the SM at three-loop level (with two weak and one strong loop) [330]. Logarithmic divergent corrections to θ start at 14th order in the weak coupling g_W [331, 332].

10.1 $F \cdot \tilde{F}$ does not matter, while $G \cdot \tilde{G}$ does

What about adding a term $F_{\mu\nu}\tilde{F}^{\mu\nu}$ to a $U(1)$ Lagrangian? This term amounts to a total divergence: it can be transformed to a pure surface integral and thus has no impact. What about QCD (or any other non-abelian gauge theory)? We have

$$G \cdot \tilde{G} = \partial_\mu K^\mu \qquad K^\mu = \epsilon^{\mu\alpha\beta\gamma} A_{i\alpha} \left[G_{i\beta\gamma} - \frac{g_S}{3} f_{ijk} A_{j\beta} A_{k\gamma} \right] \tag{10.3}$$

If one could adopt $A_\alpha = 0$ for the gluon fields at spatial infinity like for the photon field, then we would have no observable in both cases. The present discussion would amount to 'much ado about nothing'. However, that is not the case, because the topological structure of the QCD ground state,[1] as in general for non-abelian QFTs, is much more 'complex'. Under the correct boundary conditions, the amplitude of the QCD ground state can be transferred to another ground state with a non-zero surface integral. The Dirac and gauge fields in a non-abelian QFT transform as

$$\psi \to \Omega\psi \ , \quad A_{\mu;\,a} \to \Omega A_{\mu;\,a}\Omega^{-1} + \frac{i}{g_S}(\partial_\mu \Omega)\Omega^{-1} \ ; \tag{10.4}$$

$A_{\mu;\,a} = 0$ is not the only possible configuration with minimal energy. The ground state is characterized by A_μ being a *pure* gauge field: $A_\mu^{\rm vac} = \frac{i}{g_S}(\partial_\mu\Omega)\Omega^{-1}$. In the *temporal* gauge $A_{0;\,a} = 0$ one can classify the functions Ω by their asymptotic behaviour:

$$\Omega_n \to e^{i(2\pi n)} \ \text{ as } \ r \to \infty \quad n = 0, \pm 1, \pm 2, \ldots \tag{10.5}$$

The phase factor $e^{i(2\pi n)}$ can be viewed as the topological mapping of a circle unto another circle n times; i.e., it 'wraps' around the circle n times. Therefore n can be referred to as the 'winding' number. It is determined by an integral over the pure gauge fields corresponding to these ground state configurations[2]:

$$n = \frac{g_S^2}{32\pi^2} \int d^3r \, K^0_{(n)} \ , \quad K^0_{(n)} = -\frac{g_S}{3} f_{abc}\,\epsilon_{ijk}\, A^i_{a(n)} A^j_{b(n)} A^k_{c(n)} \ . \tag{10.6}$$

Transitions occur from a configuration with n_- at $t = -\infty$ to one with n_+ at $t = +\infty$.[3]

$$\nu \equiv n_+ - n_- = \frac{g_S^2}{32\pi^2} \int d\sigma^\mu K_\mu |_{t=-\infty}^{t=+\infty} = \frac{g_S^2}{32\pi^2} \int d^4x \, G \cdot \tilde{G} \ . \tag{10.7}$$

Thus there are surface integrals over the current K_μ which are *not* zero.

The true ground state is a linear superposition of the n configurations:

$$|\theta\rangle = \sum e^{-in\theta} |n\rangle \tag{10.8}$$

---

[1] 'Vacuum' is not a good choice of word in QFT: the ground state is not empty at all.

[2] The structures f_{abc} [ϵ_{ijk}] for $SU(3)_C$ [$SU(2)$] runs $a, b, c = 1, \ldots, 8$ [$i, j, k = 1, 2, 3$].

[3] Short comments might help to remind readers of subtle features of QFT. An "instanton" configuration connects ground states with $n-1$ and n and can be viewed as tunneling effects. "Instantons" are classical solutions to the equations of motions with *finite* and *non-zero actions* in QFT; the same substance with different words: solutions of the equations of motions in classical field theory on a Euclidean space-time. The word 'pseudo-particles' was used in Refs. [333] and [213], while in Ref. [334] the 'modern' word "instanton" (= 'instant-on') was used.

with θ being a real number. One gets transition between ground states $|\theta(t = \pm\infty)\rangle$

$$\langle \theta(+\infty)|\theta(-\infty)\rangle = \sum_{\nu} e^{i\nu\theta} \sum_{n} \langle (n+\nu)(+\infty)|n(-\infty)\rangle \,. \tag{10.9}$$

The 'highly complex structure' of the ground state of QCD transforms the surface term $G \cdot \tilde{G}$ into a dynamical agent and thus introduces **CP** violation into the strong interactions.[4] Thus one should use an *effective* formulation for the strong forces, see Eq. (10.2), to take into account these non-perturbative features of QCD. Even when one consider the additional term to be zero at the tree level, loop corrections would induce its presence: they induce the presence of every gauge-invariant dimension-four operator in the Lagrangian – unless protected by some other symmetry. The term $G \cdot \tilde{G}$ produces **CP** and also **P** violation, but it does *not* depend on flavor amplitudes – opposite to what we have mostly discussed above. The coupling θ is an observable and a novel one– where can we find its impact? Since the strong **CP** violating term does not change the "flavors" of the quarks (as weak interactions do), certainly it will not be the leading effect in weak decays. We expects a 'natural' scale as $\theta \sim \mathcal{O}(1)$; numbers like 0.01 are surprising. Yet θ appears extremely small in the views of high sensitivity experimental data. The reasons of its suppression are studied under the name of "strong **CP** problem".

10.2 Anomalies

Here we give an amazing example of how QFT can connect different phenomena. Our community had focused on two challenges in different directions.

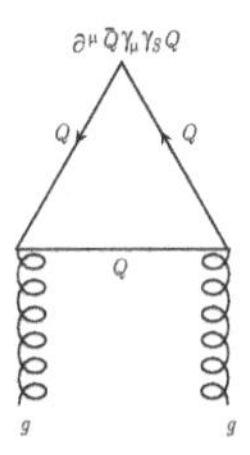

Fig. 10.1 Triangle diagram coupling an axial vertex with two vector bosons.

- Neutral states do not couple directly to electromagnetism. However, they couple at one-loop level. The decay $\pi^0 \to \gamma\gamma$ was measured with a branching ratio of $\sim 99\%$. It was described successfully by J. Steinberger in 1949 [96] during the pre-QFT era. He computed the decay from the diagram where an external pion effectively couples with two external photons through a proton loop (but not a neutron one).

[4] It is a fascinating story (at least for theorists), and serious readers should follow the discussions in Refs. [335–337].

At parton level the process can be described by the triangle diagram in **Fig. 10.1**, having a pion – a pseudo-scalar state $\partial^\mu \bar{Q}\gamma_\mu\gamma_5 Q$ at one vertex – and photons as final states g.
However, the neutral pion cannot decay into two photons if the gauge invariant *axial* vector current is conserved. The observation of this decay did not line up with the assumptions of current conservation in fashion at that time – based on current algebra that was successful in several applications, going back to Ref. [338].

- Let us consider QCD with only one family doublet made up from u and d quarks. In the limit of massless quarks – not a bad approximation with $m_u < m_d \ll \bar{\Lambda}$ – one might think that QCD possesses a *global* $U(2)_L \times U(2)_R$ symmetry. The vectorial component $U(2)_{L+R}$ is indeed conserved even after quantum corrections. The axial part $SU(2)_{L-R}$ is assumed to be spontaneously broken and leading to the emergence of a triplet of Goldstone bosons – the three pions. Those actually acquire a mass due to m_u, $m_d \neq 0$. A puzzle arises concerning the remaining axial $U(1)_A \equiv U(1)_{L-R}$: (i) It cannot represent a symmetry, since in that case one would have 'parity doubling'; i.e., hadrons had to come in mass degenerate pairs of opposite parity, which is *not* the case. (ii) If it is broken spontaneously, a fourth Goldstone boson has to exist. The only possible candidates are the neutral η and η' bosons with isospin zero [339]. Yet neither fulfills that role: an upper limit can be placed upon their masses, if they arise as Goldstone bosons; that is not satisfied even by the lighter η: $m_\eta \leq \sqrt{3}m_\pi$ does *not* hold [340]. Both these points leading to a dead end form what is called the $U(1)_A$ problem of QCD.

Both challenges summarized above are clarified referring to a general statement in QFT: once one-loop corrections are included, a *classical* symmetry may no longer be conserved. This is referred to as a "quantum anomaly".[5] It was realized that a flavor singlet axial current produces a "quantum anomaly".[6] The $U(1)_A$ axial symmetry is *not* a true symmetry of QCD, even though it is an apparent symmetry of its Lagrangian in the limit of massless quarks. The axial current $J_\mu^{(5)} = \bar{Q}\gamma_\mu\gamma_5 Q$ is no longer conserved, once one-loop corrections are included:

$$\partial^\mu J_\mu^{(5)} = \frac{\alpha_S}{8\pi} G \cdot \tilde{G} \,. \tag{10.10}$$

The assumed $U(1)_A$ symmetry was not there in the first place even for massless quarks. The diagram which generates the anomaly is a triangle diagram as in **Fig. 10.1**, with an internal fermion loop to which three external spin-1 lines are

[5]Here we talk about a crucial "quantum anomaly", based on QFT, which has impact on our understanding of fundamental dynamics. However, in HEP literature one can use the same word, 'anomaly', with another meaning: to describe a not imposing difference from SM expectations and data, usually around $2 - 3\sigma$. It may be misleading: besides, such differences both on the experimental and theoretical sides often happen without deeper reason.

[6]Weinberg suggested there is no $U(1)_A$ symmetry in the strong dynamics [340]. Basically at the same time it was pointed out that the QCD ground state is much more complex than thought before [213].

attached. One line corresponds to an axial current, while the other two to vector ones. The "triangle anomaly" or triangular fermion loop or "Adler-Bell-Jackiw (ABJ) anomaly" (after its discoverers) makes the situation 'complex' [45, 341]. On the good side: this does *not* go worse with more loops [342]. It was realized that the conservation of all three currents, as was assumed in the current algebra calculation, cannot be maintained when the model is quantized. In the absence of these symmetries, $\pi^0 \to \gamma\gamma$ is no longer forbidden [343] and happens all the time. Once this aspect of current algebra was cleared up, agreement with data was restored.

In QCD and QED one does *not* get a triangle anomaly, since they couple axial and vector gauge currents the same way. Instead the SM ($SU(3)_C \times SU(2)_L \times U(1)_Y$) leads to the triangle anomaly, which is also referred to as the chiral anomaly: axial and vector gauge fields couple differently. In the SM at one loop level, one axial and two vector gauge currents can be connected through the triangle in **Fig. 10.1**. While by itself the anomaly yields a finite result on the right-hand side of Eq. (10.10), it destroys the renormalizability of the theory. It cannot be "renormalized away", since $\epsilon_{\mu\nu\rho\sigma}$ cannot be regularized in a gauge-invariant way with dimensions less than "four". Instead this amplitude has to be neutralized by requiring that added contributions from all types of fermions in the theory yield a vanishing result. For the SM this requirement tells us: all electric charges of the fermions of a given family have to add up to zero. This imposes a connection between the charges of quarks and leptons – the number of quarks balanced against the number of leptons – yet it does not explain it.

10.3 QCD and quark masses

The $U(1)_A$ and the strong **CP** problems become further intertwined, when one includes the weak interactions. The SM tells us we have a single $SU(2)$ doublet of Higgs fields. They are connected to gauge bosons, giving them masses $0 < M^2(W^\pm) < M^2(Z^0)$, but also through Yukawa interactions to quarks and charged leptons, as described in **Sect. 4.4.1**. The physical Higgs field ϕ^0 has a ground state value $\langle\phi^0\rangle = v$. After spontaneous symmetry breaking, the quark mass terms for the up-type quarks U and down-type quarks D can be written as

$$\mathcal{L}_{\text{mass}} = v\sum_{k,l}((G_U)_{kl}\bar{U}_{k,L}U_{l,R} + (G_D)_{kl}\bar{D}_{k,L}D_{l,R}) + \text{h.c.} \tag{10.11}$$

The mass matrices $\mathcal{M}_U = vG_U$ and $\mathcal{M}_D = vG_D$ can be expressed in terms of mass eigenstates:

$$\mathcal{L}_{\text{mass}} = \bar{U}^m_L \mathcal{M}_U^{\text{diag}} U^m_R + \bar{D}^m_L \mathcal{M}_D^{\text{diag}} D^m_R + \text{h.c.} \tag{10.12}$$

So far we have no good reason to describe the pattern of the measured quark masses as due to Yukawa couplings in the SM – except it works.

However, we have not finished yet. Let us write the up-quark mass term in another form:

$$\mathcal{L}^U_{\text{mass}} = \bar{U}^m_L \mathcal{M}^{\text{diag}}_U U^m_R + \bar{U}^m_R (\mathcal{M}^{\text{diag}}_U)^\dagger U^m_L$$
$$= \frac{1}{2}\bar{U}^m[\mathcal{M}^{\text{diag}}_U + (\mathcal{M}^{\text{diag}}_U)^\dagger]U^m + \frac{1}{2}\bar{U}^m[\mathcal{M}^{\text{diag}}_U - (\mathcal{M}^{\text{diag}}_U)^\dagger)]\gamma_5 U^m \,. \quad (10.13)$$

In general, the eigenvectors of the mass matrix are complex, and the diagonal elements of $\mathcal{M}^{\text{diag}}_U$ are not necessarily real. They can be denoted by $m_k e^{i\alpha_k}$, and their phases α_i could induce **CP** violation. The imaginary parts are in the mass terms proportional to γ_5; they can be removed by performing the chiral rotation

$$U^m_k \Rightarrow e^{-\frac{i}{2}\alpha_k\gamma_5} U^m_k \ , \quad \bar{U}^m_k \Rightarrow \bar{U}^m_k e^{-\frac{i}{2}\alpha_k\gamma_5} \,. \quad (10.14)$$

The same holds for the down-quark mass term.

This would be the end of the story if QCD were invariant under this chiral transformation – but it is not:

$$\partial^\mu J^5_{\mu,k} = \partial^\mu(\bar{U}^m_k \gamma_\mu \gamma_5 U^m_k) = \frac{\alpha_S}{4\pi} G \cdot \tilde{G} + 2m_k\, i\, \bar{U}^m_k \gamma_5 U^m_k \neq 0 \,. \quad (10.15)$$

The current associated with this transformation is not conserved even for massless quarks as a consequence of the quantum anomaly we became acquainted with in **Sect. 10.2**. The chiral transformation of Eq. (10.14) changes the action S:

$$S \Longrightarrow S - \sum_k \int d^4x\, \partial^\mu J^5_{\mu,k} = S - i(\arg \det\mathcal{M}) \int d^4x \frac{\alpha_S}{8\pi} G \cdot \tilde{G} \quad (10.16)$$

$$\arg \det\mathcal{M} = \sum_k \alpha_k \,; \quad (10.17)$$

the sum runs over terms arising from U and D quark masses.

We could have started directly from the effective Lagrangian for the strong forces of Eq. (10.2). A chiral transformation can shift some of the θ terms into the mass terms, but cannot rotate them away simultaneously.

The total action remains unaffected by the *simultaneous* transformations:

$$Q_k \Rightarrow e^{-\frac{i}{2}\alpha_k\gamma_5} Q_k \ , \ m_k \Rightarrow e^{-i\alpha_k} m_k \ , \ \theta \Rightarrow \theta - \arg \det\mathcal{M} \,. \quad (10.18)$$

Thus observables depend on the *combination*

$$\bar{\theta} = \theta_{\text{QCD}} - \arg \det\mathcal{M} \equiv \theta_{\text{QCD}} - \Delta\theta_{\text{EW}} \ , \quad (10.19)$$

rather than θ or $\arg \det\mathcal{M}$ by themselves. The parameter $\bar{\theta}$ is the only real **CP** violating parameter of QCD. It shows the connection between the complex structure of the QCD ground state and the non-trivial one of Higgs dynamics.

10.4 About observable $\bar{\theta}$

We have already remarked that $G \cdot \tilde{G}$ operator is flavor-diagonal – so what? The answer is that its most noticeable impact would likely be to generate an EDM for neutrons. EDMs have been introduced in **Sect. 2.6**. Here we discuss how to probe EDMs of baryons in QFT.

Let us consider an additional Lagrangian coupling fermion and electromagnetic fields:

$$\mathcal{L}_{\rm EDM} = -\frac{i}{2}\, d\, \bar{\psi}\sigma_{\mu\nu}\gamma_5\psi F^{\mu\nu}\ . \tag{10.20}$$

This Lagrangian satisfies the low-energy properties of an EDM, since in the non-relativistic limit it produces the energy shift

$$\Delta E = \vec{d}\cdot\vec{E} + O(|\vec{E}|^2)\ , \tag{10.21}$$

when the system is placed in an electric field $\vec{E}$. The operator $\bar{\psi}\sigma_{\mu\nu}\gamma_5\psi F^{\mu\nu}$ has operator dimension "5"; then its coefficient d has usual dimension "-1" and can be calculated as a *finite* quantity. Of course, nucleon properties are notoriously difficult to calculate directly from non-perturbative QCD.

The neutron EDM – d_N – can be viewed as due to a photon coupling to a virtual proton and pion in a fluctuation of the neutron:

$$n \Longrightarrow p^*\pi^* \longrightarrow n\ \ ,\ \ n \longrightarrow p^*\pi^* \Longrightarrow n \tag{10.22}$$

That was pointed out first by Baluni [344] in a pioneering paper about the impact of $\bar{\theta}$ on the observable neutron EDM. There are actually two effective pion-nucleon couplings in a one-loop process: one is produced by ordinary QCD (conserving **P**, **T** and **CP** symmetries), while the other one is induced by $G\cdot\tilde{G}$ which is not (see **Fig. 10.2**). A rough guesstimate can be gleaned from naive dimensional reasoning: the scale for (electric or magnetic) dipole moments is set by e/M_N with M_N being the mass of the neutron; to be an EDM it obviously has to be proportional to $\bar{\theta}$ and to the ratio of (current) quark masses m_q to M_N. Thus $d_N \sim \mathcal{O}(\frac{e}{M_N}\frac{m_q}{M_N}\,\bar{\theta}) \sim \mathcal{O}(2\cdot 10^{-15}\,\bar{\theta}\,e\,{\rm cm})$.

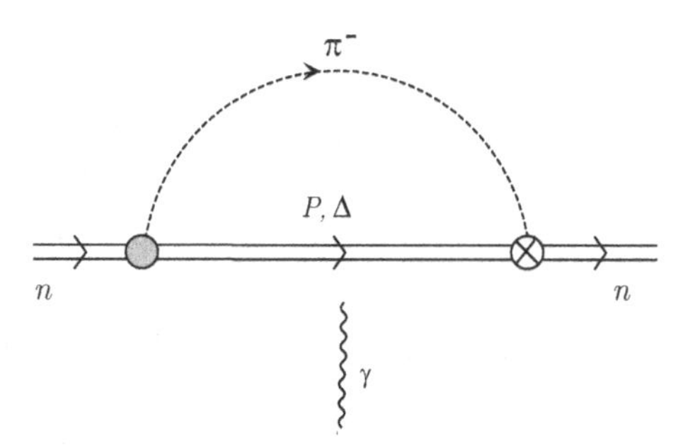

Fig. 10.2 A major contribution to the neutron EDM. The two effective vertexes are a strong vertex and a **CP** violating one (left/right, respectively).

A real estimate was given by Baluni in 1979 [344]: $d_N \simeq 2.7\cdot 10^{-16}\ \bar{\theta}\ e\,{\rm cm}$. He used MIT bag model computations of the transitions amplitudes between the neutron and its excitations. Chiral perturbation theory was also employed, leading to $d_N \simeq 5.2\cdot 10^{-16}\,\bar{\theta}\,e\,{\rm cm}$ [345]. About 10 years later, estimates remained roughly in the same range: $(4 \cdot 10^{-17} \sim 2 \cdot 10^{-15})\, \bar{\theta}\, \mathrm{e\,cm}$ [346]. Using present data, one gets from [8]:

$$d_N|_{\mathrm{PDG2020}} < 1.8 \cdot 10^{-26}\, e\,\mathrm{cm} \tag{10.23}$$

$$d_N \sim \mathcal{O}(10^{-16}\, \bar{\theta})\, \mathrm{e\,cm} \Longrightarrow \bar{\theta} < 10^{-9}\,. \tag{10.24}$$

Such a tiny upper limit[7] is a true theoretical problem (see **Chapter** 13 for more details on neutron electric dipole moment). Are there 'escape hatches'?

- If at least one of the quarks is massless (i.e., $m_i = 0 \Longrightarrow \det M = 0$), then the perfect chiral symmetry tells us that the non-physical phase α_i can be used to 'dial' $\bar{\theta}$ to zero! However, true experts (like the 'Pope of Chiral Symmetry') stated [347] that neither the u nor a fortiori the d mass can vanish: $m_d(1\ \mathrm{GeV}) > m_u(1\ \mathrm{GeV}) \sim 0.002\ \mathrm{GeV} \neq 0$ with the "running" mass evaluated at a scale of ~ 1 GeV.
- One could deny the very existence of a $U(1)_A$ problem, assuming that the 'vacuum' structures of non-Abelian gauge theories are even much more 'complex' than expected [348]. Here we are too conservative to follow those arguments.
- One can argue in favor of some unknown 'engineering' solution: $\bar{\theta}$ being the coefficient of an operator can be renormalized, in general, to any value including zero. That is technically correct; however, adjusting $\bar{\theta}$ to be smaller than $\mathcal{O}(10^{-9})$ by hand is viewed as highly 'un-natural':
  - We see no reason why θ_{QCD} and $\Delta\theta_{\mathrm{EW}}$ should practically vanish. Even if $\theta_{\mathrm{QCD}} = 0 = \Delta\theta_{\mathrm{EW}}$ were set *by fiat*, quantum corrections to $\Delta\theta_{\mathrm{EW}}$ are typically much larger than 10^{-9} and actually infinite. At which order this happens depends on the electroweak dynamics, though. Within the CKM ansatz $\Delta\theta_{\mathrm{EW}} \neq 0$ arises first at three loops, and it does not diverge before seven loops. In other models, the problem is more pressing. In models with right-handed currents or non-minimal Higgs dynamics $\Delta\theta_{EW} \neq 0$ arises already at one loop level [349, 350].
  - To expect that θ_{QCD} and $\Delta\theta_{\mathrm{EW}}$ cancel to render $\bar{\theta}$ sufficiently tiny would require fine tuning of a kind which would have to strike even a sceptic as unnatural. For θ_{QCD} reflects dynamics of the strong sector and $\Delta\theta_{\mathrm{EW}}$ of the electroweak sector.
- A "more respectable" approach would be to implement **CP** symmetry *spontaneously*, which imposes $\bar{\theta} = 0$ as the leading effect, with corrections leading to a small and *calculable* deviation from zero. This is referred to as *soft* breaking **CP** invariance in opposition to the CKM theory with its *hard* **CP** violation. A priori this approach appears as a very attractive option. On several occasions theorists have seen advantages of a *spontaneously* over a *manifestly* broken symmetry in principle. Arranging for spontaneous **CP** breaking represents a quite manageable task. For once we go beyond the minimal structure of the

[7]Ladies/Gentlemen do not quibble about $\bar{\theta} < \mathcal{O}(10^{-9})$ vs. $\bar{\theta} < \mathcal{O}(10^{-10})$.

SM: more VEVs emerge that can exhibit physical phases. There are explicit examples given with left-right symmetric models [349] or non-minimal Higgs dynamics [350]. The resulting scenarios are intriguing in their own right – at least for theorists. However, in practical realizations, tough challenges have to be faced. We give one major example:

- the *cosmic domain wall problem* raises its unpleasant head (as it does for any *discrete* symmetry). As our universe cools down to a temperature below which **CP** invariance is broken spontaneously, domains of different **CP** phases emerge. Since it is a discrete symmetry, walls have to form to separate these domains. As shown in Ref. [351], the energy stored in such walls would greatly exceed the closure density for our Universe. This cosmological disaster can be vitiated if the spontaneous **CP** breaking occurs *before* an inflationary period in our Universe's past. Then 'we' would live in a *single* domain. That means that the breaking scale of SSB has to be very high, of the order of GUT scales.
- While $\bar{\theta}$ naturally emerges to be small in these scenarios, we have to aim for truly tiny values–again, this favors a very high breaking scale.

• Let us remark that **CP** being broken spontaneously does not suffice to enforce $\bar{\theta} = 0$ at tree level. For the emerging VEVs can still contribute to "arg det $\mathcal{M}$" at tree level. Strategies have been pursued to build viable models [352]. One possibility is to allow for a complex quark mass matrix $\mathcal{M}$, but at the same time impose a special form on $\mathcal{M}$ such that "arg det $\mathcal{M} = 0$" holds at tree level. The key ingredient here is the introduction of novel super-heavy quarks. Yet at low energies, accessible to experiments so far, one recovers effectively a CKM mechanism [353, 354].
In general, one could reasonably say that the ansatz of *soft* breaking **CP** invariance flies in the face of the successes of the CKM theory with its *hard* **CP** violation. Yet, this statement sounds very conservative.

• There is another intriguing approach, where a physical quantity that is conventionally taken to be a constant is reinterpreted as a dynamical degree of freedom. This approach, and the related existence of "axions", will be discussed in details in **Chapter 11**.

Chapter 11

Axions

In **Chapter 10**, as previously in **Chapter 9**, we have shown a significant 'chink' in the SM's armor.[1] The tiny limit of $\bar{\theta} < 10^{-9}$ could hardly come by accident, and it 'begs' for an explanation. In QFTs one finds challenges all the time while producing finite results and comparing them with present data and possible future one. In case of unexpected small values, one needs an 'organizing' principle to arrange various contributions from amplitudes in a way as to render the required cancellations. To use a different word: "symmetries". The well-known and successful chiral symmetry fails to solve the "Strong **CP** violation" in presence of quark masses $m_q \neq 0$. Is it possible to invoke some other variance of chiral symmetry for this purpose, even if it is spontaneously broken? One particularly intriguing idea: an observable that is conventionally taken to be a *constant* could be re-interpreted as a *dynamic degree of freedom*; it would adjust itself to a desired value in response to forces acting upon it. One early example is provided by the original Kaluza-Klein theory [355] invoking a six-dimensional space-time manifold: two of those could be *compactified dynamically* and thus led to the quantization of electric and magnetic charges.[2]

In 1977 Roberto Peccei and Helen Quinn suggested to add a global $U(1)$ symmetry to the SM, in order to 'augment' it with an *axial* field (not another scalar field).[3] It is now referred to as Peccei-Quinn (PQ) symmetry $U(1)_{\rm PQ}$ [333]. This axial symmetry is characterized by the following properties:

- It is a symmetry of a *classical* theory.
- It is "broken spontaneously" in QFT.[4]
- It is also broken *explicitly* by non-perturbative effects, which reflect the complexity of the QCD ground state.

[1]We will discuss "Dark Matter", "Cosmology", "Baryogenesis" and "Multiverse", all outside the SM, in **Sects. 15.1, 15.3, 15.4 and 15.5**. The SM is not wrong, but it is certainly incomplete.

[2]Kaluza's theory was published in 1921 as a classical extension of general relativity, while in 1926 O. Klein gave Kaluza's theory a quantum interpretation after the 1925 revolution of Heisenberg and Schrödinger.

[3]We have used the word 'augment' since this symmetry leads to an additional neutral pseudo-scalar boson named "axion". In Ref. [334] several subtle aspects of it are pointed out.

[4]Using the words of "broken spontaneously" is somewhat misleading; it seems more appropriate to call it spontaneously *realized*; still, we follow the tradition.

The "broken spontaneously" PQ symmetry gives rise to Goldstone bosons called "axions" [334]. With $U(1)_{\rm PQ}$ being "axial", it exhibits a triangle anomaly leading to a coupling of an axion field to $G \cdot \tilde{G}$. This coupling not only generates a mass for the axion, but it crucially transforms the quantity $\bar{\theta}$ (see Eq. (10.19)) into a *dynamical* one. In other terms, it transforms the observable *number* into a *dynamical* quantity that depends on the axion field. The axion potential due to non-perturbative dynamics induces a vacuum expectation value for the axion such that $\bar{\theta} \simeq 0$ emerges; i.e., $\bar{\theta}$ relaxes *dynamically* to a value very close to zero.

The previous description is much more than just a nice yarn, as explained in some reviews by Peccei himself [337,339,346]. Here we give a very short introduction to axion dynamics. Let us consider this Lagrangian as a practice:

$$\mathcal{L}_{PQ} = -\frac{1}{4} G \cdot G + \sum_j \left[\overline{Q}_j i \gamma_\mu D^\mu Q_j - (y_j \overline{Q}_{L,j} Q_{R,j} \phi + \text{h.c.}) \right] \\ + \frac{\theta g_S^2}{32\pi^2} G \cdot \tilde{G} + \partial_\mu \phi^\dagger \partial^\mu \phi - V(\phi^\dagger \phi) \tag{11.1}$$

where we have added a spin-0 field ϕ to SM quarks Q_i and gluon fields. This Lagrangian remains invariant *classically* under the transformations:

$$\phi \overset{U_{PQ}(1)}{\longrightarrow} e^{i2\alpha} \phi, \quad Q_i \overset{U_{PQ}(1)}{\longrightarrow} e^{-i\alpha\gamma_5} Q_i \,. \tag{11.2}$$

The potential $V(\phi^\dagger \phi)$ is chosen such that the axial symmetry $U_{PQ}(1)$ is broken spontaneously by a vacuum expectation value (VEV) of ϕ:

$$\langle \phi(x) \rangle = v_{PQ} \, e^{i \langle a \rangle / v_{PQ}} , \tag{11.3}$$

where v_{PQ} is a real parameter and $\langle \bar{\theta} | a | \bar{\theta} \rangle \equiv \langle a \rangle$ denotes the VEV of the axion field $a(x)$. With Q_i acquiring mass

$$m_i = y_i \, v_{PQ} \, e^{i \langle a \rangle / v_{PQ}} \tag{11.4}$$

we obtain

$$\overline{\theta} = \theta - \sum_i \arg y_i - N_f \langle a \rangle / v_{PQ} \,, \tag{11.5}$$

where N_f is the number of families. A new important feature is that the quantity $\overline{\theta}$ depends on the axion field a through its VEV – it is no more a 'mere' parameter. In the usual scenarios for a spontaneous broken symmetry, the phase of the scalar field remains completely undetermined, which implies Goldstone bosons with zero mass.

A second new feature arises: a chiral anomaly is implemented through a term proportional to $a \, G \cdot \tilde{G}$; since it is linear in the field a, $G \cdot \tilde{G}$ acts as a non-trivial effective potential for a and the resulting dynamics determine $\langle a \rangle$. To see how it works, look at this effective Lagrangian:

$$\mathcal{L}_{\rm eff} = \mathcal{L}_{SM} + \frac{\alpha_S}{8\pi} \left[\overline{\theta} \, G \cdot \tilde{G} + \frac{\xi}{v_{PQ}} a \, G \cdot \tilde{G} \right] - \frac{1}{2} \partial_\mu a \partial^\mu a + \mathcal{L}_{\rm int}(\partial_\mu a, \psi) \tag{11.6}$$

where $\mathcal{L}_{\rm int}(\partial_\mu a, \psi)$ describes the purely derivative coupling of the axion field to the other fields ψ. The model dependent parameter ξ reflects the anomaly of the $U(1)_{PQ}$ current

$$\partial_\mu J^\mu_{PQ} = \xi \frac{\alpha_S}{8\pi} G \cdot \tilde{G} \,. \tag{11.7}$$

The effective potential for the axion field a is minimized by $\langle G \cdot \tilde{G} \rangle = 0$. One can use the fact that $\langle G \cdot \tilde{G} \rangle$ is periodic in $\bar{\theta}$ and is controlled by the combination $\bar{\theta} + \xi \langle a \rangle / v_{PQ}$ to obtain

$$\langle a \rangle = -\bar{\theta} \, \frac{v_{\rm PQ}}{\xi} \,. \tag{11.8}$$

Physical axion fields are described by a shifted $a_{\rm phys}(x) = a(x) - \langle a \rangle$ leading to

$$\mathcal{L}_{\rm eff} = \mathcal{L}_{\rm SM} - \frac{1}{2} \partial_\mu a_{\rm phys} \partial^\mu a_{\rm phys} + \frac{\alpha_S}{8\pi} \frac{\xi}{v_{\rm PQ}} \, a_{\rm phys} \, G \cdot \tilde{G} + \mathcal{L}_{\rm int}(\partial_\mu a_{\rm phys}, \psi) \,. \tag{11.9}$$

The 'offending' **P** and **T** violating bilinear term $\bar{\theta} \, G \cdot \tilde{G}$ in the strong forces has been traded in against the minimum of the axion potential of $a_{\rm phys} \, G \cdot \tilde{G}$; it leads to

$$\bar{\theta} = 0 \,. \tag{11.10}$$

Electroweak forces driving $K_L \to \pi\pi$ will move $\bar{\theta}$ away from zero, but only by an extremely tiny amount: $\bar{\theta} \sim \mathcal{O}(10^{-16})$; it means that $\bar{\theta}$ is basically zero.

Does this mean that "victory can be snatched from the jaws of defeat" with a Peccei-Quinn-type approach? It could be an amazing triumph for 'theory'. In literature one can see very active working on both theoretical and experimental sides.[5] Two short comments about PQ symmetry:
(a) Peccei and Quinn were the pioneers, as underlined by the the name of the global $U(1)_{\rm PQ}$ symmetry; even so the early team of inventors includes many more than Peccei and Quinn, like G. t' Hooft [213], S. Weinberg and F. Wilczek [334] and so on.
(b) Maybe – maybe – one could argue that it is not a truly novel idea; however, one has to open basically all of our tool box at the same time, and that is truly novel!

11.1 The dawn of axions – and their dusk?

The breaking of the global symmetry $U(1)_{\rm PQ}$ is not given for 'free', it leads to a novel (yet undiscovered) dynamical entity: the physical axion field. Its impact depends on the following parameters: its mass and its couplings to other fields [334]. From Eq. (11.6) we obtain its equation of motion:

$$-\,\partial^2 a + \partial_\mu \frac{\partial \mathcal{L}_{\rm int}}{\partial \partial_\mu a} = \frac{\xi}{v_{\rm PQ}} \frac{\alpha_S}{8\pi} G \cdot \tilde{G} \,. \tag{11.11}$$

[5] In case of success, Nobel prizes should obviously reward both experimental and theoretical efforts.

The axion is not massless due to an anomaly term, and its squared mass is

$$m_a^2 = -\frac{\xi}{v_{PQ}} \frac{\alpha_S}{8\pi} \frac{\partial}{\partial a} \langle G \cdot \tilde{G} \rangle |_{\langle a \rangle = \frac{\theta}{\xi} v_{PQ}} . \tag{11.12}$$

This expression leads to a rough estimate on dimensional grounds:

$$m_a^2 \sim \mathcal{O}\left(\frac{\Lambda_{QCD}^4}{v_{\rm PQ}^2} \right) . \tag{11.13}$$

Since we expect $v_{\rm PQ} \gg \Lambda_{QCD}$ on general grounds, we are talking about a very light boson. The question is how light would the axion be. As $v_{\rm PQ}$ goes *up*, the mass of the axion goes *down*, as do its couplings (see **Sects. 11.1.1, 11.1.2**). From the beginning of the axions era, two scenarios have been singled out, with the electroweak scale $v_{\rm EW} = (\sqrt{2}G_F)^{-1/2} \simeq 250$ GeV providing the discriminator.

(a) An energy scale $v_{\rm PQ} \sim v_{\rm EW}$ leads to $m_a \sim \mathcal{O}(1 \text{ MeV})$. These axions have been extensively searched for in accelerator based experiments and also through astrophysical observations. They are referred to as *visible* axions, because they would have been detected directly in accelerator or other lab scale experiments.

(b) An energy scale $v_{\rm PQ} \gg v_{\rm EW}$ leads to $m_a \ll 1$ MeV. Such axions could not be found in accelerator based experiments due to their tiny couplings; therefore they are called *invisible* axions. Yet that does not mean they necessarily escape detection. Actually they could be of great significance for the formation of stars, whole galaxies – and even our Universe.[6]

11.1.1 *'Visible' axions*

'Visible' axions have been searched for in accelerator experiments to no avail. The simplest scenario involves *two* $SU(2)_L$ doublet Higgs fields with opposite hypercharge.[7] They carry a $U(1)$ charge in addition to the hypercharge; this second global $U(1)$ is identified with the PQ symmetry. The anomaly induces *non*-derivative coupling of the axion a to the gauge fields [357]:

$$\mathcal{L}_{\rm anom} = \frac{a}{v} N_f \left[\left(x + \frac{1}{x} \right) \frac{\alpha_S}{8\pi} G \cdot \tilde{G} + \left(\frac{4}{3}x + \frac{1}{3x} + \frac{1}{x} \right) \frac{\alpha_{EW}}{4\pi} B \cdot \tilde{B} \right]$$

$$x = \frac{v_2}{v_1} \, , \; v = \sqrt{v_1^2 + v_2^2} \tag{11.14}$$

where N_f is the number of families, v_1 and v_2 are the VEVs of the two Higgs doublets and $B_{\mu\nu}$ is the field strength tensor of the hypercharge gauge boson field coupling to right-handed fermions: $B_{\mu\nu} = F_{\mu\nu}^{em} - \tan\theta_W \, F_{\mu\nu}^{Z^0}$. The anomaly also induces a mass for the axion [358]:

$$m_a \simeq \frac{m_\pi F_\pi}{v} N_f \left(x + \frac{1}{x} \right) \frac{\sqrt{m_u m_d}}{m_u + m_d} \simeq 25 \, N_f \left(x + \frac{1}{x} \right) \text{keV} . \tag{11.15}$$

[6] In 1983 it was pointed out [356] that the range of 0.25 TeV $\leq v_{\rm PQ} \leq 10^5$ TeV is excluded due to our understanding of stellar evolution. It has been suggested there to probe the range 10^5 TeV $\leq v_{\rm PQ} \leq 10^9$ TeV, where the latter limit includes domain walls and cosmology models. We have entered a now era, where one can talk about "TeV" rather than "GeV". In the previous century we had started to discuss jets of hadrons around few GeV.

[7] In the SM the Higgs doublet and its charge conjugate fill this role.

In the 1980s it was believed that m_a might be enhanced to even reach ~ 1 MeV, and thus the axion would have been found while decaying as $a \to e^+e^-$ very quickly. This 'lucky' event did not happen.

A more realistic scenario corresponds to $m_a < 2m_e$. In this case, an axion can decay fairly slowly into two photons:

$$\tau(a \to \gamma\gamma) \sim \mathcal{O}\left(\frac{100 \text{ keV}}{m_a}\right) \text{ s} . \tag{11.16}$$

Those axions have been looked for in beam dump experiments – without success. In a beam-dump experiment, a high-energy beam is dumped into a dense block of heavy material, in order to absorb the hadronic cascade as quickly as possible. The short decay path minimizes the background of conventional leptons from the decay of known long-lived particles, thus facilitating the search for penetrating stable particles.

Other channels where to look for axions are very rare kaon decays $K \to \pi a$. One expects a dominating contribution to $K \to \pi+$ 'nothing' from $K \to \pi a$ with the axion decaying well *outside* the detector. For long-lived axions ($m_a < 2m_e$) one expects [359]:

$$\text{BR}(K^+ \to \pi^+ a)|_{\text{th}} \sim 3 \cdot 10^{-5} \cdot \left(x + \frac{1}{2}\right)^{-2} . \tag{11.17}$$

For *two*-body FS kinematics present upper limits are [360, 361][8]:

$$\text{BR}(K^+ \to \pi^+ X^0) < 3.0 \cdot 10^{-10} , \quad \text{BR}(K_L \to \pi^0 X^0) < 3.7 \cdot 10^{-8} . \tag{11.18}$$

Eq. (11.17) does not give an accurate prediction; however, the discrepancy between expectation and observation seems conclusive. One comes to the same conclusion of not-existence of long-lived visible axions also from the absence of quarkonia decay into them: neither $J/\psi \to a\gamma$ nor $\Upsilon \to a\gamma$ has been found. As far as we know, there are no upper limits to $D^+ \to \pi^+ X^0$, $D^0 \to \pi^0 X^0$.

11.1.2 *'Invisible' axions*

Despite fruitless searches at accelerators, part of our community still thinks (in particular theorists) that axions or something with similar features collectively identified as ALP, Axion Like Particles, do exist in our Universe. Besides the attractive connections with the "Strong **CP** violation" (see **Chapter 10**), astrophysical and cosmic sources have been hypothesized.[9] Besides, axions may well be a relevant component of dark matter and many experiments search for axions assuming that is the case.

In order to fall under case (b) of **Sect. 11.1** ($m_a \ll 1$ MeV), it has been suggested to separate the $SU(2)_L \times U(1)_Y$ and $U(1)_{\text{PQ}}$ breaking scales. To that purpose, one can introduce a *complex scalar* field σ that

[8] X^0 is basically a massless and non-interacting particle.

[9] See the list in the PDG2019 Review "112. Axions and Other Similar Particles" [362].

- is an $SU(2)_L$ singlet,
- carries a PQ charge,
- possesses a huge VEV $v_{\rm PQ} \gg v_{\rm EW}$.

Since the extra boson σ is a $SU(2)_L$ singlet with zero weak hypercharge, it is immediately evident that it does not take part in the gauge symmetry of the Lagrangian. We are introducing by hand a new energy scale in the theory, $v_{\rm PQ}$, which has nothing to do with the electroweak one. The couplings of such axions to gauge bosons as well as fermions can become truly tiny. These requirements can be realized in two distinct scenarios discussed first in 1980:

(a) Only very heavy new quarks carry a PQ charge; this is referred to as the KSVZ axion [363]. The minimal version can do with a single $SU(2)_L$ Higgs doublet.

(b) Known quarks and leptons also carry a PQ charge. Two $SU(2)_L$ Higgs doublets are then required in addition to the σ field. The fermions do not couple directly to σ, yet become sensitive to the PQ breaking through the Higgs potential. This is referred to as the DFSZ axion, and it was first suggested in Ref. [364].

From current algebra one infers for the axion mass in either case:

$$m_a \sim 0.6 \text{ eV} \cdot \frac{10^4 \text{ TeV}}{v_{\rm PQ}} \,. \tag{11.19}$$

The most generic coupling of such axions is to two photons:

$$\mathcal{L}(a \to \gamma\gamma) = -\tilde{g}_{a\gamma\gamma} \frac{\alpha}{\pi} \frac{a(x)}{v_{\rm PQ}} \vec{E} \cdot \vec{B} \,; \tag{11.20}$$

$\tilde{g}_{a\gamma\gamma}$ is a $\mathcal{O}(1)$ model-dependent coefficient. Axions with such tiny masses have lifetimes easily in excess of the age of our Universe. Indeed, these axions are 'invisible' under ordinary circumstances. Yet in astrophysics and cosmology more favorable extraordinary conditions can arise, and their footprints could become visible in different settings. Most axion searches are performed exploiting the coupling to photons in different ways. In the next **Section** we summarise these techniques and the current experimental limits.

11.2 Short review of experimental axion search

The search of 'invisible' axions is a very challenging experimental task, which mostly rely on their coupling to photons, either real or virtual. Currently, the most promising techniques either search for axion-photon conversion in an external magnetic or electric field, or exploits the coupling of axions with the normal mode of an electromagnetic cavity, or also search for the possible emission of axions from nuclear reactions in lab scale or astrophysical environments.

Light-Shining-through-Walls (LSW) experiments are a general class of experiments searching for axion conversion in a large and intense magnetic field. An intense photon beam (a laser) is sent through a magnetic field and then through a thick photon-opaque but axion-transparent wall. Another magnetic field volume is located after the wall along the photon path and before a photon detector. In case axions can undergo two photon conversions in the magnetic field volumes (one before and one after the wall), a light signal should be detected through the wall, caused by the double axion conversion. This technique was first pioneered by [365] and the best current limit was given by OSQAR (Optical Search for QED Vacuum Birefringence, Axions, and Photon Regeneration):

$$g_A < 3.5 \cdot 10^{-8}\,\mathrm{GeV}^{-1} \text{ at } 95\% \text{ CL for } m_A < 0.3\,\mathrm{MeV}\,.$$

A nice curiosity: the experiment used a 9 T LHC dipole magnet [366].

Another way to probe the axion-photon coupling is to search for the rotation of the elliptical polarisation of the light passing through a magnetic field volume. The effect induced by possible axion coupling compete with genuine QED effects, which also induce dichroism and birefringence. The PVLAS experiment, after an incorrect early claim [367], has put limits which are comparable with those put by OSQAR [368].

Astrophysical sources are a key element of many experiments searching for axions, which might be produced by stars, supernovae explosions, black holes halos, or be a relic component of the Big Bang.

Stars like our Sun can produce axions in their core. In agreement with the axion-photon coupling of Eq. (11.20), a photon can be transformed into an axion. The solar axion flux can be searched for in many ways, e.g. directly by the inverse process transforming an axion (with spin 0) into a photon (with spin 1) or indirectly, by measuring solar properties and take into account the energy loss due to axion emission.

In direct searches (haloscopes) a strong magnetic field $\vec{B}$ can stimulate the conversion

$$\text{axion} \overset{\vec{B}}{\Longrightarrow} \text{photon}\,, \tag{11.21}$$

as it is sketched in **Fig. 11.1**.

An axion detection scheme, based on the "Primakoff effect",[10] uses a microwave cavity permeated by an intense magnetic field. Available microwave technology allows an impressive experimental sensitivity. No effect has been seen so far, yet these searches continue.

The most recent helioscope CAST (CERN Axion Solar Telescope) uses a decommissioned LHC dipole magnet on a tracking mount. CAST found the limit $g_A < 6.6 \cdot 10^{-11}$ GeV^{-1} at 95% CL for $m_A < 0.02$ eV [370].

The Primakoff conversion of photons to axions in the Sun is also one of the production mechanisms that contribute to the total solar axion flux. Very recently an excess of events at low energies (1–7 keV) was reported by the experiment XENON1T: one possible interpretation, among others, is that this excess could be a hint of a solar axion [371].

[10]Particles with a two-photon vertex can transform into photons in external electric or magnetic fields, an effect first discussed by Primakoff [369]. It has also been used for the measurement of the widths of neutral mesons.

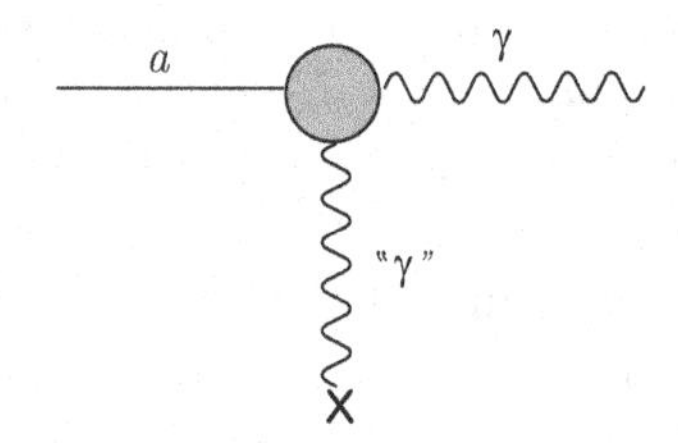

Fig. 11.1 An 'invisible' axion transforming itself into a photon, inside a magnetic field.

Axion emission could lead to energy loss and thus provide a cooling mechanism to stellar evolution. Their greatest impact occurs for the lifetimes of red giants and supernovae such as SN 1987a. The actual bounds depend on the models – whether it is a KSVZ or DFSZ axion – but mildly. Altogether, astrophysics tells us that *if* axions exist we get [372]:

$$m_a < 10^{-2}\ \mathrm{eV}\,. \tag{11.22}$$

Cosmology might provide us with a *lower* bound through a very intriguing line of reasoning. At temperatures T above $\Lambda_{\rm QCD}$ the axion is massless and all values of $\langle a(x)\rangle$ are equally likely. For $T \sim 1$ GeV the anomaly-induced potential turns on driving $\langle a(x)\rangle$ to a value which will yield $\bar{\theta} = 0$ at the new potential minimum. The energy stored previously as *latent heat* is then released into axions oscillating around its new VEV. Precisely because the invisible axion's couplings are so immensely suppressed, the energy cannot be dissipated into other degrees of freedom. We are dealing with a fluid of axions. Their typical momentum is the inverse of their correlation length, which in turn cannot exceed their horizon; one finds

$$p_a \sim (10^{-6}\mathrm{s})^{-1} \sim 10^{-9}\ \mathrm{eV} \tag{11.23}$$

at $T \simeq 1$ GeV; i.e. axions – despite their minute mass – form a very cold fluid and actually represent a candidate for cold Dark Matter. Their contribution to the density of our Universe relative to its critical value is [373]:

$$\Omega_a \sim \left(\frac{0.6\cdot 10^{-5}\ \mathrm{eV}}{m_a}\right)^{\frac{7}{6}} \cdot \left(\frac{200\ \mathrm{MeV}}{\Lambda_{\rm QCD}}\right)^{\frac{3}{4}} \cdot \left(\frac{75\ \mathrm{km/s}\cdot\mathrm{Mpc}}{H_0}\right)^{2}, \tag{11.24}$$

where H_0 is the Hubble expansion rate. For axions *not* to over-close the Universe one requires:

$$m_a > 10^{-6}\ \mathrm{eV} \;\;\leftrightarrow\;\; v_{\rm PQ} < 10^{12}\ \mathrm{TeV}\,. \tag{11.25}$$

It means that 'we' might be moving or existing in a bath of cold axions making up a significant fraction of the matter of our Universe. Ingenious suggestions have been made to search for such cosmic background axions, that often involve the process of **Fig. 11.1**.

The most sensitive experiments for cosmic background axion search are based on the possible coupling of the axion field with the normal mode of a cavity. The technique is indeed very sensitive but the need of high Q-value cavities has the obvious disadvantage that the sensitivity is limited to a narrow mass band. The existing limit provided by many such experiments (particularly, ADMX, HAYSTACK, UF and RBF) are generally better than those offered by CAST, but they are limited to specific mass ranges. For a complete review of the current limits for ALP searches made with all available techniques see Ref. [8].

11.3 Short comment about Axion Electrodynamics

As we have discussed above, the play of axions is an amazing one, but not a 'fable' or a 'mystery'.[11] Let us look at an unusual aspect underlined in the previous century by a 1987 paper [374] which studied the equations of axion electrodynamics. It has little direct connections with **CP** asymmetries, which is the main item in this book. At the first reading it can be ignored; at the third reading (we hope) one can learn by this short paper. This article begins by saying "Whether or not axions have any physical reality, their study can be a useful intellectual exercise" – at least theorists have to agree. Besides, it is not beyond the realm of possibility that the impact of axion-like fields may be realized in condensed matter systems.

The ordinary Maxwell Lagrangian leads to:

$$\vec{\nabla} \times \vec{E} = -\partial \vec{B}/\partial t \tag{11.26}$$

$$\vec{\nabla} \cdot \vec{B} = 0 \, . \tag{11.27}$$

Then one can add a term

$$\Delta \mathcal{L} = k \, a \, \vec{E} \cdot \vec{B} \, , \tag{11.28}$$

where the *axial* field a – 'the axion' – is coupled to $\vec{E} \cdot \vec{B}$ with a constant k. Thus

$$\vec{\nabla} \cdot E = \rho - \kappa \vec{\nabla} a \cdot \vec{B} \tag{11.29}$$

$$\vec{\nabla} \times \vec{B} = \partial \vec{E}/\partial t + \vec{j} + \kappa (\dot{a} \vec{B} + \vec{\nabla} a \times \vec{E}) \, ; \tag{11.30}$$

ρ and $\vec{j}$ are non-axion charge and currents. Thus there are additional terms in the charge density $\propto -\vec{\nabla} a \cdot \vec{B}$ and current density $\propto (\dot{a}\vec{B} + \vec{\nabla} a \times \vec{E})$. Since a is **P** and **T** odd, the new Maxwell Lagrangian conserves both.

Consider a special case: a *magnetic monopole* is surrounded by a spherical ball; inside for the axion field one gets $a = 0$, while outside $a = \theta \neq 0$, at large distances; i.e., a is a constant *below* the shell and *above* it, but with different values. On the other hand, the axion field has changed its value *inside* the shell. One does not need to look inside the shell, only at its effects on boundary conditions. It means that in a θ vacuum magnetic monopole becomes a "dyon"[12] with electric charge added to its magnetic charge. In an external B field axions can acquire an EDM. Again in general, θ is a "dynamical" variable.

Is this only an academic statement? No: axion fields might illuminate otherwise surprising phenomena in *solid* and/or *condensed matter* physics. Tools introduced by HEP can be of service or even applied in details there.

[11] Experimenters may see it differently – unless axions are found.

[12] In general, a "dyon" is a hypothetical particle with both electric and magnetic non-zero charges.

11.4 Our 'judgement'

It was already stated that QCD could produce **CP** violation [213], which in turn affects the neutron EDM [344]. However, there are hardly escape hatches away from non-perturbative QCD [347, 351–354]. We have no 'natural' way in the SM; i.e., with*out* huge fine tuning. Trying to resolve the strong **CP** problem has led to an impressive intellectual edifice based on an intriguing arsenal of theoretical reasoning, and has inspired fascinating experimental undertaking that are still going strong. In this chapter we have shown that there is a 'fertile' landscape built combining the thinking and working of HEP theorists and theoretical astrophysicists and observers and analysts. The fascinating story of the possible existence of axions goes back to the late 1970s. Then the search for axions entered a very different stage just before the end of the second century, namely with the connection with Dark Matter (that we discuss in **Sect. 15.1**). To make it short: the life of axions started with 'dawn', then 'dusk' and finally 'renaissance'.[13] In our view axions are a wonderful example of fundamental dynamics. One starts with a novel idea, but its confirmation needs new detectors and unusual analyses in different direction with long time commitment.[14]

[13]See Ref. [362] for further details.

[14]A cynic might summarize differently: why we certainly do not know the solution, we cannot even be sure that there is a problem in the first place. The situation about axions might remember a reader about an often told story of a French officer from the period of 'Enlightenment' (or the 'Age of Reason'): he was overheard praying before a battle: 'Dear God – in case you exist – save my soul if I have one.'

Chapter 12

Moments for the charged leptons

The dynamics for charged leptons e and μ is simple: they are truly elementary, and there is no experimental hint that one needs to go beyond the SM. No **CP** violation is expected in μ decays. A more intriguing case is represented by the τ leptons, where the SM basically gives zero **CP** asymmetries *except* for $\tau^+ \to \bar{\nu} K_S \pi$ decays. In this case the **CP** asymmetry is predicted to be $O(10^{-3})$ due to $\bar{K}^0 - K^0$ oscillations [327, 328], as explained in **Sect. 9.7.**[1]

In this **Chapter** we discuss instead **CP** and **T** asymmetries that are *in*dependent of flavor physics. Since the discovery of **CP** violation in 1964, there have been relevant experimental efforts to find non-zero values for EDMs of leptons, neutron, atoms and molecules. As EDMs violate both parity and time reversal symmetries (see **Sect. 2.6**), their values yield a mostly model-independent measure of **CP** violation in Nature (assuming **CPT** invariance). Hence EDMs are very interesting quantities to study: if non-zero values were observed, a novel source of **CP** violation, completely *independent* of flavor dynamics, would be required because no SM contribution is expected [375].[2] Limits on EDMs place strong constraints upon the amount of **CP** violation that extensions of the SM may allow. A new source of **CP** violation is one essential ingredient to lead to a successful electroweak baryogenesis to account for the baryon asymmetry in our Universe (see **Sect. 15.4**).

In this **Chapter** we discuss EDMs for charged leptons, and in **Chapter 13** EDMs for baryons. Yet in our search for ND we have more 'arrows' in our 'quiver': *magnetic moments* of charged leptons, which we will discuss first. It was one of the first successes of QFT with QED, although it has no direct connection with **CP** or **P** violation.

[1] Is there a hint of ND in BaBar data [329] – maybe? The theoretical prediction is found to be about 2.8σ away from the BaBar measurement.

[2] In the SM EDMs are generated at three loop level as a minimum [376], so they size is effectively zero.

12.1 Anomalous *magnetic* moment of charged leptons

Let us consider the impact of an external magnetic field (as described by off-shell photons) on charged lepton. Applying to the electromagnetic current the "Gordon identity" in relativistic QM

$$\bar{u}(p_2)\gamma^\mu u(p_1) = \bar{u}(p_2)\left(\frac{p_2^\mu + p_1^\mu}{2m} + \frac{i\sigma^{\mu\nu}q_\nu}{2m}\right)u(p_1) \tag{12.1}$$

one gets (from the second term on the right side) a spin magnetic moment $\vec{\mu}_S$ in the non-relativistic limit for *fundamental* charged fermions:

$$\vec{\mu}_S = g\frac{q}{2m}\vec{S} = g\frac{q}{2m}\frac{\vec{\sigma}}{2}\,. \tag{12.2}$$

The charged lepton interacts via both its charge and its magnetic moment. A guess based on classic electrodynamics suggests a gyro-magnetic ratio $g = 1$, but Dirac pointed out [377] that the solutions of 'his' equation lead to $g = 2$ for electrons.[3] The same holds for μ and τ leptons.

One can calculate deviation from the value $g = 2$ of the gyro-magnetic ratio due to quantum loop effects for charged leptons in the SM. One of the signature success of QFT was the analytical result for the 1-loop QED contribution,[4] which gives a tiny difference a, as pointed out after World War II [378]:

$$a|_e \equiv \frac{1}{2}(g-2) = \frac{1}{2}\frac{\alpha}{\pi} + ... \simeq 0.00116 \tag{12.3}$$

This value depends on the QED at the lowest scale: $\alpha(0) \sim 1/137$. Comparison of experiment and theory tests the SM at its quantum loop level. Presently, the estimate of the value of the (QED tenth-order) electron anomalous magnetic moment is

$$a|_e = 1159652182.032(720) \cdot 10^{-12}\,. \tag{12.4}$$

The most accurate measurement of $a|_e$ thus far has been carried out using a cylindrical Penning trap [379].

$$a|_e = 1159652180.73(28) \cdot 10^{-12}\,. \tag{12.5}$$

The difference between experiment and theory is thus

$$\Delta a|_e = a|_e(\exp) - \mathrm{a}|_e(\mathrm{SM}) = (-1.30 \pm 0.77) \cdot 10^{-12}\,; \tag{12.6}$$

After an improved determination of the fine structure constant α [380], the same comparison leads to a slightly different value [381]

$$\Delta a|_e = (-0.87 \pm 0.36) \cdot 10^{-12}\,. \tag{12.7}$$

The result above represents a 2.4σ discrepancy between experiment and theory, while the previous result (12.6) represented a 1.7σ effect.

[3] Pauli commented that Dirac: "... with his fine instinct for physical realities, he started his argument with*out* knowing the end of it". Our community faces similar situations now.

[4] This is probably the most celebrated result by J. Schwinger, and $\alpha/2\pi$ is engraved on his tombstone at Mount Auburn Cemetery in USA.

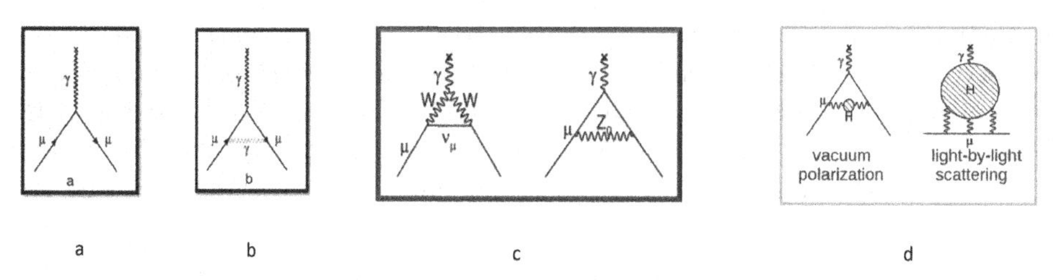

Fig. 12.1 Some of the diagrams contributing to $a|_\mu$ calculation. The tree level vertex (a) with first order QED (Schwinger term)(b), lowest order weak (c) and hadronic (d) contributions.

12.1.1 $a|_\mu = (g-2)/2|_\mu$

PDG2020 data show a truly precise result for the value of $g-2$ for muons:

$$a|_\mu(\text{exp}) = (11\,659\,209 \pm 6) \cdot 10^{-10} \tag{12.8}$$

It is an amazing achievement by the BNL E821 experiment [382]!

The SM deals with the challenge of computing quantum loop corrections by filling the following blanks:

$$a|_\mu(\text{SM}) = \frac{1}{2}\frac{\alpha}{\pi} + ... \left(\frac{\alpha}{\pi}\right)^2 + ... \left(\frac{\alpha}{\pi}\right)^3 + ... \left(\frac{\alpha}{\pi}\right)^4 + ... \left(\frac{\alpha}{\pi}\right)^5 ... \tag{12.9}$$

Obviously the leading source is QED with an excellent record [383–388]:

$$\begin{aligned} a|_\mu(\text{QED}) &= \frac{1}{2}\frac{\alpha}{\pi} + 0.765857425(17)\left(\frac{\alpha}{\pi}\right)^2 + 24.050\,509\,96(32)\left(\frac{\alpha}{\pi}\right)^3 \\ &\quad + 130.879\,6(63)\left(\frac{\alpha}{\pi}\right)^4 + 752.2(1.0)\left(\frac{\alpha}{\pi}\right)^5 + ... \\ &= (116\,584\,718.92 \pm 0.03) \cdot 10^{-11}\,, \end{aligned} \tag{12.10}$$

where the small error results mainly from the uncertainty in α. There are about 12670 diagrams for the QED contributions, which is known at 5 loops. We have to admire both the courage of the 'Cardinals' of QED and their achievements. Beyond QED there are two other classes of SM contributions:

$$a|_\mu(\text{SM}) = a|_\mu(\text{QED}) + a|_\mu(\text{EW}) + a|_\mu(\text{Had})\,. \tag{12.11}$$

Some of the contributing diagrams are drawn in **Fig. 12.1**. The impact of electroweak dynamics with $W^\pm$, Z^0 and Higgs fields gives [389]

$$a|_\mu(\text{EW}) = 153.6\,(1.0)\,\cdot 10^{-11} \tag{12.12}$$

from a full 2-loop numerical evaluation. The main challenges come from loops of intermediate states of hadrons which give [390, 391]

$$a|_\mu(\text{Had, LO}) = 6939\,(39)\,(7)\,\cdot 10^{-11} \tag{12.13}$$

$$a|_\mu(\text{Had, N(N)LO}) = 19\,(26)\,\cdot 10^{-11} \tag{12.14}$$

at leading order. The hadronic terms can be divided into contributions of diagrams containing hadronic self-energy insertions in the photon propagators and the hadronic light-by-light contribution. While the former can be computed from the low energy data on e^+e^- annihilation using a dispersive integral [392], the estimate of the latter relies on models [391, 393]. Summing these contributions one gets (PDG2020)

$$a|_\mu(\mathrm{SM}) = (116\,591\,830\,(1)\,(40)\,(26) \cdot 10^{-11} \tag{12.15}$$

where the uncertainties are due to the electroweak and low-order and higher-order hadronic contributions, respectively. Theorists have many reasons to be proud of their achievements after hard work!

One can compare the results of Eq. (12.8) and Eq. (12.15):

$$\Delta a|_\mu = a|_\mu(\exp) - \mathrm{a}|_\mu(\mathrm{SM}) = (261 \pm 63 \pm 48) \cdot 10^{-11}\ ; \tag{12.16}$$

where the uncertainties come from the experiment data and from theory predictions. There is a discrepancy of 3.3 times the combined 1σ error between experiment and theory results.

This is a good example of how, in a 'complex' landscape, one needs excellent tools from both the experimental and theoretical sides. We are looking at the work of two groups, experimenters and theorists: they knew what they were going after, and they spent a lot of their lives dedicated to find their 'goals'.[5]

Fig. 12.2 The Muon g-2 magnet moving toward Wilson Hall at Fermilab (July 26, 2013) [Fermilab].

Our community has continued to probe $a|_\mu$ and whether one can find the impact of ND. It is 'natural' for ND to leave its footprint there, *not* just a 'fashion'. We have candidates for 'explaining' the difference between the SM prediction and data, like supersymmetry (see **Chapter 14**) or scenarios involving "dark photons" (see **Sect. 15.1**). The well-known and 'mature' detector of the E821 experiment has

[5]The landscape is much more 'complex' for baryons, since they are bound states of quarks and gluons: $g_{\mathrm{proton}} \simeq -5.5857$ and $g_{\mathrm{neutron}} \simeq 3.8261$. Refined quark models test that, but hardly go beyond that.

been moved from BNL to the Fermilab [394]. It is an achievement by itself; see the pictures in [395], one is reported in **Fig. 12.2**. The BNL-E821 equipment, together a new muon accumulator ring and other improvements, is allowing a new measurement of a_μ at Fermilab [396], with expected gain in statistical and systematic uncertainties.

Very recently the Muon g-2 Collaboration, using the BNL-821 equipment with an improved muon accumulator, has published [397] a new value: $a|_\mu(\text{exp}) = (11659204.0 \pm 5.4) \cdot 10^{-10}$. This value is 3.3σ higher than SM prediction and is in excellent agreement with that of the previous E821 experiment at Brookhaven National Laboratory. Although the statistics is comparable with the measurement made at BNL, it makes sense to combine them in order to improve the precision. When combined with previous BNL result, the new measurement yields:

$$a|_\mu(\text{exp}) = (11659206.1 \pm 4.1) \cdot 10^{-10}$$

leading to

$$a|_\mu(\text{exp}) - a|_\mu(\text{SM}) = (25.1 \pm 5.9) \cdot 10^{-10}$$

with a significance of 4.2 σ (see Fig. 4 in Ref. [397]). The foreseen research program may allow to reduce the statistical error by a factor 2, offering the possibility of important discoveries.

A novel detector has being built at J-PARC in Japan [398]; it should be able to probe both $a|_\mu$ and the EDM for a muon at the same time.

12.1.2 $a|_\tau = (g-2)/2|_\tau$

The measured value of the muon anomalous magnetic moment shows a sizable difference with the SM prediction. It has triggered a huge interest in measuring also the anomalous magnetic moment of the τ. In a large class of theories beyond the SM, the parameter $a|_\tau = (g-2)/2|_\tau$ should be more sensitive to ND than the muon one: the new contributions to the anomalous magnetic moment of a lepton ℓ of mass m_ℓ are expected proportional to m_ℓ^2/Λ^2, where Λ is the ND scale, giving an enhancement of a factor of $m_\tau^2/m_\mu^2 \sim 283$.

SM prediction for a_τ relies on the same ingredients as for a_μ [399]:

$$a|_\tau(\text{SM}) = a|_\tau(\text{QED}) + a|_\tau(\text{EW}) + a|_\tau(\text{Had}) = 117721(5) \cdot 10^{-8}\,; \qquad (12.17)$$

$a|_\tau(\text{QED})$ is computed up to three loops [400], $a|_\tau(\text{EW})$ up to two loops [399, 401] and $a|_\tau(\text{Had})$ includes both the leading order and higher order hadronic contributions [392, 399].

The very short lifetime of the τ lepton ($2.9 \cdot 10^{-13}$ s) makes practically impossible the determination of a_τ by measuring the τ spin precession in a magnetic field, like in the electron and muon cases. Instead of spin precession experiments, high energy accelerator experiments have been done which include pair production of τ leptons. The best current experimental limit

$$-0.052 < a|_\tau(\text{exp}) < 0.013 \qquad (12.18)$$

at 95% confidence level has been derived from the DELPHI measurement at the Large Electron-Positron (LEP-2) collider of the total cross section for $e^+e^- \to e^+e^-\tau^+\tau^-$ [402]. This is well above the SM prediction, but nevertheless it provides a model-independent bound on ND contribution $a|_\tau(\text{ND}) \sim -0.018 \pm 0.017$. A re-analysis using various measurements of the $e^+e^- \to \tau^+\tau^-$ cross section, the transverse τ polarization asymmetry at LEP and SLD as well as the $W \to \tau\nu_\tau$ decay width from LEP and Tevatron in an effective field theory framework has led to a much stringent constraint on ND:

$$-0.007 < a|_\tau(\text{ND}) < 0.005$$

at 95% confidence level [403].

There is hardly a chance to measure a_τ close to the SM predictions as done for a_μ. Yet it is not a waste of time to discuss whether one can go down where one could find the impact of ND. At least our community would learn novel lessons about strong dynamics. As shown for instance in Ref. [404], the prospects of more precise measurements of $e^+e^- \to e^+\gamma^*\gamma^*e^- \to e^+\tau^+\tau^-e^-$ at future machines such as the proposed CLIC (Compact Linear Collider) and CEPC (Circular Electron Positron Collider), or at Belle II, would certainly allow for a remarkable improvement in the existing experimental bound on a_τ.

Several new methods have been suggested to determine a_τ. One can use the precise measurements of $\tau^- \to \nu_\tau\ell^-\bar{\nu}_\ell\gamma$ ($\ell = e, \mu$) decays [405] data from high-luminosity B factories. With radiative leptonic τ decays one can take advantage of the radiation zero of the differential decay rate which occurs when (in the τ rest frame) the final lepton ℓ and the photon are back-to-back, and ℓ has maximal energy [406]. Or one can use the decay $B^+ \to \tau^+\nu_\tau$ which would produce polarized τ leptons. Their spin can be precessed in a bent crystal and their final polarization can be measured through the angular distribution of the daughter lepton in the decays [407]. The channeling of a short-lived polarized particle through a bent crystal has already been tested successfully [408] to measure the magnetic moments of short-living baryons such as the Σ^+ hyperon [409]. Or one can measure $e^+e^- \to \tau^+\tau^-$ to determine a_τ by studying the angular distributions of the decay products of polarized τ using unpolarized electron beams [410]. A study has already been done for a future flavor factory showing that with 10^{10} $\tau^+\tau^-$ pairs the estimated sensitivity is of order 10^{-6} [411].

12.2 Electric dipole moments (EDMs) for leptons

EDMs have been introduced in **Sect. 2.6**. One describes EDMs for leptons with an operator in the Lagrangian

$$\mathcal{L}_{\text{EDM}}(x) = -\frac{i}{2}\, d_\ell\, \bar{\psi}_\ell(x)\sigma_{\mu\nu}\gamma_5\psi_\ell(x)F^{\mu\nu}(x)\,. \tag{12.19}$$

Since this operator has dimension 'five', its dimensional coefficient d_ℓ can be calculated as a finite quantity. In the non-relativistic limit, $\mathcal{L}_{\text{EDM}}$ describes the interaction of the EDM with an electric field.

EDMs for leptons are excellent hunting regions for ND. In the SM, the **CP** violating non-renormalizable interaction of (12.19) is generated by virtual effects. The CKM phase in the quark sector can induce a lepton EDM via a diagram with a closed quark loop, but a non-vanishing result appears first at the four loop level [412]. Electrons may also acquire an EDM through the **CP** violating QCD parameter $\bar{\theta}$. In both cases the SM gives a basically zero value, at the most [413]: $d_e(\mathrm{SM}) \sim 10^{-38}$ e cm.

In the search for EDMS of charged leptons, a 'natural' goal would be to find $d_e \neq 0$, since we can probe many sources (atoms and molecules), where electrons are not 'free'. However, it may not be the better 'road': maybe, maybe one can find EDMs first with $d_\mu \neq 0$ and/or $d_\tau \neq 0$. The mechanisms that generate an EDM in muons or tau leptons are identical to those in electrons. The EDM of charged leptons acquired through the two SM mechanisms scales with the inverse ratio of the masses [413]. We expect to scale up the SM prediction according to the factors $m_\mu/m_e \sim 200$ and $m_\tau/m_e \sim 3600$. Furthermore, their values might be enhanced by novel features of ND.[6]

12.2.1 *Experimental limits on d_ℓ*

The Schiff's theorem, discussed in **Sect. 2.6.4**, applies to the relativistic moving electrons in atoms with *partial shielding* for atoms with several protons. However, there are special cases, namely with alkali metals, in particular Cesium with its single valence electron, where a huge *enhancement* factor emerges [418]

$$d_{\mathrm{Cs}} \simeq 100 \cdot d_{\mathrm{e}} \tag{12.20}$$

Experiments to search for the electron EDM attempt to measure energy shifts arising from the interaction of an internal atomic or molecular electric field with a bound electron, a relativistic effect which can be greatly enhanced in some electronic states. The present best limit comes from ACME collaboration, where novel technology has been applied to a (polar) molecule (thorium monoxide), giving [419]:

$$|d_e| \leq 0.11 \cdot 10^{-28}\,\mathrm{e\,cm}\,. \tag{12.21}$$

Improved results are always welcome, but the true goal is *not* to get smaller limits. It is rather to develop new ideas and/or tools and/or analyses to *find* an electronic EDM. In these experiments one measures the dynamics of electrons not as free states, but as bound states with baryons. Once an EDM was found, we have to think about the information the data would give us about the connection of electrons and baryons, and of their EDMs. Aside the intrinsic leptonic EDM d_e, there are the interactions due to the coupling of an electron to the nucleus, and interference between them. Elaborate atomic, molecular and nuclear calculations are required to account for these contributions.

[6]Digressing from the topics of this **Chapter**, we mention that Okun tackled some aspects of EDMs of neutrinos (with a right-handed helicity) [414, 415]. There are experimental constraints for neutrino EDMs, coming from indirect measurements involving e^+e^- collisions [416, 417].

The latest upper limit on the muon EDM has been obtained by using spin precession data from the muon g-2 storage ring at Brookhaven National Laboratory (BNL) [420]. It reads

$$|d_\mu| \leq 0.19 \cdot 10^{-18}\ \mathrm{e\,cm}\,. \tag{12.22}$$

at 95% CL. As mentioned above, there is a novel technology to probe both d_μ and $a_\mu = (g-2)/2|_\mu$ at the same time with the new g-2/EDM experiment at J-PARC in Japan [398]. The plan is to reach the EDM for the muon with $\mathcal{O}(100)$ more sensitivity than the BNL experiment; d_μ might show the best 'road' to ND 'soon'.

While strong experimental limits have been placed on the electron [419] and muon [420] EDMs giving stringent constraints on ND [375, 421], the measurement of the tau EDM suffers the same difficulty as the measurement of its anomalous magnetic moment: the tau lifetime is too short! The best current limit comes from the Belle collaboration that has searched for **CP** violating effects in the process $e^+e^- \to \gamma^* \to \tau^+\tau^-$ using triple momentum and spin correlations of the decay products. The squared spin density matrix for τ-pair production is given by the following sum [417]:

$$\mathcal{M}^2_{\rm prod} = \mathcal{M}^2_{\rm SM} + |d_\tau|^2\mathcal{M}^2_{d^2} + \mathrm{Re}(\mathrm{d}_\tau)\mathcal{M}^2_{\rm Re} + \mathrm{Im}(\mathrm{d}_\tau)\mathcal{M}^2_{\rm Im}\ , \tag{12.23}$$

with the SM contribution $\mathcal{M}^2_{\rm SM}$, the squared EDM contribution $|d_\tau|^2\mathcal{M}^2_{d^2}$ and the interference between them as $\mathrm{Re}(d_\tau)\mathcal{M}^2_{\rm Re}$ and $\mathrm{Im}(d_\tau)\mathcal{M}^2_{\rm Im}$. The initial electron and positron are assumed to be unpolarized and massless particles. The interference terms contain the following combinations of spin-momentum correlations:

$$\mathcal{M}^2_{\rm Re} \subset (S_+ \times S_-)\cdot\hat{k}\ \ ,\ \ (S_+ \times S_-)\cdot\hat{p}, \tag{12.24}$$

$$\mathcal{M}^2_{\rm Im} \subset (S_+ - S_-)\cdot\hat{k}\ \ ,\ \ (S_+ - S_-)\cdot\hat{p}, \tag{12.25}$$

where $S_\pm$ is a $\tau_\pm$ spin vector, $\hat{k}$ and $\hat{p}$ are the unit vectors of the τ^+ and e^+ momenta in the center of mass of the system, respectively. These terms are **CP**-odd since they change sign under a **CP** transformation; $\mathcal{M}^2_{\rm Re}$ is also **T**-odd, while $\mathcal{M}^2_{\rm Im}$ is **T**-even. The analysis uses 26.8 million τ pairs (29.5 fb^{-1}) accumulated with the Belle detector at the KEKB accelerator. The following limit at 95% confidence level is obtained on the scale of (10^{-16} e cm) [417]:

$$-0.22 < \mathrm{Re}\,(d_\tau) < 0.45\ \ ,\ \ -0.25 < \mathrm{Im}\,(d_\tau) < 0.08\,. \tag{12.26}$$

The limit on the real part of d_τ comes from a **CP**-odd and **T**-odd observable, while for constraining the imaginary part of d_τ a **CP**-odd and **T**-even observable is used.

The τ lepton EDM can be measured at Belle II with polarized electron beams at energies near and on top of the Υ resonances. Having a polarized beam will allow, in the same way as for a_τ, to set limits on these observables using only the decay products of a single polarized τ. The upper limit sensitivity for the real part of d_τ has been estimated to be $\mathrm{Re}\,(d_\tau) \sim \mathcal{O}(10^{-19})$ with 50 ab^{-1} at Belle II [422, 423]. It is also possible to set upper limits on d_τ from the precision measurements of the electron EDM, by calculating its contribution into the electron EDM [424].

12.2.2 *Weak dipole moments of the τ lepton*

By analogy with the couplings to the photon, one can define two weak dipole moments of the τ lepton, the weak magnetic term a_τ^W and the **CP** violating weak electric term d_τ^W, through the effective Z couplings [425]:

$$\mathcal{L}_{\rm wdm}^Z = -\frac{1}{2\,\sin\theta_W\cos\theta_W} Z_\mu \bar{\tau} \left[i a_\tau^W \frac{e}{2m_\tau} \sigma^{\mu\nu} q_\nu + d_\tau^W \sigma^{\mu\nu} \gamma_5 q_\nu \right] \tau \,. \tag{12.27}$$

Experimentally, some bounds have been derived considering the process $e^+e^- \to \tau^+\tau^-$ at LEP [426–428] and studying the correlations between the spins of the two τ in the final state as proposed in Refs. [429–432]. The current best experimental limits come from the ALEPH collaboration at CERN that has obtained at 95% confidence level [426]:

$$\begin{aligned} &\left|{\rm Re}\left(a_\tau^W\right)\right| < 1.1 \cdot 10^{-3}\,, \quad \left|{\rm Im}\left(a_\tau^W\right)\right| < 2.7 \cdot 10^{-3} \\ &\left|{\rm Re}\left(d_\tau^W\right)\right| < 0.50 \cdot 10^{-17} {\rm e\,cm}\,, \quad \left|{\rm Im}\left(d_\tau^W\right)\right| < 1.1 \cdot 10^{-17} {\rm e\,cm}\,. \end{aligned} \tag{12.28}$$

Here also the experimental limits are well above the SM prediction [429, 430] giving a lot of room for improvement on these limits in the CEPC project [433]. This would allow to constrain ND appearing in these quantities. Moreover, by combining these constraints with the ones coming from the electroweak precision observables and the parity violation measurements, we should be able to constrain theories beyond the SM that predict new couplings of leptons to the Z boson, such as the left-right symmetric models.

Similarly, a weak magnetic coupling $W\tau\nu_\tau$ of the form [425]

$$\mathcal{L}_{\rm wdm}^W = -\frac{g}{2\,\sqrt{2}} i \frac{\kappa_\tau^W}{2m_\tau} W_\mu \left[\bar{\tau}\sigma^{\mu\nu} q_\nu (1-\gamma_5)\nu_\tau\right] + {\rm h.c.}\,, \tag{12.29}$$

has been studied by the DELPHI collaboration at LEP [434], considering the $e^+e^- \to \tau^+\tau^-$ decay. It led to a limit at 90% confidence level for the anomalous tensor coupling κ_τ^W [434]:

$$-0.096 < \kappa_\tau^W < 0.037\,. \tag{12.30}$$

By considering, within an effective field theory framework, various measurements, as $\sigma(e^+e^- \to \tau^+\tau^-)$, the transverse τ polarization asymmetry and $\Gamma(W \to \tau\nu_\tau)$ from LEP, SLD and Tevatron, a more stringent bound has been given at the 95% confidence level [403]:

$$-0.003 < \kappa_\tau^W < 0.004\,. \tag{12.31}$$

In lieu of a conclusion, we remark that as we go from the simplest to the more complex systems, i.e. from leptons to baryons to atoms to molecules, the possible mechanisms that can generate an EDM increase. Hence a statistically significant measurement in one case may not be enough to understand the origins of its EDM, thereby requiring measurement of EDM in multiple systems to parse out the contributions from various underlying mechanism. In the next **Chapter** we will discuss EDMs for baryons.

Chapter 13

EDMs for baryons

In **Sect. 2.6** we have discussed the impact of Quantum Mechanics on EDMs, while in **Sect. 12.2** we developed the tools one gets from QFT and discussed how to apply them to leptons. The SM gives basically zero contributions to the EDMs of leptons and the same is true for baryons. Yet the description of EDMs for baryons is quite different from the case of leptons. It is much more 'complex', and in our opinion there are therefore more chances to find non-zero EDM values from ND. It should be pointed out, however, that even if a non-zero value for a baryon EDM is found, it would not automatically lead to a clear understanding of the underlying dynamics. To establish EDM $\neq 0$ is, of course, a 'golden goal', but it would probably be just the beginning of a novel journey toward the ND and its features.

In this **Chapter** we describe a possible impact of ND on EDMs of baryons, namely with classes of non-minimal Higgs dynamics or right-handed currents. There is one class – super-symmetry (SUSY) – that we discuss with some details in **Chapter 14**. If somebody does *not* love tough challenges – like going up steep mountains in the fog – that person is in the wrong business. If no ND has been found in our world yet, a person should feel that is just bad luck and try again. One has to perform careful analyses, requiring alliances among atomic, nuclear and high energy physics.[1]

Some comments are at order. A nonzero EDM necessarily involves the breaking of **P** and **CP** symmetries. In order to characterize the possible sources of particle EDMs, effective Lagrangians are usually constructed to include both the short and long distance sources of **CP** violation. In **Sect. 3.4** we introduced an *effective* Lagrangian[2]:

$$\mathcal{L}_{\text{eff}} = \mathcal{L}_{\text{dim } 4} + \mathcal{L}_{\text{dim } 5} + \mathcal{L}_{\text{dim } 6} + \mathcal{L}_{\text{dim } 8} + \dots \, . \tag{13.1}$$

The relevant hadronic operators responsible of EDMs are identified by simple requirements of gauge invariance and hermiticity. We 'paint' the landscape with quark

[1] It is well known that Napoleon had said: 'Being lucky is part of the job description for generals.' It seems to be true also for theorists working on fundamental dynamics. On the other hand, Napoleon was not an expert on how to produce true alliances. Instead the Habsburg family had a long history to avoid wars. What was their secret? Marry and often!

[2] In that section we focused on $\Delta F_{\text{lavor}} \neq 0$ forces; yet *overall* EDMs are flavor independent – $\Delta F_{\text{lavor}} = 0$.

and gluon fields, ignoring third family because of its mass:

$$\mathcal{L}_{\rm eff}^{\bf CP\ odd} = -\frac{i}{2}\left[\sum_{i=u,d,s,c} \frac{a_{i;(5)}}{\Lambda}\,\bar{q}_i\sigma\cdot F\gamma_5 q_i + \sum_{i=u,d,s,c} \frac{b_{i;(5)}}{\Lambda}\,\bar{q}_i\sigma\cdot G\gamma_5 q_i\right]$$
$$+ \sum_{i,j=u,d,s,c} \frac{c_{ij;(6)}}{\Lambda^2}\,(\bar{q}_i q_i)(\bar{q}_j i\gamma_5 q_j) + \mathcal{L}_{\rm glue}^{\bf CP\ odd} \qquad (13.2)$$

with $\sigma\cdot F \equiv \sigma_{\mu\nu}F^{\mu\nu}$, $\sigma\cdot G \equiv \sigma_{\mu\nu}t^a G^{\mu\nu,a}$ and t^a are representation matrices of $SU(3)_C$ [435, 436]. On the first line there are operators of dimension "5" with quarks and gauge bosons; on the second line there are operators of dimension "6" with quark fields plus an additional Lagrangian with only 'outside fields' of gluons:

$$\mathcal{L}_{\rm glue}^{\bf CP\ odd} = \bar{\theta}\,\frac{\alpha_S}{8\pi}G\cdot\tilde{G} + c_{(6)}\,G^2\cdot\tilde{G} + c_{(8)}\,G^3\cdot\tilde{G} + ... \qquad (13.3)$$

The operators of this Lagrangian have dimension "4", "6", "8" and so on; G and $\tilde{G}$ denote $G^{\mu\nu,a}$ and $\tilde{G}^{\mu\nu,a}$. The coefficients $c_{(6)}$, $c_{(8)}$ ... are calculable as finite numbers in terms of the weak parameters, in contrast to the parameter $\bar{\theta}$. The contribution of operators independent of $\bar{\theta}$ can be determined, for instance, by 'naive dimensional analysis'–that is easy to apply, but also gives easily the wrong results [435, 436].

13.1 Neutrons, deuterons and beyond

The level of experimental precision achieved in EDM searches has improved dramatically since the early work of Purcell and Ramsey [23], and has been broadened to many atomic and nuclear quantities. The EDMs of sub-atomic particles, as the neutron or Λ^0, and of several atoms and molecules have been measured to vanish, to often remarkably high precision. The experiments that currently provide the best constraints on ND are the EDM of the neutron and the atomic EDMs of thallium and mercury, paramagnetic and diamagnetic atoms, respectively. There are two excellent articles about EDMs from baryons [436]. Their limits have to be updated; however, one can learn about the tools to describe its dynamics. Some comments are in order.

- In the SM the situation for the neutron EDMs is unusual. It has been realized that they can*not* be produced with two-loop diagrams: one needs three-loops including gluon exchanges as pointed out first by Shabalin [437]. Even the three-loop contributions are suppressed, as discussed with more details in Ref. [438].
- A nonzero permanent EDM requires **P** and **CP** violation. The cost of **P** violation can be roughly estimated in terms of the product of Fermi's constant $G_F \simeq 1.2\cdot 10^{-5}$ GeV^{-2} times the square of the axial decay constant of the pion, $F_\pi \sim 92.2$ MeV. The latter represents the order-parameter of the spontaneous breaking of chiral symmetry of QCD at low energies. The dimensionless product scales therefore as $G_F F_\pi^2 \sim 10^{-7}$. In the SM, once the θ term is absent, the sole source for **CP** violation is the CKM mechanism. The CKM generated

CP violation is flavor-violating, while the EDMs are flavor-diagonal. Hence one expects a further suppression factor $G_F F_\pi^2 \sim 10^{-7}$ to undo the flavor violation. Summarizing: the starting scale is given by the **CP** and **P** conserving (magnetic) moment of the nucleon, which is of the order of the nuclear magneton $\mu_N \sim 10^{-14}$ e cm, while the suppression factors are a factor $\sim 10^{-3}$ (as in kaon **CP** violation) and the two suppression factors $G_F F_\pi^2 \sim 10^{-7}$. Hence we have that the EDM of the nucleon cannot be larger than

$$|d_N| < 10^{-17} \cdot \mu_N \sim 10^{-31}\, e\,\text{cm} . \tag{13.4}$$

This result agrees in magnitude with complete SM calculations [439, 440].

- In the previous century $SU(6)$ wave functions were mostly used in relations with "constituent quarks" u and d.[3] Obtaining the contribution of quark EDMs to the EDM of the nucleon then amounts to evaluating the relevant Clebsch-Gordan coefficients, and leads to

$$d_N \simeq \frac{1}{3}(4d_d - d_u) \ , \ d_P \simeq \frac{1}{3}(4d_u - d_d) . \tag{13.5}$$

  In estimating the contributions of the physics beyond the SM this relation works generally well, and puts stringent bounds on the **CP** violation sources in the underlying model.
- A neutron [proton] EDM can*not* be described only by single quark dynamics: the wave functions of quarks combine to produce them.
- Many examples of ND produce non standard one-loop diagrams or at least two-loops ones.

13.1.1 *Direct contributions*

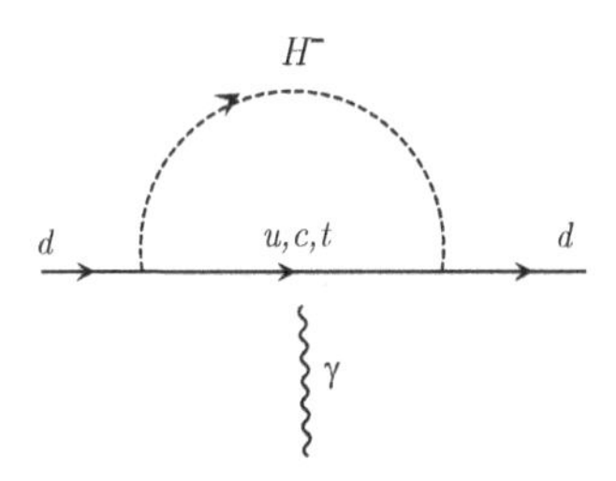

Fig. 13.1 One loop contribution to EDM from single d quarks with Higgs exchange.

A non-minimal Higgs sector is a fertile ground for producing an EDM. One looks at diagrams from ND and thinks about their possible impact. Direct contributions are given by diagrams where the exchange of a charged Higgs generates an EDM for

[3] $SU(6)$ wave functions were used to classify groups of baryons (and mesons). They used global $SU(3)_{\text{flav}} \times SU(2)_{\text{spin}}$. One associated a non-relativistic wave function to the nucleon which includes three constituent quarks and allows for the two spin states of each. It cannot go for accuracy, but that is not the goal here.

individual quarks, as in **Fig. 13.1**. For example, one predicts $d_N \simeq \frac{1}{3}(4d_d - d_u) \simeq \frac{4}{3}d_d$ in models with natural flavor conservation [441] where $d_u \ll d_n$. A neutral spin-0 meson has been discovered as predicted by the SM, while neither a charged Higgs boson nor any other neutral one has been found yet.

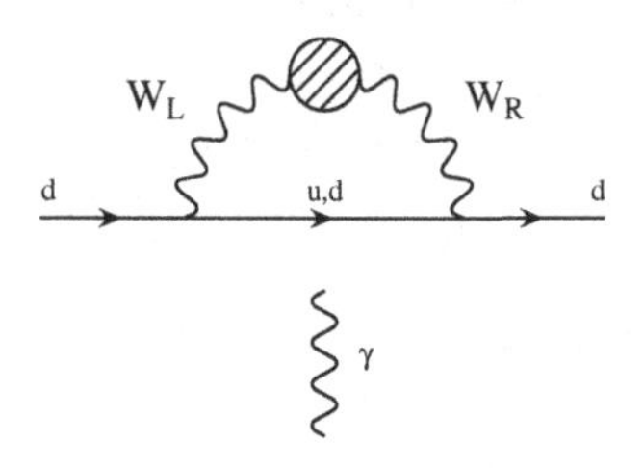

Fig. 13.2 Left-right exchanges for d quarks.

Likewise for left-right symmetry models one uses $d_N \simeq \frac{1}{3}(4d_d - d_u) \simeq \frac{4}{3}d_d$. The leading contribution to the EDMs is coming from one-loop diagrams, as shown in **Fig. 13.2**. Thus in left-right models we can expect to obtain relatively large effects since in the SM the EDM is produced only at the three-loop level. Even in case left-right currents contribute little to the observed **CP** violation in kaons, they could still generate values as high as $d_N \sim \mathcal{O}(10^{-28})$ e cm. Are those values within 'striking distance' of on-going or scheduled experiments? Not really, but they are not so far. The most recent result is that obtained in 2020 at the Paul Scherrer Institute in Switzertland [442], obtained with a refined implementation of the Ramsey technique of separate oscillating fields [23] and a large flux of ultra-cold neutrons. The best experimental limit obtained so far is:

$$d_N < 1.8 \cdot 10^{-26} \quad \text{e cm} \qquad (90\%\,\text{C.L.})\,.$$

This limit is two orders of magnitude above the theoretical expected value quoted above. While certainly challenging, covering this two orders of magnitude gap might not be impossible. As we will see below, other mechanisms, including CKM effects in the SM model, predict values which are not easy to reach in the near future.

13.1.2 *(Color-)electric dipole moments*

In several cases, as in the SM, cancellations among different diagrams, mainly due to the unitarity of the CKM matrix, suppress low order loop contributions of a single free quark to the EDM of the neutron. This problem can be circumvented taking into account the multiparton content of the neutron in calculating the EDM, that is, instead of dealing with a single free quark, considering more quarks simultaneously. Once it was realized that di-quark effect could provide the leading contributions, many such mechanisms were found. One possibility is given by the so-called quark colour dipole moment operator shown in **Fig. 13.3**: it represents an electroweak two-loop effect and leads to $d_N|_{\rm ND} \sim \mathcal{O}(10^{-26})$ e cm [443–445]. It is close to

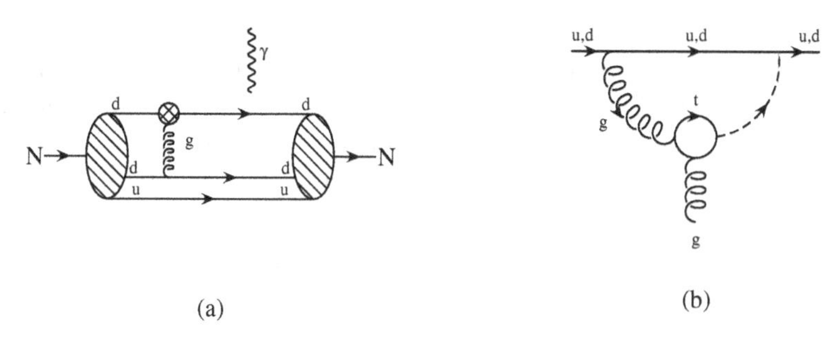

Fig. 13.3 In (a), the neutron EDM is generated by a quark colour magnetic moment; in (b) the diagram giving rise to a quark colour dipole moment – the blob in (a) – is shown.

the present upper bound reported in Ref. [442]. Two-loop amplitudes can produce sizable contributions to neutrons with gluon exchanges, as discussed in Ref. [445].

As far as the electric dipole moment of electrons d_e is concerned, one obtains a two-loop contribution by replacing the gluons in the colour-electric dipole operator by electroweak bosons and attaching it to a lepton. It has given $d_e \sim \mathcal{O}(10^{-27})$ e cm [435,446–448,450]. The ACME collaboration has given limit on electron EDM [449]: $|d_e| < 8.7 \cdot 10^{-29}$ e cm.

13.1.2.1 *'Pure' gluon dynamics*

Let us consider the effective Lagrangian $\mathcal{L}_{\text{glue}}^{\mathbf{CP}\text{odd}}$ in Eq. (13.3), after setting $\bar{\theta} = 0$. Diagrams which give rise to these operators in models based on a non-minimal Higgs sector are shown in **Fig. 13.4**. The study of this Lagrangian is not of academic interest only [435, 450]. Non-minimal Higgs models can produce sizable neutron EDM while contributing very little to $K_L \to \pi\pi$ due to the emergence of the $G^2\tilde{G}$ operator. Besides, the operator $G^2\tilde{G}$ is induced in different classes of models for **CP** violation.

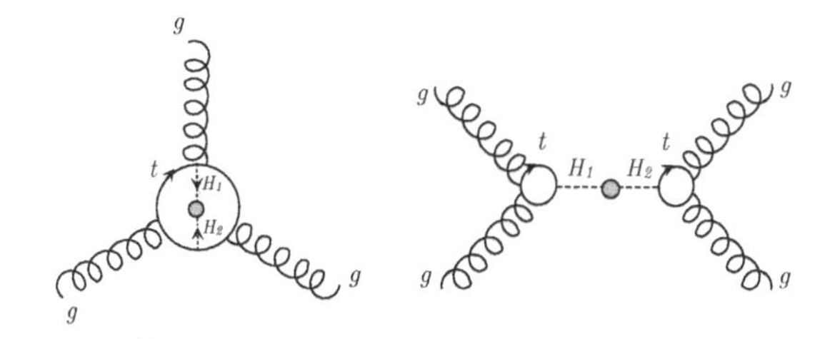

Fig. 13.4 Diagrams giving rise to $G^2 \cdot \tilde{G}$ and $G^3 \cdot \tilde{G}$ operators.

The leading term of this Lagrangian is $c_{(6)}\eta_{\text{QCD}}\tilde{O}_{(6)}$, where $\tilde{O}_{(6)} = \frac{\alpha_S}{4\pi} f_{abc} G^a_{\mu\nu} G^{b;\nu\rho} \tilde{G}^c_{\rho\mu}$ is the dimension "6" operator. Its coefficient $c_{(6)}$ carries mass dimension -2 and can be calculated within a specific dynamical model by integrating out the heavy fields in the loop diagrams.

The parameter $\eta_{\rm QCD}$ denotes the radiative QCD corrections that emerge when the operator $\tilde{O}_{(6)}$ is evaluated at a lower scale than the one provided by the internal heavy fields. The QCD radiative corrections suppress rather than enhance the operator $G^2\tilde{G}$. Can we estimate the size of its matrix element, that depends crucially on non-perturbative QCD? It has been suggested in Ref. [435] to assume

$$\frac{\langle N|(g_S^3/16\pi^2)f_{abc}G^a_{\mu\nu}G^{b;\nu\rho}\tilde{G}^c_{\rho\mu}|N\rangle}{\langle 0|(g_S^3/16\pi^2)f_{abc}G^a_{\mu\nu}G^{b;\nu\rho}G^c_{\nu\mu}|0\rangle} \sim \frac{\langle N|(\alpha_S/4\pi^2)G^a_{\mu\nu}\tilde{G}^{a;\mu\nu}|N\rangle}{\langle 0|(\alpha_S/4\pi^2)G^a_{\mu\nu}G^{a;\mu\nu}|0\rangle} \tag{13.6}$$

QCD sum rules based on low energy collisions normalized at ~ 1 (GeV)^2 yield:

$$\langle 0|(\alpha_S/4\pi^2)G^a_{\mu\nu}G^{a;\mu\nu}|0\rangle \sim 3\cdot 10^{-3}(\mathrm{GeV})^4 \tag{13.7}$$

$$\langle 0|(g_S^3/16\pi^2)f_{abc}G^a_{\mu\nu}G^{b;\mu\rho}G^c_{\nu\rho}|0\rangle \sim 4\cdot 10^{-4}(\mathrm{GeV})^6\ ; \tag{13.8}$$

$g_S^3G^2\cdot\tilde{G}$ and $g_S^3G^3$ operators possess non-trivial anomalous dimensions in contrast to $\alpha_S G^2$ and $\alpha_S G\cdot\tilde{G}$. One can compare SM predictions [445] with most recent experimental data [442]

$$d_N|_{\rm CKM} < 10^{-31}\ \mathrm{e\,cm} \tag{13.9}$$

$$d_N|_{\rm experim} < 1.8\cdot 10^{-26}\ \mathrm{e\,cm} \qquad (90\%\,\mathrm{C.L.})\ . \tag{13.10}$$

On the other hand ND could produce values of d_N in the range of $\mathcal{O}(10^{-28})$ e cm, while it might give impact on the measured ϵ'/ϵ_K.

At first it appears that contributions due to quark interaction with *neutral* Higgs bosons can be ignored since – in the absence of FCNC – they are proportional to the cube of the u and d (current) quark masses. Yet they induces a $G^3\cdot\tilde{G}$ term that could conceivably generate d_N as large as $\mathcal{O}(10^{-25})$ e cm! The reason for this relatively large effect is very interesting [451]: this contribution is related – via the anomaly – to the nucleon mass rather than the current quarks m_q; thus it will not vanish in the chiral symmetry limit $m_q\to 0$.

13.1.3 *Deuteron*

As the neutron, the deuteron is attractive to search for hadronic EDMs because of its relatively simple structure–it is the simplest nucleus. The deuteron consists of a weakly bound proton and neutron in a predominantly $^3S_1$ state, with a small admixture of the D-state. In the world of nuclear dynamics it is the simplest system in which the **P**- and **T**-odd nucleon-nucleon interaction contributes to an EDM; furthermore deuteron properties are well understood, so we expect reliable calculations.

The general form of the interaction, based on discrete symmetry considerations and discussed in Ref. [452], contains ten **P**- and **T**-odd meson-nucleon coupling constants for the lightest pseudo-scalar and vector mesons π, η, ρ and ω. This interaction induces a P-wave admixture to the deuteron wave function, which leads to an EDM. Since the proton and neutron that make up the deuteron could have EDMs themselves, the situation is 'complex', although the deuteron is weakly bound. It has been suggested in Ref. [452] to disentangle the one- and two-body contributions to the EDM:

$$d_{\mathcal{D}} \simeq d_{\mathcal{D}}^{(1)} + d_{\mathcal{D}}^{(2)} \tag{13.11}$$

The two-body component is predominantly due to the polarization effect, while the one body contribution is simply the sum of the proton and neutron EDMs. These contributions are described by *hadronic* loop diagrams in the world of mesonic and baryonic degrees of freedom. Reproducing only the pion dependence, one finds [452]:

$$d_P = -0.05\bar{g}_\pi^{(0)} + 0.03\bar{g}_\pi^{(1)} + 0.14g_\pi^{(2)} + ... \tag{13.12}$$

$$d_N = +0.14\bar{g}_\pi^{(0)} - 0.14\bar{g}_\pi^{(2)} + ... \tag{13.13}$$

$$d_{\mathcal{D}} = +0.09\bar{g}_\pi^{(0)} + 0.23\bar{g}_\pi^{(1)} + ... \tag{13.14}$$

One can estimate [453]

$$d_D < -10^{-16}\,\bar{\theta}\,e\,\mathrm{cm}\,, \qquad \frac{d_D}{d_N} = -\frac{1}{3}\,, \tag{13.15}$$

since $\bar{\theta}$ contributes differently to the neutron and the deuteron.

13.2 Lessons learnt so far and future perspectives

We have learnt that probing EDMs of single baryons or hadrons gives us a fertile hunting range for the existence of ND and its features. However, we need very experienced hunters. Here we have just 'painted' the landscape; a committed reader can find experimental details in Chapter 9 of Ref. [453].

Even if no non-zero values of EDMs have been found yet, it is a 'hot' item.

- New *experimental* tools have been applied, others more tested or even novel ones suggested. Likewise for analyses in different regions of fundamental dynamics: atoms, leptons, baryons, light nuclei (like deuteron) and heavy deformed nuclei, molecules and solid-state samples. Such analyses requite small (by modern scale) collaborations; on the other hand their members have to commit for a long-term effort.
- In the world of quarks and gluons one can distinguish three classes of contributions to EDMs. In pictures:
  (a) EDMs with single quarks, see **Fig. 13.1** and **Fig. 13.2**;
  (b) color-electric-dipole moments, see **Fig. 13.3**,
  (c) gluon dynamics, see **Fig. 13.4**.
  Many more diagrams giving contributions to EDMs arise from SUSY, as we will discuss in **Chapter 14**.
- It is obvious there are different 'cultures' between HEP (based on quarks and gluons with strong forces) and Hadrodynamics/nuclear physics based on

hadrons.[4] In the world of hadrons one can 'paint' diagrams with 'constituent' quarks (hardly including gluons). However, when we discuss EDMs of baryons, it only matters when the first *non-zero* value will be found.

Factors of two or three are not a problem before we find non-zero values for EDMs. To find non-zero values 'close' to the present limits or even differing by factors of 10–100 would be wonderful. Miracles can happen – rarely.

[4]On can see it right away by checking what is conventionally indicated as "Standard Model" in the two areas. The same of course applies to other different 'cultures', as cosmology, astrophysics, etc.

Chapter 14

Super-symmetry

> Super-symmetry: a theory with a great future – in its past?

The SM is based on the non-abelian gauge dynamics $SU(3)_C \times SU(2)_L \times U(1)$. As a first step, the symmetry provides gauge bosons, quarks and leptons that are massless; then one adds Higgs dynamics that couples with $SU(2)_L \times U(1)$ gauge bosons, but also with quarks and charged leptons to produce their masses. Higgs dynamics can do the 'job', never mind whether it is a piece of art or not.[1] No rationale more 'compelling' than simplicity has emerged for limiting to minimal Higgs dynamics or close to it; thus one has to look beyond that, although no experimental hint, so far, suggests that this is indeed necessary.

In a quantum field theory there is always some sensitivity to very high scales of physics due to ultraviolet (UV) renormalization in one-loop corrections and more, namely the UV cut-off; usually, it is a very mild one expressed by a logarithmic dependence. However, the situation is *very different* for *scalar* fields, see **Sect. 4.7**, since

$$\Delta M_\phi^2 \propto \Lambda_{\rm UV}^2 + \dots .$$

Renormalization introduces a *quadratic* infinity in the scalar mass, which has to be removed through the renormalization process by adding counter-terms to the Lagrangian. Yet physicists tend to view such a situation as very contrived because it requires extreme *fine tuning.* In their view $\Lambda_{\rm UV}^2$ is *not* merely a mathematical 'place holder' for an infinity. It parametrizes the sensitivity of observables – in this case the scalar mass.

The 'landscape' has changed when 'the' Higgs state has been found in 2012 at CERN with $M_{H^0} \simeq 125$ GeV [4,8]; it is consistent with what the SM can produce, if one does not 'care' about huge fine tuning. A particularly disturbing feature is that this mass is of the order of $v \simeq 246$ GeV, namely as the breaking part of the electroweak (EW) gauge symmetry. It must be highly cognizant of whether new

[1] It may not be a piece of art by itself, but it could be as a part of a larger 'landscape'.

layers of dynamics enter at much higher scales, to achieve the needed cancellation. A 'natural' scale would be close to Planck scale $\sim \mathcal{O}(10^{19})$ GeV or 'at least' to GUT scales $\sim \mathcal{O}(10^{15})$ GeV. Often this is referred to as the 'gauge hierarchy problem' of the SM: the Higgs sector has been perceived as the product of effective, yet unsatisfying theoretical 'engineering'.

Three types of scenarios had been suggested to provide more appealing frameworks (for some theorists):

- Higgs fields are 'composites' rather than 'elementary' ones, and represent an effective description of underlying dynamics like, e.g., in 'technicolor' models.
- In "Little Higgs Models" the Higgs is a pseudo-Goldstone boson resulting from a spontaneously broken approximate symmetries, which would explain its natural lightness. Little Higgs Models do not solve the gauge hierarchy problem – they 'delay the day of reckoning' to a higher scale [454]. They provide scenarios where ND quanta can be produced at LHC collisions with*out* creating conflicts with present electroweak constraints; at the same time they introduce fewer additional parameters than 'Extra Dimension' scenarios, see **Sect. 15.3**. In order to avoid stringent electroweak constraints, one subclass of those models introduces a discrete symmetry called "T-parity" [455], under which the new particles are odd and can therefore contribute only at the "loop" level. As a consequence a set of six T-odd "mirror" quarks needs to be introduced that are organized into three families. While constraints from flavor dynamics are not part of their motivation, this latter class of models is *not* of the Minimal Flavor Violation (MFV) variety. They could connect findings in *pp* collisions at high energies with flavor dynamics in the decays of strange, charm and beauty hadrons. They have the potential to be of practical use due to the relatively paucity of their new parameters: ten in the quark flavor sector with three **CP** violating phases. Little Higgs models have been strongly constrained by LHC run-2 data.

There is a third very elegant theoretical scheme that provides a natural habitat for scalar fields – namely *Super*symmetry (SUSY). Since we consider it so attractive, we will describe and analyze in more details despite the fact that no evidence of it has been seen so far in the data.

We view SUSY as a 'landscape', not just a class of models, which we regard as a positive feature, rather than a negative one.

- From the beginning of the project of *pp* collisions at CERN a main motivation was to find SUSY states *directly*. There were good arguments (and still are) that its scale is around 1 TeV. Yet analyses of the huge amount of data at the LHC *pp* collisions provided by ATLAS and CMS collaborations have *not* seen any impact of ND and in particular of SUSY. Another Higgs state might be the first step to find SUSY. It is still too early to give up *direct* searches. For run-3 LHC will be upgraded to operate at a center-of-mass energy of 14 TeV;

future dataset from High-Luminosity LHC (HL-LHC) after 10 years of operation is expected to represents an increase of a factor of forty. It will provide unprecedented opportunities to search for SUSY fields with masses of $\mathcal{O}(1\,\text{TeV})$.[2]
- There is another way to find SUSY: its *indirect* impact on weak transitions due to loop diagrams, in particular about **CP** asymmetries and EDMs.

There is one statement about SUSY that is beyond any doubt: if it exists, it is broken by large mass gap between 'ordinary' fields and their SUSY partners on the scale of hundreds of GeV. Our community has probed SUSY at run-1 of LHC at 7 and 8 TeV, run-2 at 13 TeV and will probe it later at run-3 and beyond (we hope).[3] That is a blow to the professional pride of many theorists; yet we should accept this lesson in humility – and keep working on SUSY rather than abandon it.

SUSY was 'found' from two different 'roads':

- Fron the papers of Neveu, Schwarz and Ramond in 1971 a new anti-commuting gauge symmetry arose [456].[4]
- In 1971 Golfand and Likhtman were looking for a generalization of the usual space-time algebra and found the super-Poincare algebra [457]. In 1972 Volkov and Akulov found a nonlinear supersymmetric theory [458]. In 1974 Wess and Zumino wrote down the first four-dimensional quantum field theory "action" [459]. Colleagues have continued to work on that wonderful idea in different eras and scenarios right away [460] and later, including Dark Matter analyses [461–463].

SUSY is discussed in details by M. Kaku in his 1993 book "Quantum Field Theory" [464], M. Shifman in his 1999 two books of "ITEP Lectures on Particle Physics and Field Theory" [465], in PDG2019 Review of Particle Physics about Supersymmetry [466, 467] and in Section 19 of Ref. [12], which we follow closely.

---

[2]Finding SUSY fields around a few TeV is not its strongest point by far.

[3]Severe constraints have been posed on SUSY classes, especially on the the minimal or close to minimal versions. Still, SUSY appears an exciting challenge as before, since for theorists there is *no* reason that our Universe is a building based on such SUSY classes – obviously our experimental colleagues could have different views. From the lack of any experimental evidence, it seems quite likely that, if SUSY is a real feature of nature, its realisation is far from being 'minimal'. In any case, for simplicity, we give in the following a short introduction to SUSY mostly based on the Minimal Supersymmetric Standard Model (MSSM).

[4]Just one example that pioneers of fundamental dynamics are not good with words: the title of the published paper of Neveu and Schwarz – leading to superstring – is "Factorizable dual model of pions". Young readers may not understand the meaning of most of these words. We just recall here that the birth of string theory originated from a first attempt to understand and describe hadrons and particularly their confinement.

14.1 The virtues of SUSY

SUSY is the ultimate symmetry for a local QFT: in its local form it forms a bridge to gravity, and it alleviates several vexing theoretical problems.

- The Coleman-Mandula theorem [468] proved that all possible symmetry groups $\mathcal{S}$ in local quantum field theories can be expressed as the direct product of the Poincare group $\mathcal{P}$ and an internal symmetry group G:

$$\mathcal{S} = \mathcal{P} \otimes G \,. \tag{14.1}$$

  No symmetry connects states of different spins; in the language of mathematics: each irreducible representation of $\mathcal{S}$ contains only states of the *same* spin.
  There was a loophole in the proof of this theorem: it admitted only *bosonic* operators as generators of a symmetry.[5] It was realized later that *fermionic* operators can also generate a symmetry group and that they relate fermions and bosons to each other. This symmetry could 'outwit' Coleman-Mandula theorem.[6] Thus it represents the "ultimate" symmetry of a local quantum field theory and was aptly called *Super*symmetry.
- Once SUSY is implemented as a *local* gauge symmetry, coordinate covariance becomes local as well – i.e., *general relativity* has to emerge. That is why local SUSY is referred to as *supergravity.*
- SUSY also forms an integral part of superstring theories. It also gives a successful framework for gauge coupling unification and predicts a potential candidate for Dark Matter.
- SUSY *per se* does not solve the problem of extreme fine tuning in Higgs forces, but it makes it more tractable: when a Higgs potential is chosen such that it yields the required large ratio to tree-level, one can invoke SUSY to stabilize this ratio against quantum corrections. This is referred to as one of the *non-renormalization theorems* of SUSY: light scalars become *natural* by riding piggy-back on the shoulders of light fermions. It can be achieved by cancelling Λ^2_{UV} terms in the scalar mass quantum corrections against those due to fermion ones. The connection between fermions and bosons is both unique and essential.
- With the *quadratic* Higgs mass renormalization removed by SUSY, electroweak symmetry breaking can be induced *radiatively*, if the top quark is sufficiently massive that its Yukawa coupling g^y_t dominates the renormalization of the Higgs mass: $\partial M^2_\phi/\partial \log\mu = 3(g^y_t)^2 m^2_t/8\pi^2 + ...$; this would happen around $m_t > 160$ GeV. This feature had been noted in 1982 [164]; such a mass was perceived by most of our community as extravagantly high. Well – top quarks were discovered in 1995 from the data of $\bar{p}p$ collisions at Fermilab with $m_t \sim 173$ GeV. It was a true prediction.

[5]That is a general lesson on the impact of theorems: they are based on assumptions; therefore we should *not* stop thinking. History shows that this lesson, as obvious as it is, is often forgotten.

[6]There might be a parallel between symmetry and (classical) music: 'L'Infedelta Delusa' (Joseph Haydn) = 'Deceit Outwitted'.

In SUSY models each field has to possess a super-partner; i.e. a field whose spin differs by half a unit. For manifest SUSY the super-partner must be mass degenerate.[7] Since not a single super-partner has been observed, we know SUSY breaking has to be implemented one way or another.

The simplest class of SUSY is called $N = 1$ and connects fields of bosons and fermions with their spins of n and $n + \frac{1}{2}$ [469]. One can also discuss SUSY with $N = 2$, 3 and 4, describing particles of spin $0, \frac{1}{2}, 1, \frac{3}{2}, 2$ [470].

14.2 Introduction to SUSY

We can be brief about the general structure of SUSY– details can be found in excellent reviews such as, e.g., Refs. [465–467, 471–473]. Going beyond generalities we sketch the minimal supersymmetric standard model (MSSM)[8] and use it as a reference point. SUSY has a long history (on the scale of quantum field theories): the "A Supersymmetry Primer" [472] first appeared in 1997, while its Version 7 is from the beginning of 2016. While present data have not shown any impact of SUSY, there are excellent reasons *not* to give up, as we have seen in **Sect. 14.1**.[9]

In SUSY to remove lepton and baryon number violating interactions one can introduce a new multiplicative quantum number – the R parity – assigning a value $+1$ to ordinary fields (quarks, leptons, gauge bosons, Higgs, graviton) and value -1 to their superpartners (squarks, sleptons, gauginos, higgsinos, gravitinos).

Superspace and superfields, although not strictly necessary, are elegant tools for analyzing the formal structure of SUSY in a succinct way. The superfield formalism allows us to treat all superpartners as a single field (or superfield). Scalars and fermions related by supersymmetry correspond to different components of a single superfield like spin up and spin down states are different components of a single fermion. The simplest class of SUSY combines spins $J = 0$ and $J = 1/2$ as *chiral* or *matter* superfields and $J = 1/2$ and $J = 1$ as *gauge* or *vector* superfields (and for $J = 2$ graviton with $J = 3/2$ gravitino).

The superfield content of MSSM is given in **Table 14.1** with*out* gravitino and graviton. Each field of the SM is extended into a superfield. These superfields are denoted by their R $= +1$ component. At least two *distinct* Higgs superfields H_1 and H_2 are needed: H_1 with isospin $\frac{1}{2}$ and hypercharge $-\frac{1}{2}$, while H_2 with isospin $\frac{1}{2}$ and hypercharge $+\frac{1}{2}$; otherwise chiral anomalies arise due to higgsino loops. Thus one gets not only 'the' neutral "Higgs" as established, but four more at least: two more neutral ones, and two charged ones.

---

[7] We do not discuss *non*linear realizations of SUSY where the super-partner can be composites.

[8] It is the 'minimal' extension of the SM field content which makes it supersymmetric.

[9] The composer Mozart had vowed to wed his future wife and write a solemn Mass if she could recover from a serious illness; she did. He had not finished this 'project' that he passed away, thus the *Great Mass in C minor* was left in a fragmentary state. Some experts of SUSY might see an analogy there: SUSY is an amazing project, but we still have no real idea where to go.

Table 13.1: Superfields of MSSM with $i = 1, 2, 3$ as family index.

superfield	color	isospin	hypercharge	lepton number	baryon number
Q_i	3	$\frac{1}{2}$	$+\frac{1}{6}$	0	$\frac{1}{3}$
U_i^c	3	0	$-\frac{2}{3}$	0	$\frac{1}{3}$
D_i^c	3	0	$+\frac{1}{3}$	0	$\frac{1}{3}$
L_i	0	$\frac{1}{2}$	$-\frac{1}{2}$	+1	0
E_i^c	0	0	+1	−1	0
H_1	0	$\frac{1}{2}$	$-\frac{1}{2}$	0	0
H_2	0	$\frac{1}{2}$	$+\frac{1}{2}$	0	0

Non-gauge interactions are introduced through a superpotential G coupling up to three super-fields together:

$$G = \mu H_1 H_2 + y_{ij}^u Q_i H_2 U_j^c + y_{ij}^d Q_i H_1 D_j^c + y_{ij}^l L_i H_1 E_j^c \ . \tag{14.2}$$

The dimension*less* numbers $y_{ij}^{u,d,l}$ contain the Yukawa couplings of the ordinary fermion fields (among other things). With superfields and mixing parameter μ carrying dimension 'one' (in mass units), G gets dimension 'three'.

Constructing a SUSY model requires five steps: (i) adopt a gauge group; (ii) choose superfields with the appropriate gauge quantum numbers; (iii) formulate the gauge interactions; (iv) construct the superpotential with them; (v) implement SUSY breaking. The last step provides the greatest challenge; it is also the one that obscures the intrinsic elegance of SUSY.

One might invoke SUSY to shield Higgs masses against getting *quadratically* renormalized. It has been suggested that the extreme fine tuning of Higgs dynamics could be overcome if the spectrum of novel SUSY states starts around the TeV scale. A *spontaneous* breaking of SUSY could be achieved through a conventional Higgs mechanism – yet that would lead to a phenomenological conflict in the relationship between fermions and sfermions masses. The sum rule given by the super-trace

$$\mathrm{Str}\, M^2 = \sum_J (-1)^{2J} M_J^2 = 0 \ ; \tag{14.3}$$

trivially holds in manifest SUSY, but it is *not* modified by this spontaneous symmetry breaking. As the sum of all fermion masses squared has to equal that of all boson masses squared, it predicts that some squarks and sleptons must be light: some SUSY fields should have already been found.

To avoid this phenomenological conflict, one can adopt a pragmatic approach and add terms to the Lagrangian that break SUSY explicitly, yet 'softly':

$$\mathcal{L}_{\text{effect}} = \mathcal{L}_{\text{SUSY}}(d \leq 4) + \mathcal{L}_{\text{soft}}(d \leq 3) \ . \tag{14.4}$$

where all dimension-4 operators obey SUSY. It can be seen an 'engineering' device to generate higher masses for all SUSY partners.[10] A more ambitious approach

[10]The words 'softly broken' is a polite way to say we do not understand what is happening here.

requests that SUSY is breaking 'dynamically'. We have actually in mind that ND beyond MSSM dynamically generates such soft SUSY breaking. One might envision a hidden sector characterized by very high scales where SUSY breaking originates, which is transmitted to the MSSM through flavor blind interactions. Such an ansatz effectively disassociates the scale at which the intrinsic SUSY breaking occurs – Λ_{SSB} – from the one that gives rise to different flavors – $\Lambda_{\rm flav}$ – with $\Lambda_{SSB} \ll \Lambda_{\rm flav}$. SUSY breaking masses emerge as flavor singlets; squark masses are basically degenerate up to small corrections of the order of the corresponding quark masses. There can be at least two quite different scenarios:

- SUSY breaking is *gravity* mediated and enter $\mathcal{L}_{\rm soft}(d \leq 3)$ through soft mass terms:
$$M_{\rm soft} \sim \mathcal{O}\left(\frac{\Lambda_{SSB}^2}{M_{\rm Planck}}\right) \; ; \tag{14.5}$$
it suggests $\Lambda_{SSB} \sim \mathcal{O}(10^{11})$ GeV leading to $M_{\rm soft} \sim 1$ TeV.
- It could be *gauge* mediated instead, with the soft mass terms being generated by virtual exchanges of messenger fields from the hidden sector which has $SU(3)_C \times SU(2)_L \times U(1)$ couplings:
$$M_{\rm soft} \sim \mathcal{O}\left(\frac{\alpha_{\rm gauge}}{4\pi}\frac{\Lambda_{SSB}^2}{M_{\rm mess}}\right) \; ; \tag{14.6}$$
$\alpha_{\rm gauge}$ denotes the gauge coupling of the messengers fields and $M_{\rm mess}$ their typical scale. It turns out that $M_{\rm mess} \sim \Lambda_{SSB} \sim \mathcal{O}(10^5)$ GeV or so is quite a 'natural' value; i.e., SUSY breaking occurs at much lower scales than in the gravity-mediated case.

Of course, these two scenarios are not in a competition for accuracy. Summarizing, one starts from the 'hidden' sector of dynamical fields that do *not* carry SM quantum numbers. SUSY can be broken there by, say, fields acquiring a vacuum expectation value. This breaking is then 'mediated' through a sector of 'messenger' fields that carry non-trivial SM quantum numbers to produce a supersymmetric SM with the required breaking. This 'mediation' has to occur in a 'flavor blind' way; otherwise it would be hard to understand why SUSY has not manifested itself in heavy flavor transitions. There are several ways known for achieving this goal, and we have mentioned some characteristics belonging to the classes of 'anomaly mediation' and 'gauge mediation' [474–476]. Yet they seems to require a great deal of theoretical finesse, and they are not particularly 'robust'.[11]

Let us illustrate by a simple example how to produce a sizable non-zero value on the right side of the Eq. (14.3), that would solve the phenomenological problem expressed by this equation. When SUSY is implemented as a *local* symmetry, it requires a spin-3/2 fermion – called the gravitino – as the super-partner for the

[11]However, one should remember that often unusual new ideas are not 'robust' from the beginning.

spin-2 graviton. One manifestation of SUSY breaking is the emergence of a finite gravitino mass leading to

$$\text{Str}\, M^2 = \sum_J (-1)^{2J} M_J^2 = m_{3/2}^2 \,, \tag{14.7}$$

since the graviton remains massless. It is expected $m_{3/2} > 100$ GeV.

We illustrate these general remarks by one example of soft SUSY breaking (see Eq. (14.4)) generated at some high scales in a flavor blind scenario. The Lagrangian is given by:

$$\begin{aligned}\mathcal{L}_{\text{soft}} = m_{3/2}^2 \sum_i |A_i|^2 &+ A\, m_{3/2}[y_{ij}^u \tilde{Q}_i h_2^* \tilde{U}_j^c + y_{ij}^d \tilde{Q}_i h_1 \tilde{D}_j^c \\ &+ y_{ij}^l \tilde{L}_i h_1 \tilde{E}_j^c + \text{h.c.}] + B\, m_{3/2}\, \mu\, h_1 h_2 \\ &+ \frac{1}{2} M(\lambda_1\lambda_1 + \lambda_2\lambda_2 + \lambda_3\lambda_3) \,;\end{aligned} \tag{14.8}$$

A_i denote scalar fields in general, $\tilde{Q}$, $\tilde{U}$, $\tilde{D}$, $\tilde{L}$ and $\tilde{E}$ scalar parts of superfields of quarks and leptons, $h_{1,2}$ the scalar ones of the superfields H_1 and H_2 and $\lambda_{1,2,3}$ gaugino fields for the gauge groups $U(1)$, $SU(2)_L$ and $SU(3)_C$, respectively. The low energy effective Lagrangian is then obtained by running the parameters down to the electroweak scale.

In addition to the SM parameters we have five new classes of dimensional parameters in the MSSM: μ describing the mixing between the two Higgs superfields plus four more representing SUSY breaking: gravitino and gaugino masses ($m_{3/2}$ and M), a higgsino mixing term ($B\, m_{3/2}$) and quark-squark-higgsino coupling ($A\, m_{3/2}$).[12] Besides, the requirement that the EW symmetry breaking occurs at the right scale implies that $4M_W^2/g^2 = v_1^2 + v_2^2$, and we have a new parameter $\tan\beta \equiv v_2/v_1$.

It is crucial to keep probing the fundamental dynamics of the Higgs sector with more and more refined tools: the Higgs sector in SUSY is more complex that in the SM, even in the MSSM.[13] In data from run-1 and run-2 of LHC the ATLAS and CMS collaborations have not found charged Higgs states or other neutral ones.

There is no deep reason why SUSY should enter our world as a MSSM; still one can use it as a reference point. As we have seen, SUSY is able to solve the Higgs fine-tuning problem by imposing a space-time symmetry between bosons and fermions. However, another fine-tuning problem exists. In the MSSM the expression for the mass of the gauge boson Z contains terms which are naturally of the order of the SUSY breaking scale. Thus, in order to obtain the measured value of the Z mass, there must be a cancellation between these terms, if they are much larger than m_Z. If the required cancellation is large, small changes in the SUSY parameters will result in a widely different value of m_Z, in which case the considered spectrum is fine-tuned. Since many years it is claimed that to have a natural (i.e. no fine-tuned) version of the MSSM, one expects SUSY particles with masses that lie around the

[12] The number of parameters can be reduced to four in the simple cases where $B = A - 1$.

[13] The common *bon mot* (or 'jest'): half the SUSY states has already been found. Yet R parity assigns a value $+1$ for ordinary fields in the SM; thus one should find Higgs fields beyond the SM.

TeV scale. However, to date, there has not been any evidence for the existence of SUSY particles at these scales. An example of way out could be if the required cancellation among different terms is by a lucky chance provided by the fundamental theory underlying the MSSM, e.g. string theory. It has also been pointed out [477] that fine-tuning might not be as sizable as expected because of different boundary conditions at the GUT scale. These examples are interesting, but we would not bet on them.

14.2.1 *Squark mass matrices of one class of MSSM*

The superpartners of the left- and right-handed quarks are referred to as left- and right-handed squarks, although the latter (being scalars) possess no chiral structure. Squark masses are greatly affected by soft SUSY breaking. Let us consider the term in $\mathcal{L}_{\text{soft}}$ proportional to the trilinear scalar coupling A which mixes left- and right-handed squarks; see Eq. (14.8), where we can read off the squark mass matrices. For *Down*-type squarks:

$$\tilde{V}_{D-\text{mass}} = (\tilde{D} \quad \tilde{D}^{c*}) \begin{pmatrix} \tilde{M}^2_{DLL} & \tilde{M}^2_{DLR} \\ \tilde{M}^{2\dagger}_{DLR} & \tilde{M}^2_{DRR} \end{pmatrix} \begin{pmatrix} \tilde{D}^* \\ \tilde{D}^c \end{pmatrix} \tag{14.9}$$

$$\tilde{M}^2_{DLL} = \left(m^2_{3/2} + (v_1^2 - v_2^2)\left(\frac{g'^2}{12} - \frac{g^2}{4}\right)\right)\mathbf{1} + \mathcal{M}_D\mathcal{M}_D^\dagger$$

$$\tilde{M}^2_{DRR} = \left(m^2_{3/2} + (v_1^2 - v_2^2)\frac{g'^2}{6}\right)\mathbf{1} + \mathcal{M}_D^\dagger\mathcal{M}_D$$

$$\tilde{M}^2_{DLR} = (A^* m_{3/2} + \mu^* \tan\beta)\,\mathcal{M}_D\ ; \tag{14.10}$$

$\mathcal{M}_D$ represents the mass matrix for *Down*-type quarks. We write the three families of down squarks as six component vectors with $\tilde{D} \equiv (\tilde{d}_L, \tilde{s}_L, \tilde{b}_L)$ and the other three components given by $\tilde{D}^c$, where $\tilde{q}_L^c = \tilde{q}_R^*$. Likewise for their analogue in the *Up*-type squark mass matrix.

In this class of MSSM, the leading contribution to the squark masses, namely the gravitino mass $m_{3/2}$, is flavor independent, originating the super-GIM mechanism. If $\mathcal{M}_D$ is diagonalized by $\mathcal{M}_D^{\text{diag}} = \mathbf{U}_L^D \mathcal{M}_D \mathbf{U}_R^{D\dagger}$, we can make the unitary transformation

$$\begin{pmatrix} U_L^D e^{i\phi_A} & 0 \\ 0 & U_R^D e^{-i\phi_A} \end{pmatrix} \begin{pmatrix} \tilde{M}^2_{DLL} & \tilde{M}^2_{DLR} \\ \tilde{M}^{2\dagger}_{DLR} & \tilde{M}^2_{DRR} \end{pmatrix} \begin{pmatrix} U_L^{D\dagger} e^{-i\phi_A} & 0 \\ 0 & U_R^{D\dagger} e^{i\phi_A} \end{pmatrix} \rightarrow \begin{pmatrix} \tilde{M}^2_{DLL} & \tilde{M}^2_{DLR} \\ \tilde{M}^{2\dagger}_{DLR} & \tilde{M}^2_{DRR} \end{pmatrix} \tag{14.11}$$

where the phase factor

$$\phi_A = \arg(Am_{3/2} + \mu\,\tan\beta) \tag{14.12}$$

is introduced to absorb the phase into the off-diagonal matrix elements. We get

$$\tilde{M}^2_{DLL} = \left(m^2_{3/2} + (v_1^2 - v_2^2) \left(\frac{g'^2}{12} - \frac{g^2}{4} \right) \right) \mathbf{1} + (\mathcal{M}_D^{\text{diag}})^2$$

$$\tilde{M}^2_{DRR} = \left(m^2_{3/2} + (v_1^2 - v_2^2) \frac{g'^2}{6} \right) \mathbf{1} + (\mathcal{M}_D^{\text{diag}})^2$$

$$\tilde{M}^2_{DLR} = \left(|A| m_{3/2} + \mu^* \frac{v_1}{v_2} \right) \mathcal{M}_D^{\text{diag}} \,. \tag{14.13}$$

While this mass matrix mixes left- and right-handed squarks, it is still *diagonal* in the flavor space. Eq. (14.13) holds at a large unification scale. In much lower energy region $\sim M_W$ and below, the couplings acquire radiative corrections [478]; they induce corrections to $\tilde{M}^2_D$ due to Yukawa couplings of *Up*-type quarks.

Although the details depend on the SUSY models, qualitatively new features emerge here due to the radiative corrections. Since in general $\mathcal{M}_D$ and $\mathcal{M}_U$ cannot be diagonalized by the same unitary transformations, $\tilde{M}^2_D$ and $\mathcal{M}_D$ can*not* be diagonalized simultaneously. Expressing the squark-quark-gluino coupling defined for *flavor* eigenstates in terms of *mass* eigenstates could reveal the emergence of *flavor changing neutral currents* (FCNC): the flavor structure is still controlled by the CKM matrix. Usually in MSSM one can start with flavor blind SUSY breaking at high energy scales, and effective dynamics evolves down to lower energies.

14.2.2 *Beyond MSSM*

Above we have discussed situations in 'flavor blind' MSSM. However, 'minimality' is not a clear virtue or even unambiguously defined in SUSY.[14] Scenarios with $\Lambda_{\text{flavor}} \sim \Lambda_{SSB}$ are particular intriguing, since they may bring a connection between SUSY breaking and the physics of flavor generation. The 'communication' between high and low energy domains no longer proceeds in a flavor blind manner even before radiation corrections are included. Soft breaking terms can be expected to exhibit a non-trivial flavor structure; for example, the mass terms for the third family squarks might be quite different than for the first two ones; or the mass matrices for the *Up*- and *Down*-type squarks might follow different pattern.

In general, FCNC can arise that are not suppressed by a super-GIM mechanism; the resulting phenomenology is less conservative and potentially much richer than in MSSM scenarios. This opens the door to a rich and multi-layered phenomenology; yet still there is little to say concretely beyond these generalities.

One can parameterize SUSY contributions *beyond* MSSM to the flavour changing processes by employing the so called Mass Insertion (MI) approximation. The advantage of this approach is that it allows to treat such contributions in a model independent way without resorting to specific assumptions about the SUSY flavor

[14] In fact, to specify completely a model, it is necessary to fix the soft breaking terms: this amounts to more than a hundred parameters at the electroweak scale–luckily enough, most of this enormous parameter space is already ruled out by phenomenological constraints.

structures. One chooses a basis for the fermion and sfermion states where all the couplings of these particles to neutral gauginos are flavour diagonal, leaving all the sources of FC inside the off-diagonal terms of sfermion mass matrix. These terms are denoted by $(\Delta_{ij})_{AB}$, where $A, B = (L, R)$ and $i, j = 1, 2, 3$ indicate chiral and flavour indices respectively. One expands the squark propagators in powers of $(\delta_{ij})_{LL}$, $(\delta_{ij})_{LR}$, $(\delta_{ij})_{RL}$ and $(\delta_{ij})_{RR}$, where $\delta \equiv \Delta^2/\tilde{m}^2$ with $\tilde{m}$ being an appropriate mass scale like the average squark mass. In other words: while in this treatment the quark-squark-gluino vertex conserves flavor, there is a mass insertion in the squark propagator that is *not* flavor diagonal – $\delta_{ij} \neq 0$ for $i \neq j$ – and can also mix left- and right-handed squarks. Within MSSM the parameters δ_{ij} contain an effective super-GIM filter reducing them below the 10^{-2} level or so without fine tuning. In non-minimal SUSY models one only gets limits $|\delta_{ij}| < 1$ as 'painted' in Ref. [479]. On one hand, such representation allows us to express phenomenological bounds in a model-independent way; on the other hand, it is feasible to evaluate the parameters $(\delta_{ij})_{AB}$ in a given model. In the second decade of 2000s the situations has changed: we have much more data and have to go after accuracy! The MI approximation is still popular in the experimental literature, since one can summarize present data. However, the situation has become more 'complex' and we are not sure that now and in the future probing squark propagators is the best way to find the impact of SUSY.

14.2.3 *Present limits on SUSY masses*

Of course, data are the referees of fundamental dynamics, but often it takes time to understand the information they provide. There is a long history on probing SUSY; we can find direct or indirect impacts of SUSY – and in principle they are connected. Since the turn of the millennium the most important development has been the harvest of high quality data from the BaBar and Belle experiments at the B factories (with assistance from the Fermilab CDF and D0 analyses). Now we are in new era with the LHC experiments from *pp* collisions at CERN, BESIII in Beijing and Belle II in Japan. In the run-1 and run-2 of LHC no signs of *direct* impact of SUSY has been found so far. It has provided significant constraints on SUSY models: gluino production has been probed by inclusive searches until about 2.3 TeV, first and second generation squarks in the range of about 1 to 1.9 TeV and third generation ones in the range of about 600 GeV to 1.2 TeV; sleptons at scales around 700 GeV and EW gauginos at scales around 400–1100 GeV. These limits are not conclusive, since they depend on the assumptions made on the underlying SUSY spectrum. Rather than from direct searches for masses of sparticles, model-dependent interpretations of allowed SUSY parameter space could come more easily from global SUSY fits, combining them with indirect constraints from low-energy experiments, flavor physics, high-precision EW results, and astrophysical data.

14.2.4 *Gateways for* CP *asymmetries*

SUSY models in general introduce a host of possible sources of **CP** violation through a superpotential[15] or soft breaking terms or both. They generate phases that at low energies can surface on both charged and neutral currents couplings and in Higgs interactions. Their pattern depends on the specifics of the dynamics driving SUSY breaking. Even simple SUSY models have many potential sources of **CP** violation. For example, one can look at the MSSM scenario described by Eq. (14.8), and make the following comments: (i) Hermiticity of the potential requires $m^2_{3/2}$ to be real; we can do that by absorbing its phase into the definition of A and B. (ii) The gaugino mass term M can be chosen real by adjusting the phases of the gaugino fields λ_i. (iii) The squark-quark-gluino coupling will be affected by such a choice of the gaugino phase. (iv) The phase associated with a linear combination of A and μ can be detected by experiments.

Usually in MSSM one can start with flavor blind SUSY breaking generated at high energy scales, and the effective dynamics evolves down to low energies in a way that

- FCNC could emerge even in the *strong* sector described by gluino coupling;
- FCNC arise *radiatively* and thus are reduced considerably in strenght;
- in general they do *not* conserve **CP** and **T** symmetry;
- **CP** violation occurs also in flavor *diagonal* transitions.

With the increased layers in the dynamical structure there are many additional gateways through which the impact of (broken) SUSY can be seen indirectly in **CP** asymmetries and on rare transitions.

14.3 SUSY impact on flavor dynamics

At the moment the wealth of experimental information on rare decays of strange and beauty mesons seems well described by the CKM dynamics of the SM – with some discrepancies ('anomalies').[16] Hence "the" question is where we can find the impact of SUSY and of what entity. The strong constraints set by the first interpretation of the LHC data have a price: they switch the attention towards simplified models with more freedom in parameter spaces – and obvious limitations.

14.3.1 *SUSY contributions to* $\Delta S \neq 0$ *transitions*

In the SM roughly 70% of the measured $K_L - K_S$ mass difference is described by the real parts of the box diagrams, where the contributions from charm exchanges are by

[15]The irreducible phase contained in the Yukawa couplings of quarks that is present already in the SM is embedded there.

[16]Not everybody is convinced that what we see are really 'anomalies', that is anomalous behaviours with respect to the SM; more data and more analyses are crucial.

far dominant [480,481]; the remaining percentage is attributed to long distance contributions. The control of the theoretical uncertainty due to the non-perturbative estimate is essential here, also in view of the accurate experimental measurement (PDG2020) $\Delta m_K = m_{K_L} - m_{K_S} = 3.483(6) \cdot 10^{-12}$ MeV. The theoretical uncertainty is much larger than the experimental one, being around 30%. The calculation of the kaon mass difference is one component of LQCD programs based on the evaluation of matrix element of bilocal operators. The actual dominant systematic error is due to its discretization effects [482–485]. The value of Δm_K is truly tiny on the scale of Λ_{QCD}. This FCNC process could be therefore an excellent one in which to search for the effects of ND.

CP violation is described mostly by *imaginary* parts of box diagrams through the parameter ϵ_K, whose uncertainties are smaller. In principle, an impact of SUSY could 'hide' there; however, we would not bet on that.[17]

An example of a new diagram contributing to the neutral kaon mass matrix is the leftmost of **Fig. 14.1**; a complete set can be found in Ref. [486]. The matrix elements of the full set of $\Delta S = 2$ operators beyond the SM have been computed in LQCD [487–489]; their results can be used to set additional constraints on several SUSY models.

The situation is quite different for $\Delta S = 1$ processes, in particular the ones leading to *direct* **CP** asymmetry. From the experimental analysis of $K_L \to \pi\pi$ one gets $\text{Re}(\epsilon'/\epsilon_K)|_{\text{PDG2020}} = (1.66 \pm 0.23) \cdot 10^{-3}$. On the theoretical side, it has been suggested [117] that there is a decent chance to find ND there, and SUSY is an obvious candidate for that, see **Fig. 14.1**. However, there is still an open debate on the values predicted by the SM, as discussed in **Sect. 7.2,**

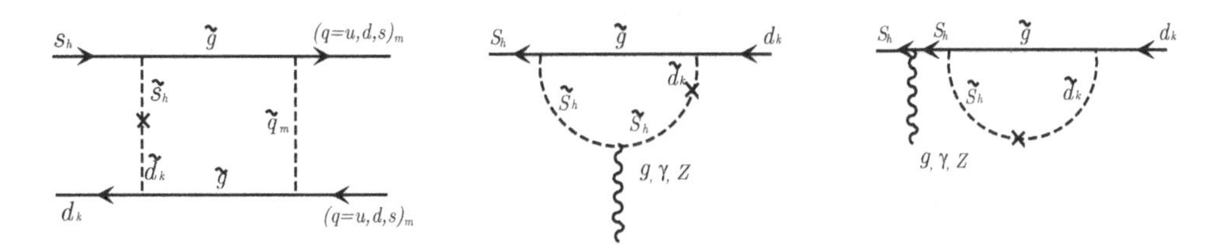

Fig. 14.1 Examples of SUSY diagrams in kaon physics.

14.3.1.1 CP *asymmetry for strange baryons and SUSY*

CP violation can be probed in transitions of strange baryons, in particular hyperons. They can be polarized which allows for a simultaneous measurement of angular distributions of hyperons and anti-hyperons and to test **CP** symmetry directly. This has been done for the process $e^+e^- \to J/\psi \to \bar{\Lambda}\Lambda$ recently measured by the BESIII experiment. To determine the asymmetry parameters the decay chain

[17] Even if one can draw SUSY diagrams, it is possible they mostly cancel each other.

$e^+e^- \to J/\psi \to \bar{\Lambda}\Lambda \to [\bar{p}\pi^+][p\pi^-]$ was analyzed, see **Sect. 7.7**:

$$\alpha_-(\Lambda \to p\pi^-)|_{\rm PDG2020} = +0.732 \pm 0.014$$
$$\alpha_+(\bar{\Lambda} \to \bar{p}\pi^+)|_{\rm PDG2020} = -0.758 \pm 0.012 \,. \tag{14.14}$$

It would be a miracle to find **CP** asymmetry as a few$\times\mathcal{O}(10^{-3})$. However, the LHCb collaboration could probe $pp \to [\bar{\Lambda}\Lambda]_{J/\psi} + X$ in runs-3 and -4 and find it at the level of $\mathcal{O}(10^{-4})$. It could be originated by ND, and SUSY is an obvious candidate.

14.3.2 *Possible impact of SUSY on $K \to \pi\bar{\nu}\nu$*

In **Sect. 6.5.10** we have discussed ultra-rare kaon decays. These processes are considered to be one of the most powerful probes of ND. The SM predicts non-zero values with very small uncertainties [165, 190]:

$$\mathrm{BR}(K^+ \to \pi^+\bar{\nu}\nu)|_{\rm SM} = (8.39 \pm 0.30)\cdot 10^{-11}\left[\frac{|V_{cb}|}{40.7\cdot 10^{-3}}\right]^{2.8}\left[\frac{\phi_3/\gamma}{73.2^0}\right]^{0.74}$$
$$\mathrm{BR}(K_L \to \pi^0\bar{\nu}\nu)|_{\rm SM} = (3.36 \pm 0.05)\cdot 10^{-11}\left[\frac{|V_{cb}|}{40.7\cdot 10^{-3}}\right]^{2}\left[\frac{\phi_3/\gamma}{73.2^0}\right]^{2} .$$

The information that can be extracted from precise measurements of the two neutrino modes turns out to be very useful in restricting the SUSY parameter space. In the MSSM with MFV, these decays remain to a large extent SM-like, with small deviations (typically within 10%). In the MSSM beyond MFV these decays can be modified independently and are unique probes of flavor violation in the up-squark sector. Several motivated frameworks exist that could receive sizable contributions with respect to the SM [490]. Future challenges come from the experimental analyses:

$$\mathrm{BR}(K^+ \to \pi^+\bar{\nu}\nu)|_{\rm data} = (10.6^{+4.0}_{-3.4}|_{\rm stat} \pm 0.9_{\rm sys})\cdot 10^{-11}$$
$$\mathrm{BR}(K_L \to \pi^0\bar{\nu}\nu)|_{\rm data} < 300\cdot 10^{-11} .$$

14.3.3 *SUSY contributions to $\Delta B \neq 0$ and $\Delta C \neq 0$ transitions*

B and D decays have always been foremost in testing the limits of the SM. Presently, examining processes that might not respect lepton flavor universality is one major endeavors at the LHC and other experiments because of some discrepancy in B decays data with SM predictions. That concerns an enhancement of the charged-current interaction $b \to c\tau\nu$ with respect to the tree-level induced SM amplitude, and a deficit in neutral-current transition involving $b \to s\ell^+\ell^-$ at one-loop level. The theoretical challenge to devise an UV complete model that accommodates all other low-energy observables has proven to be difficult; however, there are a handful of proposals, including SUSY models (see for example Refs. [491, 492]).

Interesting processes from the point of view of SUSY are the decays $B^0_{(s)} \to e^+e^-$ and $B^0 \to \mu^+\mu^-$. They only proceed at second order in weak interactions in the SM, but may have large contributions from supersymmetric loops, proportional to $(\tan\beta)^6$. Global analyses of these and other rare exclusive B meson decays, do not exclude the possibility of ND [493]. However, severe constraints are imposed by present data, and even more so in the future, with the expected increase of precision data in the next years.

Non-minimal SUSY can also give significant contributions to mass differences of $B_{(d/s)}$, although the accurately measured ratio $\Delta M_{B_d}/\Delta M_{B_s}$ is fully consistent with the SM prediction. By introducing large new weak phases it can also enhance **CP** violation in $B^0 - \bar{B}^0$ oscillations and modify the **CP** asymmetries in $B_d \to \psi K_S$, $B_d \to \pi^+\pi^-$, $B_s \to D_s K$ and $B_s \to D_s \bar{D}_s, \psi\phi$, which involve oscillations [494].

In **Sect. 7.4.2** we have discussed Dalitz plots for $B^+ \to K^+\pi^+\pi^-/K^+K^+K^-$ and $B^+ \to \pi^+\pi^+\pi^-/\pi^+K^+K^-$, in particular for direct **CP** asymmetries. Dalitz plot analyses can be useful to identify patterns of ND as SUSY.

Direct **CP** asymmetry has been established for the first time in $\Delta C \neq 0$ decays [174], see **Sect. 7.5.1**: $\Delta A_{\mathbf{CP}} = A_{\mathbf{CP}}(D^0 \to K^+K^-) - A_{\mathbf{CP}}(D^0 \to \pi^+\pi^-) = -(1.54 \pm 0.29) \cdot 10^{-3}$. The possibility that the value of $\Delta A_{\mathbf{CP}}$ is driven by ND effects has been investigated in different models, including MSSM, where the dominant SUSY contribution has been assumed to come from loops involving gluinos and up-squarks [495].

14.4 Indirect impact of SUSY on $(g-2)/2|_{\rm lept}$?

In **Sect. 12.1.1** we have discussed a sign of a possible impact of ND in $(g-2)/2|_\mu$ expressed by the gap $a|_\mu(\exp) - a|_\mu(\mathrm{SM}) \simeq (25.1 \pm 5.9) \cdot 10^{-10}$ with a significance of 4.2σ [398]. Both experimenters and theorists have applied a large part of their best tool boxes to reach the present sensitivity in $(g-2)/2|_\mu$. It has been suggested that SUSY can account for the discrepancy, for instance through the EW sector of the MSSM, consisting of charginos, neutralinos and scalar leptons, escaping direct searches at the LHC because of rather small production cross-sections [496].

14.5 SUSY EDM for baryons and charged leptons

As said before, by probing EDMs we are not going after accuracy, but rather exploiting the fact that the SM can*not* produce a 'background' here.[18] Thus measurable values of EDMs of hadrons and leptons give us a rich landscape to establish ND in known matter and beyond. The SM predictions for EDMs of electron and neutron are extremely small, because the first nonvanishing contributions arise at higher–loop level. The SUSY contributions arise already at one–loop level, thus the electron and neutron EDMs are well suited to yield important information

[18] Maybe, maybe somebody will find a non-zero EDM due to $\bar{\theta}$ from 'true' QCD.

about SUSY models and can considerably restrict the allowed parameter regions. In general (a) One can get 1-loop corrections due to non-minimal Higgs structures; (b) One can get SUSY 1-loop corrections with squarks, sleptons, gauginos.

Let us address SUSY corrections in the case of baryons. A tight experimental constraint can be placed on ϕ_A, the phase of the trilinear coupling A in $\mathcal{L}_{\rm soft}$ in Eq. (14.12), as it generates a neutron electric dipole moment already at the one-loop level. The diagram in **Fig. 14.2** gives, using the static approximation $d_N \simeq 2\,d_d + d_u$ and the MI approximation

$$d_N = -\frac{4}{27}\frac{e\alpha_s}{\pi}\frac{M_{\tilde{g}}}{M_{\tilde{q}}^2}\mathrm{Im}\,[2(\delta_{11}^D)_{LR} + (\delta_{11}^U)_{LR}]\,F_1(x), \tag{14.15}$$

$$F_1(x) = \frac{1}{(1-x)^3}\left(\frac{1+5x}{2} + \frac{2+x}{1-x}\ln x\right), \quad x = \frac{M_{\tilde{g}}^2}{M_{\tilde{q}}^2}, \tag{14.16}$$

where $M_{\tilde{g}}$ and $M_{\tilde{q}}$ denote the gluino and average squark mass, respectively. Since the dipole operator is chirality changing, a complex mixing between left and right handed squarks is needed to produce an effect: $\mathrm{Im}(\delta)_{LR} \neq 0$. The experimental bound on d_N yields for $M_{\tilde{g}} \simeq M_{\tilde{q}}$

$$\mathrm{Im}\,(\delta_{11}^D)_{LR},\ \ \mathrm{Im}\,(\delta_{11}^U)_{LR} \leq \text{few} \times 10^{-6}. \tag{14.17}$$

Within MSSM $(\delta_{11}^D)_{LR} = (\delta_{11}^U)_{LR} = A \cdot m_{3/2} m_d / M_{\tilde{q}}^2$; i.e., assuming $m_{3/2} \sim M_{\tilde{q}} \sim$ 300 GeV as a benchmark, we infer from Eq. (14.17)

$$\arg(A) \leq \mathcal{O}(10^{-2}). \tag{14.18}$$

While the super-GIM mechanism goes a long way towards explaining the minute size of $\mathrm{Im}(\delta_{11})_{LR}$, the phase of A is small for $M_{\tilde{q}}$ below 1 TeV. It is then tempting to conjecture the intervention of some symmetry to yield $\arg(A) \simeq 0$. In that case there is basically only one source for **CP** violation as in the SM, namely the irreducible KM phase (in addition to $\overline{\theta}_{QCD}$), yet it can enter through a new gateway – the renormalized squark mass matrices. In non-minimal models *without* an effective super-GIM filter, the need for a symmetry suppressing new phases in the low energy sector appears compelling. Of course, one may hope that any improvement in experimental sensitivity for EDMs might finally reveal a non-vanishing result!

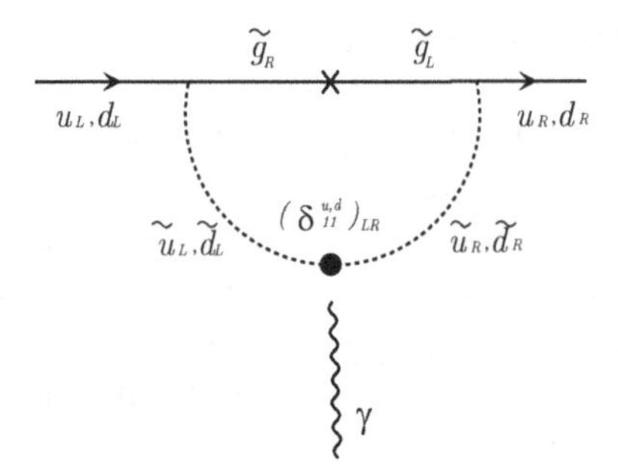

Fig. 14.2 Feynman diagram which gives the neutron electric dipole moment.

The ACME collaboration has recently announced a new constraint on the electron EDM from measurements of the ThO molecule [419]:

$$|d_e| < 1.1 \cdot 10^{-29}\, e\,\text{cm} \tag{14.19}$$

This is about an order of magnitude better than previous bounds and is a powerful constraint on **CP** violating new physics: even ND generating the EDM at two loops is constrained at the multi-TeV scale. This bound is interpreted in 1-loop probing sleptons above 10 TeV, and in 2-loops probing multi-TeV charginos or stops.

In MSSM, when one describes the EDMs for e, μ and τ, the dominant contribitions are given by one-loop diagrams with off-shell fields of sleptons and gauginos rather than squarks and gluinos as in the case of the neutron EDM [497]. Studies of the electron EDM in SUSY commenced in the eighties [498], and a variety of scenarios have been shown to predict interesting EDMs, including stops [499], electroweakinos [500], split SUSY [501] and SUSY beyond the MSSM [502].

14.6 The pundits' résumé

After this crash course in SUSY technology and phenomenology, we can evaluate **CP** violation in these models in a slightly more informed and considerably more refined manner.

(1) Indeed, SUSY is truly a symmetry 'sans-pareille' – like no other. To realize many of its attractive features we have to formulate it as a local theory – SUGRA – embedded in a GUT scheme. It is not an obvious symmetry. It connects bosons and fermions with different mass scales and forms an integral part of string theories.
(2) Our community has searched direct evidence for SUSY at the scale of around 1 TeV through to productions of novel states. They have not been found yet, but our community continues. There are also several 'roads' to find *in*direct evidence. The scales involved could be around 10–100 TeV, or more, for fans of SUSY. Since we have not found novel states yet, our lack of understanding of SUSY breaking is the main problem. Theoreticians like challenges.
(3) SUSY many new dynamical layers can support a plethora of additional sources of **CP** violation in the form of complex Yukawa couplings and other Higgs parameters, including those driving SUSY breaking.
(4) Through quark–squark–gluino couplings, flavor changing neutral currents mediating even **CP** violation enter the nominally strong dynamics.
(5) In MSSM only two observable phases emerge: one is the usual KM phase having its origin in the misalignment of the mass matrices for *Up*- and *Down*-type quarks; it also migrates into the squark mass matrices and controls the **CP** properties of the quark–squark–gluino couplings. The other one is ϕ_A, see Eq. (14.12), reflecting soft SUSY breaking; it is severely restricted by the experimental bound on the neutron EDM, which makes it irrelevant for $K_L \to \pi\pi$ decays.

(6) Yet once we enter the vast regime of non-minimal SUSY models the 'floodgates' open for additional sources of **CP** asymmetries:

- Observable effects are likely or at least quite conceivable for the EDMs of neutrons, deuterons, heavy nuclei, electrons, muons and taus.

Beyond that, completely different scenarios can occur:

(a) MSSM: the same **CP** asymmetries arise in b decays as with the SM implementation of the CKM ansatz, although the $B^0-\overline{B}^0$ oscillations are different.
(b) Non-minimal SUSY: large effects emerge in b decays, due to $\Delta B = 2$ dynamics modified by SUSY with sizable deviations from the CKM expectations.

(7) This does not mean, though, that an interpretative chaos of 'everything goes and nobody knows' will rule. For within each scenario there are numerous non-trivial correlations among the **CP** observables, rare decay rates, the CKM parameters and gross features of the sparticle spectrum.
(8) The SM predicts small **CP** violation effects for charm hadrons in SCS decays and basically zero for DCS ones. In the theorists' view the latter ones give a wonderful hunting region for ND; we need only much more data. Indeed, the LHCb collaboration has established direct **CP** asymmetry [174] in charm decays opening a new era.
(9) The SM predicts accurately very rare transitions for $K^+ \to \pi^+\bar{\nu}\nu$ and $K_L \to \pi^0\bar{\nu}\nu$. Investigations of $K^+ \to \pi^+\bar{\nu}\nu$ have been performed using the data collected by the NA62 experiment at CERN during Run 1 (2016–2018). This result constrains ND models, including SUSY, that can predict large enhancements previously allowed by the measurements published by the E787 and E949 BNL experiments. Further optimization of the analysis strategy is expected to reduce significantly the uncertainty in the measured branching ratios. The upcoming analyses might show impact of ND above the SM expectations. So far, no candidates outside backgrounds. New more precise results are also expected from KOTO experiment, which so far has studied the rare decay $K_L \to \pi^0\bar{\nu}\nu$ with the dataset taken at the J-PARC in 2016, 2017, and 2018.
(10) It is natural to ask if SUSY yields DM candidates or if it helps us to understand how many Universes exist. We discuss this **Sect. 15.2.2** and **Sect. 15.5**.
(11) Last but not least, finding SUSY directly or indirectly can be an important step to establish string theories to include gravity in the description of fundamental dynamics.

We have said about the virtues of SUSY. Yet – we wonder if SUSY is an example of a 'Faustian bargain with the devil'.[19]

[19]Two of the co-authors are super-fan of operas. One interesting analogy is represented by "The Damnation of Faust" (by Hector (!) Berlioz), since the aging scholar Faust has not found his goal.

Chapter 15

CP violation in modern cosmology

'A creative artist works on his next composition because he was not satisfied with his previous one.'

D. Shostakovich

In this book, we have mostly focused on **CP** asymmetries and rare decays in the SM, the theory of the fundamental structure of the *known* matter. Still, nowadays more than ever, astronomical discoveries are driving the frontiers of elementary particle physics beyond the SM, and, conversely, our knowledge of elementary particles is driving progress in understanding the Universe and its contents. In the "Prologue" we already stated that in the standard cosmological framework *known* matter gives us $\sim$ 4.5% of our Universe, while Dark Matter (DM) $\sim$ 26.5% and Dark Energy provide the rest.[1] As discussed in **Sect. 4.6**, the SM assumes that neutrinos are massless; yet due to their oscillations we have learnt that they are not massless. Thus the SM is incomplete even for known matter.

The goal of this **Chapter** is not to find solutions to the numerous challenges that the modern Cosmology is facing, but, more realistically, to give some suggestions and explain their strengths and weakness. Or to use different words: to 'paint' the 'landscape'. Indeed, it is quite possible that we are still *completely missing* what our Universe is telling 'us'.

15.1 Connections with Dark Matter

Dark Matter is a hypothetical form of matter different from what we are made of, that gives off no light, is not coupled to color and seems to produce the gravitational pull leading to the formation of large-scale structures in our Universe.

It presumably consists of one or more unidentified particles. Astronomical evidence in favor of the existence of DM has been found at all scales, in particular by studying the rotation curves of galaxies, the dynamical properties of galaxy clusters, the formation and evolution of large scale structures, and the cosmic microwave

[1]At seminars speakers usually talk about the ratio of *known* matter vs. DM, which is $\sim$ 15%.

background. At all these very different scales the need of large amount of DM is confirmed and, despite strong efforts, no theoretical picture is able to explain all data consistently by any other mean than assuming the existence of DM. Alternative theories try to explain data by modifying gravity [503] or modifying the dynamics of celestial bodies at small velocities. Even when partially successful, they mimic DM effects on a limited range of scales, but fail globally, and especially at the largest scales, for example in explaining the combined observation of weak lensing and X-ray emission in systems such as merging clusters as 1E 0657-558 (the "Bullet" cluster) [504] and in predicting the observed anisotropy power spectrum of the CMB. The recent observation of a neutron star-neutron star collision by means of gravitational waves in connection with electromagnetic radiation has put further strong constraints on many such theories [505].

The idea of DM required several decades to be accepted. The interested reader may see Refs. [506] and [507] and references therein for more details and for an historically precise account of this idea. For a general nice overview see also Ref. [508].

Astronomical observation provides very few information about DM, although crucial to direct experimental searches. What we know with relative certainty is:

- The DM candidate(s) must be stable or have a lifetime of $O(10^{10})$ y or more.
- DM matter candidates are neutral or at most have an electric charge that is a tiny fraction of e ($q_{DM} < 10^{-4} - 10^{-7}\, e$, depending on DM candidate mass). Best limits so far come from evaluating the impact of DM charge to CMB structure through photon coupling, and particularly to the position of the first acoustic peak in the CMB spectrum [509].
- The average DM energy density in the Universe is about 0.3 GeV/cm^3; it should be underlined that local density around the Earth (where experiments are done) can be different because of Milky Way dynamics and history. The most recent studies of the local DM density, derived from the data taken by the Gaia satellite, is in the range (0.4–1.5) GeV/cm^3 [510], compatible with the average, but possibly higher.
- A substantial fraction of DM must be "cold", i.e. it must have been already non-relativistic at the time of recombination and CMB formation. A fraction of "hot" (relativistic) or "warm" DM is compatible with existing data.
- The dark matter candidate, even assuming it is a single particle, can have a very wide range of masses. Astronomical observations provide solid information only about energy density through gravitational effects. If the candidate is a fermion, its mass is bounded from below by the Pauli exclusion principle (the so called Tremaine-Gunn limit [511]). Using data from velocity dispersion in dwarf galaxies, a lower limit of 70 eV for a single fermion candidate is obtained [512]. If it is a boson, the only lower limit comes from the fact that the Compton wavelength of super-light massive particles can become large enough to erase small scale structures. Observations put a lower limit of the order of 10^{-22} eV [513]. It should not be forgotten, however, that none of this limit applies

if DM is a mixture of several particles, each contributing to a fraction of total DM energy density. Upper limits are even weaker than lower limits. The only indication comes from the tidal effects of the DM halo on small systems such as globular clusters or galactic disks. The most stringent limits force the DM candidate (assuming it is a single candidate) to be lighter than about 5 solar masses.

The astronomer Fritz Zwicky (often credited to be the initiator of this field[2]) already postulated in 1933 that our Universe was produced by DM [514]! A reader might be interested to the personal history of this astronomer [515].[3] For a rich account of the history of DM and the role of Zwicky see also [507].

From the theoretical perspective, there are excellent candidates for DM, and in the future new ideas will come forward. We list four main theoretically well motivated candidates: DM "Axions", DM "SUSY", "Asymmetric Dark Matter" and "Forbidden Dark Matter".[4] Furthermore, the world of DM could be described as a 'cocktail'.[5] We do not 'know' how DM is produced beyond gravity. The landscape has become more 'complex' with the third millennium. History has often shown that novel technology appears to enhance our understanding of fundamental dynamics. Even if the original goal was not reached, one might find another good goal. A large number of experiments are searching for DM candidates using several techniques and assuming different physical properties for the possible DM candidate(s). **Section 15.2.1** summarises the current status of experimental searches.

15.2 DM searches: brief overview

The struggle to identify DM is an extremely active field of research.[6] DM research is made by means of three very different and complementary techniques:

a) by direct production, using accelerators;
b) by direct detection, using suitable underground detectors;
c) by indirect detection, searching for particular radiation from space that may originate from DM annihilation and/or decay.

Even a short review of the whole topic goes well beyond the scope of this book. However, given the extreme importance of the field, we provide a short account

[2]In many occasion he chose not to be polite; thus at his time he did not get fair credit for his achievements.

[3]The author of Ref. [515] claimed that 'Zwicky is celebrated mainly as the father of dark matter'.

[4]A reader of this book should not be surprised at least by our choice of the first three ones. We remind that axions were discussed in **Chapter 11** and SUSY was introduced in **Chapter 14**.

[5]Our Universe = known matter + $\text{'cocktail'}|_{\rm DM}$ + Dark Energy. That our Universe has three sources is not a true problem; the challenge is that the three ones are not tiny.

[6]The reader may get a good feeling by looking at the agenda of IDM 2020 (https://indico.cern.ch/event/766367/).

of most recent results and future programs. As of today, we have no confirmed evidence of DM except the DAMA signal, which we will briefly discuss below.

Experiments at accelerators search for new particles in high energy collisions or by means of dedicated beam dump experiments. At LHC the search is mostly done by assuming some specific interaction model and seeking specific signature in the data. Typical assumptions include one of the following options [516]:

- DM particles are produced together with Standard Model particles. One looks for an energetic SM particle recoiling against the invisible DM system. The signature is given by the strongly imbalanced momentum;

- DM mediators are produced and decay to pair of SM particles, typically quarks– then one searches for bumps in the two-jets invariant mass spectrum. A bump in the two jets or two leptons invariant mass distributions or an excess in their angular distribution, produced by a dark matter mediator, should appear;

- DM production through the Higgs portal, assuming that the Higgs boson can decay into the DM channels.

It is clear from this list that all searches are heavily model dependent. Besides, it should be underlined that any discovery through these methods, though sensational for fundamental physics in general, would not prove that DM has been discovered, not being able to demonstrate that at least one DM candidate is stable over time scales compared to the age of the Universe.

The direct search experiments seek for DM signal assuming it has some kind of interaction with matter. It is nowadays a vivid research field, with several tens of experiments running or under construction around the world. Each underground laboratory has a DM search program. Detectors must be located underground to shield from cosmic rays and environmental radiation and must be further shielded against residual radiation present in underground laboratories, chiefly γ photons and neutrons. Besides this, the materials with which the detector is made of must be as radio-pure as possible, being natural radioactivity of isotopes present inside the detector even in tiny sub-traces a major source of background.

Most experiments are designed to search for the recoil of a target nucleus against the DM candidate, assuming that the latter has a finite elastic (or possibly also inelastic) cross section. Some experiments have sensitivity also to various other kind of electromagnetic interactions, but not all. The target nuclei can be either bosons or fermions, allowing in some cases spin-dependent cross section. The reader can look at Ref. [517] and references therein for a review of recent experimental limits, bearing in mind that all limits are model dependent, i.e. they assume some specific interaction model. We summarise here the main numbers, also shown in **Fig. 15.1**:

- For a generic weakly interactive massive particle (WIMP) with mass above 5 GeV, the best cross section limit assuming a model where the cross section is spin-independent is that of XENON1T experiment [518] $\sigma < 4.1 \cdot 10^{-47}\text{cm}^2$ on Xe mixture.

- For lower WIMP masses, the best limits come from DarkSide-50 [519] experiment ($\sigma < 1 \cdot 10^{-41}$ cm^2 in the range 1.5–5 GeV/c^2, Ar target) and from the CRESST-II [520] ($\sigma < 1 \cdot 10^{-39}$ cm^2 in the range 1–1.5 GeV/c^2, $CaWO_4$ target).
- Best limits for spin-dependent cross sections (depending on an odd number of either protons or neutrons) are obtained, respectively by the PICO60 [521] ($\sigma < 3.2 \cdot 10^{-41}$ cm^2 on F target) and again by Xenon1T [522] on Xe target ($\sigma < 6.3 \cdot 10^{-42}$ cm^2).

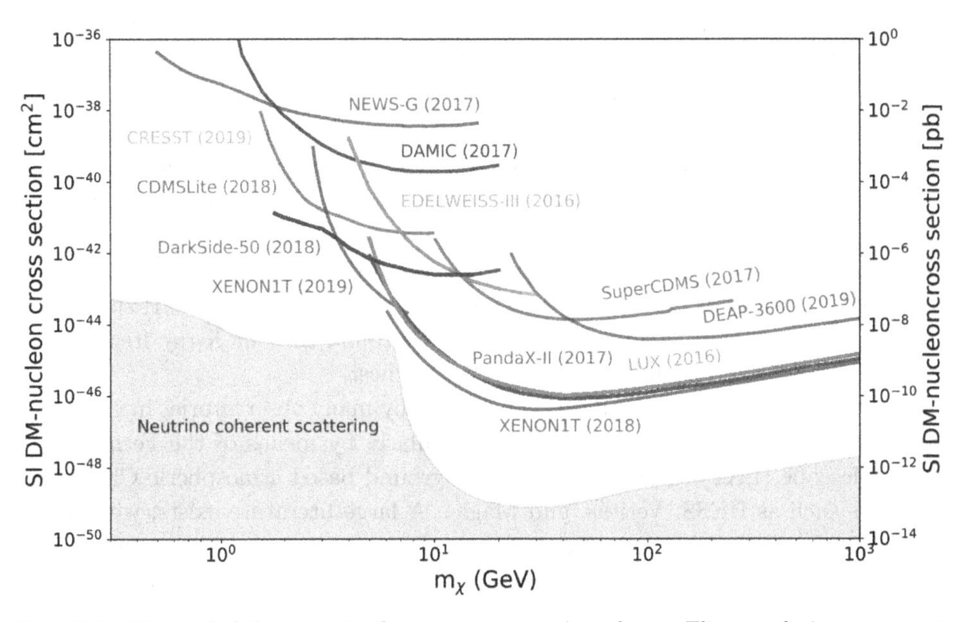

Fig. 15.1 The excluded region in the mass-cross section plane. These exclusions assume a 'standard' spin-independent WIMP-nucleus interaction model. Courtesy of PDG 2020 [8]. See references therein for details about experiments quoted in the plot.

The only experiment that, so far, has claimed a signal consistent with being produced by the direct interaction of dark matter particle into a NaI scintillator detector is Dama [523]. The Dama experiment, located at Laboratori Nazionali del Gran Sasso, follows a different approach with respect to other searches. Particularly, they make no assumption about the DM-detector interaction model (not applying, therefore, any selection cut except electronics noise rejection) and search for the annual modulation induced in the counting rate by the annual revolution of the Earth around the Sun [524]. The Earth velocity adds to the Sun's velocity in the Galactic frame, modulating with one year period the DM flow through the detector. Dama has taken data with two different detectors and through several technological upgrades, always confirming a clear and not ambiguous modulation signal [525] in the region 1–6 keV (electron recoil equivalent). No experiment has so far succeeded to confirm it, but no experiment has actually achieved the required large exposure (several 100 kg $\times$ y) and the required radio-purity to be able excluded it. The scientific debate is still in progress. No known interaction mechanism is able to reconcile Dama result neither with experimental limits nor with theory, but all limits are made using assumptions about the interaction models, which might prove to be wrong, and the theoretical expectations about DM are not more than guesswork. The future will tell us who is right. 'Se son rose, fioriranno'.[7]

DM particles, unless they share the very unique feature to be coupled to gravitational interactions only, should give measurable effects in astronomical observations, either through DM-DM annihilation into SM particles or DM-decay into SM particles. Many observations have been and are being carried forward to search for astronomical signals into photons, neutrinos, and anti-nuclei, especially anti-protons, anti-deuterium and anti-helium particles.[8]

High energy photons are an excellent probe. Any final state either through annihilation or decay should produce photons, e.g. via bremsstrahlung or neutral pions, gauge bosons or H^0 decay. On several energy regions the signal can exceed the astrophysical known background and, in case of DM decay to 2 photons, a specific mono-energetic line may appear in the spectrum. The maximum sensitivity is obtained by looking at carefully selected systems, such as nearby dwarf spheroidal galaxies, which have a small astrophysical background at γ or X-ray frequencies, and are therefore an optimal choice for DM searches.

A photon excess signal has been searched for by many observatories in the range from a few GeV up to TeV scale, and particularly by means of the Fermi Large Area Telescope (LAT) satellite [526] and by ground based atmospheric Cherenkov detectors such as HESS, Veritas, and Magic. A large literature exists, with many observations and signal claims. None of them, however, lacks possible astrophysical explanations and we do not report them here. The interested reader can again look at Ref. [517] and references therein.

The search of anti-matter particles is another very active field, made particularly by the former Pamela satellite and now chiefly by the AMS-02 observatory on the International Space Station. While the signal expected on Earth suffers from serious uncertainties [527], the observation of positrons, anti-protons, anti-deuterium and anti-helium fluxes in substantial excess with respect to astrophysical expectations would have a tremendous impact [528], not necessarily for DM search only.

A positron excess was first observed by Pamela [529] and then confirmed with much higher sensitivity by AMS-02 [530]. However, several astrophysical explanations [531] are invoked to explain it without DM. The anti-proton spectrum measured again by Pamela [532] and then with much higher precision by AMS-02 [533] also shows a small excess between 10 and 20 GeV, and above 100 GeV, which might indeed be interpreted as signal of DM annihilation. However, other explanations based on systematic errors in the calculated expected spectrum related to the

[7] An Italian saying, literally meaning: "If they are roses, they will blossom".

[8] The signal should exist also in nuclei but it is generally accepted that astrophysical backgrounds give no hope on those channels.

transport of charged particles in the Galaxy are probably more convincing [534]. Anti-nuclei such as anti-deuterium and anti-helium could also form as a result of DM annihilation or decay. The search of these anti-nuclei is part of AMS-02 experimental program. Specific detector designs have been developed to single out low-energy cosmic-ray anti-nuclei (e.g. $E < 0.25$ GeV/nucleon in the case of the General Antiparticle Spectrometer, or GAPS [535]).

15.2.1 *Finding axions in DM*

Axions can enter as DM – i.e., 'Renaissance' axions might be crucial actors on the stage of our Universe. The 'usual' axions have to be in this window:

$$10^{-5}\ \text{eV} \leq m_a|_{\text{'suggested'}} \leq 10^{-2}\ \text{eV}\ . \tag{15.1}$$

A publication has appeared with a courageous title 'Lattice QCD for Cosmology' [536]. It determines the axion mass using lattice QCD, assuming that these particles are the dominant component of DM. QCD and the EW theory determine the evolution of our early Universe. The key quantities of the calculation are the equation of state of the Universe and the temperature dependence of the topological susceptibility of QCD, which are notoriously difficult to calculate. The authors extend the analyses of equation of state in 2+1+1 flavor LQCD (up/down, strange and charm quarks) to equation of state in 2+1+1+1 by adding axions '... to describe the evolution of our Universe from temperatures several hundreds of GeV to the MeV scale'. They include '... known effects of the electroweak theory and give the effective degree of freedoms' and 'calculate the topological susceptibility up to the few GeV temperature region'. The axion mass can be expressed in terms of the topological susceptibility hence the authors 'predict the DM axion's mass in a post-inflation scenario ... as an initial condition of ... Universe'. Indeed, it shows courage.

The mass of the axion can be described in terms of its decay constant f_a: $m_a|_{\text{DM}} \sim 5.7 \cdot (10^{11}\text{GeV}/f_a) \cdot 10^{-5}$ eV [537]. For a decay constant larger or equal to 10^9 GeV, axions are dominantly produced non-thermally in the early Universe and hence contribute to the so-called cold DM, with small velocity dispersion [537]. Although the expected theoretical mass range is somewhat smaller, there is a number of experimental projects looking to DM axions in the vast range of 10^{-12} eV $\leq m_a|_{\text{DM}} \leq 10^{-4}$ eV. This is a good example that experimenters listen to theorists, but do not totally trust them.

We can 'paint' the landscape with tools from quite different dynamics:

(a) Nuclei could interact with a 'background' of DM axions (if they exist). They would acquire time-varying **CP**-odd nuclear moments. The proposed 'Cosmic Axion Spin Precession' experiment might probe the region of 10^{-12} eV $\leq m_a|_{\text{DM}} \leq 10^{-9}$ eV [538].

(b) A 'dielectric haloscope' consists of a mirror and several dielectric disks placed in an external magnetic field and a receiver in the field-free region [539].

(c) In the presence of a DM axion a large toroidal magnet acts as an oscillations current ring, whose induced magnetic flux can be measured by an external pickup loop inductively coupled to a SQUID magnetometer. It might show potential sensitivity to axion-like DM with 10^{-14} eV $\leq m_a|_{\rm DM} \leq 10^{-6}$ eV [540]. Likewise for the Axion Dark Matter Experiment (ADMX) with $1.9{\cdot}10^{-6}$ eV $\leq m_a|_{\rm DM} \leq 3.7{\cdot}10^{-6}$ eV [541].

(d) Microwave cavity experiments have been used already at test runs with some data leading to $3 \cdot 10^{-5}$ eV $\leq m_a|_{\rm DM} \leq 3 \cdot 10^{-4}$ eV [542] and 10^{-6} eV $\leq m_a|_{\rm DM} \leq 10^{-4}$ eV [543].

(e) The CERN Axion Solar Telescope (CAST) has searched for $a \to \gamma$ conversion in a strong magnetic field [544], see **Fig. 11.1** above. It gives leading limit of $g_{a\gamma} < 0.66 \cdot 10^{-10}$ GeV^{-1} on the axion-photon coupling strength for $m_a < 0.02$ eV. Its innovations are pathfinders for possible future axion helioscopes.

Let us highlight two very captivating proposals. The first one is named QUAX (QUaerere AXion)[9] [545]. The authors proposed to study the interaction of the *cosmological* axion with the spin of fermions (electrons or nucleons). The Solar System is *moving* through the galactic halo, and the Earth is effectively *moving* through the cold DM cloud surrounding the Galaxy; thus an observer on Earth will see such axions as a *wind.* In particular, the effect of the axion wind on a magnetized material can be described as an effective oscillations radio frequency field with frequency determined by m_a and amplitude related to f_a. This detection scheme is sensitive only to DFSZ axion models [364]. Some of these ideas go back to theorists long time ago [546]. By 2020 the QUAX experiment has given only limits [547].

The second proposal is to try "Unifying Inflation with the Axion, Dark Matter, Baryogenesis and the Seesaw Mechanism" [548]. This paper aims at providing a consistent picture of particle physics and cosmology up to the Planck scale. Baryogenesis proceeds via leptogenesis, that we discuss in **Sect. 15.4.4**. At low energies this model extend the SM by including seesaw-generated neutrino masses, plus an axion of about 10^{-4} eV, which solves the strong **CP** problem and accounts for the DM.

We do not bet that most of those projects (or proposals) will reach their objectives, but it is a wonderful challenge and our community should keep trying.

15.2.2 *SUSY candidate for DM*

We have three classes of SUSY candidates for DM, gravitinos, sneutrinos and neutralinos, with spins "3/2", "0" and "1/2", respectively. We give a few comments:

- Gravitinos are the superpartner of gravitons and appear in any supersymmetrization of gravity. However, there is hardly a chance to observe them

[9] In Latin 'quaerere' means 'seek for'; the authors of Ref. [545] are Italians except one. The first author is a well-known theorist, while the others are experimentalists.

since they only have a weak gravitational coupling to matter. The interactions of gravitinos are completely fixed by SUSY; since they belongs to the gravity sector, all couplings are suppressed by the Planck scale.
Gravitinos are often a nuisance in cosmology: since in general they are unstable towards decays into lighter superparticle, their lifetimes are 'naturally' long on cosmological scales. If they are not sufficiently heavy, they decay during or after Big Bang Nucleosynthesis [549].

- Sneutrinos are more suitable candidates, as mentioned in Ref. [550]. A long-held paradigm is that most of cold DM is made up of some weakly interacting massive particles (WIMPs). The WIMPs arise naturally in a large number of models included SUSY. In SUSY it is understood that WIMPs must be the Lightest Supersymmetric Particle (LSP), which is stable, and the lightest sneutrino is a LSP contender. However, most (but not all) regions of sneutrino parameter space are ruled out by WIMP direct-detection experiments [551].
- The lightest neutralino $\tilde{\chi}^0$ (without charge) in R-parity SUSY remains an excellent DM candidate, as detailed in the 1996 Report [550], and twenty years later by Ref. [552]. The neutralino is most likely the LSP in a MSSM. It is stable, weakly interacting, and it has become over time the paradigm of WIMP DM.
While searching for yet undiscovered WIMPS it is understood that one needs at least that (a) WIMPs exist, (b) are not baryonic (c) leave footprints in astrophysics, nuclear physics and particle physics and (d) can be detected directly or indirectly.
However, that is not enough in our view. It is worth pursuing candidates in a full-fledged theory, which includes solutions to other open issues in particle physics. It happens for SUSY WIMPs, specifically for the neutralino, which arises naturally in the theory.

15.2.3 *Asymmetric DM*

The idea of asymmetric DM (ADM) is based on the observation that the present day mass density of DM is about a factor of five higher than the density of visible matter, rather than the default expectation of many orders of magnitude. The similarity in these observed densities suggests a relation between dark and visible matter in their physics and cosmological history and even a common origin. The present day density of visible matter is expected to be due to the baryon asymmetry of the universe, and, according to the ADM hypothesis, the present-day DM density could similarly be due to a DM particle-antiparticle asymmetry. There are several models of asymmetric DM, where the DM asymmetry may or may not be related to the baryon-antibaryon asymmetry of visible matter [553]. These models have explored ADM in different astrophysical contexts, for instance in the Sun and other stars. If DM interacts with nucleons, it can scatter off the nuclei in stars, lose energy and get captured in their interiors. The accretion of DM in stars can have observable consequences. Data show a disagreement between helio-seismological observables in our Sun and predictions of the solar models computed with the latest surface abundances. It has been suggested that most of these challenges might be solved by the presence of asymmetric DM coupling to nucleons as the square of the momentum exchanged in the collisions [554]. Of course, any possible conclusion depends also on the standard *solar models*.

15.2.4 *Forbidden DM*

DM may be a thermal relic that annihilated into *heavier* states in the early Universe [555]. One may assume that DM dominantly annihilates into heavier particles; such annihilations are called forbidden channels, because they vanish at zero temperature. One example: $\bar{\psi}\psi \to xy$ with $m_x + m_y > 2m_\psi$, where we take the DM to be a Dirac fermion ψ, which is neutral under SM quantum numbers and charged under a hidden $U(1)_d$ gauge symmetry that is broken at low energies. It decays into x and y, which are dark photons. Dark (or hidden or secluded) photons are the $U(1)_d$ gauge bosons. Forbidden dark matter refers to the class of models where the DM relic abundance is dominantly set by forbidden channels. How can that happen? Forbidden channels can proceed at finite temperature at the early Universe, due to thermal tail with high velocity ψ's. If DM is a thermal relic, its energy density today depends on its annihilation rate $\langle \sigma v \rangle$ in the early Universe [556]:

$$\Omega_{\rm DM} h^2 \sim 0.1 \frac{(20 \text{ TeV})^{-2}}{\langle \sigma v \rangle} \,. \tag{15.2}$$

The observed DM abundance – $\Omega_{\rm DM} h^2 \sim 0.1$ – could be produced if DM annihilates with a weak scale cross section. Models of DM lighter than $\sim$ 100 MeV are highly constrained by the Cosmic Microwave Background, but forbidden DM manages to evade these constraints because annihilations shut off at later times and low temperatures.

15.2.5 *Rich landscape*

Our community may be able to show the DM existence by correlations with known matter with weak dynamics in many ways, and maybe even understand, at least approximately, its features as suggested by theorists. (a) One step in this direction is to continue to use the well-known charged kaon beams at CERN to measure possible impact of DM, for example by analyzing the decay $K^\pm \to \pi^\pm +$ missing energy [557]. It has been assumed that the missing energy can be carried by a massive dark photon $\tilde{\gamma}$. (b) Furthermore one can focus on *massless* $\tilde{\gamma}$ in $K^+ \to \pi^+\pi^0\tilde{\gamma}$ in subtle ways [558]: it can interact with SM fields only through higher dimensional operator under an unbroken dark $U(1)$ gauge symmetry, typically suppressed by the mass scales; thus this transition must proceed through short-distance effects.

The NA62 experiment at CERN will provide a sample of 10^{13} K^+ with hermetic photon coverage and good missing energy resolution. Thus one gets a probe for massless dark photon, when massive dark-photon channels are nonviable.

Experimenters often work for producing tools to go *beyond* the limits the theorists had 'suggested' before. That is the way to make progress in fundamental dynamics on different levels. Some of these classes of DM will disappear 'soon'.

15.3 Extra dimensions and cosmology

Fig. 15.2 'Dream' of Constantine I, by Piero della Francesca, from the fresco cycle, the History of the True Cross, 1459-66, Basilica di San Francesco, Arezzo, Italy.

Paintings describe the real world in two dimensions. In the Middle Ages that was obviously accepted– then the 'Era of the Renaissance' and the 'Age of Discovery' came, and with them the challenge of going beyond. One should see the wonderful painting by the Italian Piero della Francesca[10] in the Basilica of San Francesco in Arezzo (Italy). One of the co-authors (IIB) has visited it thrice at least, in particular for the painting in **Fig. 15.2**.[11] At a first sight one can think it describes

[10]He was trained as a mathematician, worked on linear perspective creating illusions of depth on a flat surface and applied it to painting. He wrote De Prospectiva pingendi (On the Perspective of painting) the earliest Renaissance treatise solely devoted to the subject of perspective.

[11]It shows Constantine I (called the 'Great') sleeping just before a crucial battle outside of Rome; he dreams of a sign in the one he could win next day.

the situation in a sequence of two-dimensional slides – but then one realizes that it shows the impact of the third dimension close to the higher corner on the left side: an angel moves through three space dimensions.[12] Can one see the analogy with 'Extra Dimensions (and Cosmology') we discuss in this **Section**?

In **Chapter 14** and **Sect. 15.1** we have studied possible impact of SUSY and DM. So far without successes, but our community has to continue investigating. Yet one could take another route, namely looking for 'extra dimensions'.

15.3.1 *Extra dimension crash course*

There have been speculations that our Universe consists of more than the three space and one time dimensions. At first sight, such a speculation might be seen as a 'non-starter', since the theories that successfully describe 'nature' are based on four dimensions. Kaluza and Klein addressed this question in the 1920s in an intriguing way [355]. They considered general relativity in five dimensions and suggested that the fifth one has been compactified or 'curled' up – possibly due to quantum corrections – into a closed circle with a radius so tiny that it could not be resolved experimentally and even its existence could have not been inferred. Considering only fields that do *not* depend on coordinates in the fifth dimension allows to decompose the metric tensor in five dimensions into one in four dimensions plus a four-dimensional vector field that acts like the vector potential of electrodynamics. In other words: 'general relativity' in five dimensions leads to the usual photon field coupled to gravity in four dimensions. While general coordinate invariance in five dimensions is a purely geometric concept, it induces general coordinate invariance in four dimensions plus local gauge invariance; the latter is usually viewed as an internal rather than geometric symmetry. Such a unification of Maxwell's and Einstein's theories leaves some other footprints. With the fifth dimension having the topology of a circle, fields must satisfy periodic boundary conditions in the fifth dimension: $\psi(x_5) = \psi(x_5 + 2\pi R)$ with R denoting the radius of the fifth dimension. With Newton's constant the only dimensional quantity available, we have $1/R \sim 1/\sqrt{G_N} = M_{\rm Pl}$. Such fields can be expanded as follows:

$$\psi(x_5) = \sum_{n=0}^{\infty} \psi_n e^{inx_5/R} \, . \tag{15.3}$$

While the zero modes ψ_0 can be identified with the known fields, there is a whole 'Kaluza-Klein' (KK) tower for such a field with their masses given in units of $M_{\rm Pl} \sim 10^{19}$ GeV $\sim (10^{-33}$ cm$)^{-1}$.

This bold proposal soon ran into road blocks that could not be overcome. Later it turned into the tale of the 'sleeping beauty' and the 'charming prince': in this case extra dimensions were re-awakened by the emergence of super-string theories as a candidate for a 'theory of everything' (TOE). These theories combine the concept

[12] and 'higher'?

of higher dimensions with that of internal symmetries to have anomalies cancelled that otherwise would destroy the viability of the theory; i.e., they have abandoned the program of reducing all internal symmetries to a general coordinate invariance in higher dimensions. In exchange super-string theory gives gives an almost unique answer to the question how many dimensions are realized in 'our' Universe. Thus a dynamical implementation of new physics can arise in an unusual setting, namely in models with extra (space) dimension. They lend themselves to a treatment with the formalism of effective field theory, discussed in **Sect. 3.4**.

Subsequently speculations about extra dimensions morphed into different versions not necessarily connected with super-strings. The idea of 'large' extra dimensions – compared to the Planck length $\sim 10^{-33}$ cm – was invoked to 'solve' the gauge hierarchy problem. It was argued that the huge ratio between the Planck scale $\sim 10^{19}$ GeV and the electroweak scale is an artifact of our basic misconception about why gravity is so weak. The tiny value of $G_N = 1/M_{\rm Pl}^2$ can be seen as a geometric effect: gravity's strength is 'diluted', since gravitational fields unlike all other SM fields get dissipated into extra space dimensions. Denoting by $M_{\rm QG}$ – 'QG' stays for *quantum gravity* – the scale induced by quantizing gravity and by n the numbers of compactified extra dimensions, one obtains for the Planck mass as inferred from measuring the classical gravitational potential:

$$M_{\rm Pl}^2 = M_{\rm QG}^{n+2}(2\pi R)^n \ . \tag{15.4}$$

If there were one or two extra dimensions ($n = 1, 2$), then the observed validity of Newtonian gravity rules out a scale of $M_{\rm QG} \sim 1$ TeV by a long shot. Yet for $n \geq 3$, $M_{\rm QG} \sim 1$ TeV becomes a tenable scenario and one that for practical reasons cannot be probed through searching for deviations from Newton's gravity.[13]

Nice introductions to the theories with extra dimensions can be found in three TASI lectures [559]. There is also a nice book about warped dimensions; its goal is to reach a large audience [560]. It should be interesting also for scientists, especially Chapters IV-VI. The ADD [561] and RS [562] models differ in the topology assumed for the extra dimensions, namely 'flat' in the former and 'warped' in the latter case. The ND effects are due to the exchanges of the graviton's KK tower. While its members are only gravitationally coupled, they are so densely populated in mass that the overall production rates for KK gravitons and black holes can be sizable as can be contributions to particle production due to virtual KK graviton exchanges. Yet no new sources to flavor violation arise.

One should note that a priori there is nothing special about 1 TeV scale in this context. It is picked as a reference scale, since we know that $M_{\rm QG}$ cannot be significantly lower; for otherwise one should have seen the footprints of such extra dimensions in, say, hadronic collider experiments at Fermilab Tevatron.

[13]For $n = 3$ one gets $M_{\rm QG} \sim 1$ TeV and $R \sim 10^{-6}$ cm. While the intrinsic scale of QG and of the weak forces no longer exhibit an 'unnatural' hierarchy, no good answer is offered to why R exceeds its 'obvious' scale $1/M_{\rm QG} \sim 10^{-17}$ cm.

Then the interpretive 'floodgates' opened:

- The ansatz of 'universal' extra dimensions (UED), as exemplified by the ACD model [563], opened up those dimensions to the 'masses'; i.e., they allow all SM fields to access to the new dimensions. In such an ansatz one *no longer* undertakes to explain the feebleness of gravity as a geometric effect as sketched above, and the inverse radius R is related to the mass scale of the KK towers of the various SM fields. In return one can obtain many new sources of flavor and **CP** violation. UED models have another qualitatively new feature: 'KK parity' can be defined as a conserved quantity allowing KK states to be produced only in pairs; i.e., at low energies they can contribute only through loop effects. This lowers the bounds on their masses as inferred from phenomenological constraints in KK contributions. A value of $1/R$ as low as 250 GeV is still allowed leading to a potentially exciting phenomenology at the LHC. The ACD version of UED models will affect rare K and B decays in a sizable and noticeable way, albeit in a minimal flavor violation fashion [564]. It does not provide a deeper reason, but it is practical.
- The topology of the extra dimensions can be even more 'complex', as in the 'split-fermion' approach [565]: different fermion fields (quark vs. lepton fields, left-handed vs. right-handed fields) are described by Gaussian wave functions centered at different locations in the extra dimension(s). The overlap between such Gaussians is very small, unless their locations are very close to each others. Splitting quarks and lepton fields provides a very efficient way to suppress proton decay. Separating the chiral components of the quark flavors by different distances can naturally lead to vastly different mass terms.
  Yet at the same time the KK towers of the gauge fields can generate FCNC in K, B and D decays that are not particularly suppressed and thus could lead to sizable deviation from SM predictions with the resulting phenomenology not of the minimal flavor violation variety. Constraints derived from present data depend much on details of such models like whether the extra dimensions is 'flat' or warped'; the latter allows for a ND scale as low as a few TeV [566].

While primarily motivated by attractive particle physics features, the extra dimension framework has evolved not only to give a different viewpoint on the basic symmetries of our Universe, but to address also specifically cosmological problems as the entropy, flatness and homogeneity ones. In the title of **Sect. 15.3** we added the words '... and cosmology' – it is 'natural' to think about connections of extra dimensions and cosmology. Constraints on extra-dimensional models arise not only by collider experiments (dominated by LHC results), but from astrophysical and cosmological considerations, as well as tabletop experiments exploring gravity at sub-mm distances.

Studying extra-dimensional theories with gravitational waves provides a new way to constrain extra dimensions. On 11 February 2016, the LIGO and Virgo Scientific Collaborations announced that they detected, directly, a transient gravitational wave (GW) signal on 14 September 2015, which was named GW150914 [567]. Based on these data and several subsequent GW events, a new field of action has been rapidly developing. The detection of GWs provides also a new way to obtain constraints on extra-dimensional theories. For example, in some extra-dimensional theories, the number of extra dimensions could affect the amplitude attenuation of GWs, or could affect the size of the shortcut that a gravitational signal takes in the bulk. Some limits on the parameters of extra-dimensional models based on the existing data have been obtained already (see Ref. [568]).

15.3.2 *The pundits' call*

'Generic' versions of ND anticipated for the 1 TeV scale should have impacted B, K and D decays in a discernible way. Yet past experience should have taught us that 'Nature' has a tendency not to follow our notions of what is 'natural'. Thus it is premature to argue, for instance, that ND quanta can be found at LHC only of the minimal flavor violation variety. In this framework one conjectures that the anticipated ND by itself might be 'flavor neutral'; i.e., the dynamics driving all flavor-nondiagonal transitions – including **CP** breaking effects – are related to the known structure of the SM Yukawa couplings. The principle of minimal flavor violation represents a classification scheme rather than a theory or even a class of theories. It is more likely to be realized approximately than exactly. Future heavy flavor studies could reveal manifestations of ND beyond the direct reach by LHC collisions; i.e., with masses well above 10 TeV – in particular if it is not of the minimal flavor violation variety. It also appears premature to enforce the argument that the origin of **CP** violation is geometrical in nature, even if new sources of **CP** violation are generated in some models with extra dimensions.

The realm of models with extra dimensions does not like a unitarian state. It rather resembles a loose federation between several 'tribes' held together by little more than just the symbol of extra dimensions. It is not surprising then that those 'tribes' at times co-operate and at others times compete with each others. On the other hand postulating extra dimensions is a radical departure from conventional model building. As such it might lead us out of the 'dead end' in our efforts to decode the presumably profound message that 'Nature' has given us through the family structure.

We follow the words of our colleague L. Sehgal from Aachen: we view the ansatz of extra dimensions as an 'imagination stretcher'. Some see SUSY as a solution in search of a problem. We think this quote is a more appropriate characterization of the concept of extra dimensions.[14]

[14] The term "extra dimensions" is usually meant as extra *space* dimension; it is intriguing to speculate about extra *time* dimensions.

15.4 Baryogenesis in our Universe

The word "cosmology" makes an entrance when we analyze our Universe as a whole. It has already been discussed in different contexts, for instance when analyzing the impact of DM in **Sect. 15.1**. In this **Section** we have a 'personal' interest, namely understanding the huge asymmetry in matter vs. anti-matter in the Universe we live in.[15] Our Universe seems to have started from a thick soup of (anti)quarks, (anti)leptons and gauge bosons. The collisions were just too energetic for any nuclear binding to survive, but later (on the cosmology time scale), as the Universe expanded and hence cooled, nuclei could form and survive, and matter became a collection of nuclei of hydrogen, helium, electrons and their neutrinos. 'Today' our Universe contains nuclei mostly of atoms of hydrogens and heliums, a few per cent of heavier atoms, hardly of any known anti-atom (and anti-protons). Let us emphasize three main cosmological challenges:

(1) One of the most intriguing goal of "Big Bang" cosmology was to understand nucleosynthesis: to produce the observed abundances of nuclei in the Universe as dynamically generated rather than dialed as input values. Before the 1960s, it was postulated that all elements in our Universe were produced either in star interiors or during supernova explosion, but quantitative estimates, especially of the helium abundance in our Universe, were in contrast with experimental observations. The nuclear community has now met successfully this challenge for the light nuclei, that, according to Big Bang Nucleosynthesis (BBN), are originated in the first stages of our Universe after the Big Bang.[16] The matter-antimatter asymmetry is a crucially important parameter for BBN.

(2) Can we 'understand' the difference between matter vs. anti-matter in our Universe? Let us start from a basic quantity, the baryon number density, namely the difference in the abundances of baryons vs. anti-baryons:

$$\Delta n_{\rm Bar} \equiv n_{\rm Bar} - n_{\overline{\rm Bar}} \; . \tag{15.5}$$

From BBN and observations on cosmic microwave background radiation, one gets

$$r_{\rm Bar} \equiv \frac{\Delta n_{\rm Bar}}{n_\gamma} \sim \text{few} \, \cdot \, 10^{-10} \quad , \tag{15.6}$$

where n_γ denotes the number density of photons in the cosmic background radiation. We conclude that in our Universe as a whole:

$$0 \neq \frac{n_{\rm Bar}}{n_\gamma} \simeq \frac{\Delta n_{\rm Bar}}{n_\gamma} \sim \mathcal{O}(10^{-10}) \; , \tag{15.7}$$

where n_γ denotes the number density of photons in the cosmic background radiation. This relation summarizes experimental observations that can be qualitatively expressed as: (a) our Universe is not empty; (b) our Universe is almost empty;

[15]'Matter' is the 'stuff' 'we' are built off. Due the **CPT** invariance matter and anti-matter have the same mass and width. Collisions of 'matter' vs. 'anti-matter' lead to enhanced energies due to a loss of equal numbers of 'matter' and 'anti-matter'.

[16]More working is still needed to solve open issues on nucleosynthesis with the heavier nuclei.

(c) there are practically no primary anti-baryons. All of these statements are inferred from our knowledge of "known" matter.

(3) On the other side, in the past century[17] our community has learnt that our Universe may be very different and even more complex than thought before: hardly any part of our Universe is empty on the cosmology distance scale! There are two new 'actors': DM and Dark Energy, whose existence we have found out only due to gravity so far. We seem to have only $\sim 4.5\%$ of known matter and $\sim 26.5\%$ of DM, as discussed in **Sect. 15.1**.

It is interesting to ascertain if discussions about baryogenesis can give us information about DM in our Universe.

15.4.1 *Ingredients for baryogenesis in known matter*

If in the part of the Universe close to us 'matter' dominates, it is conceivable that other 'neighborhoods' exist where antimatter dominates. Such a Universe would be formed as a patchwork quilt, with matter and anti-matter dominated regions, being on the whole matter-antimatter symmetric. There are several reason why this scenario is widely held to be very unlikely. No mechanism has been found by which a matter-antimatter symmetric Universe following a Big Bang evolution can develop sufficiently large regions with non-vanishing baryon number. While there will be statistical fluctuations, they can be nowhere nearly large enough. Likewise for dynamical effects: baryon-antibaryon annihilation is by far not sufficiently effective to create pockets with the observed baryon number. Moreover, the observed isotropy at large scale of cosmic microwave background confirms that in our local area all visible galaxies consist of matter. If there were distant galaxies made of anti-matter, those would be in contact with matter at their boundaries and should produce gamma rays between these domains that would show up in cosmic microwave background. Then antimatter in this kind of Universe should be well separated from matter. There is another possibility, which is also very unlikely, that the observed baryon asymmetry can be obtained from a hot Big Bang as the result of a small excess of baryons over anti-baryons. Indeed, the number density of surviving baryons can be estimated [569]

$$n_{\rm Bar} \sim \frac{n_\gamma}{\sigma_{\rm annih}\, m_N\, M_{\rm Pl}} \sim 10^{-19}\, n_\gamma \tag{15.8}$$

where $\sigma_{\rm annih}$ denotes the cross-section of nucleon annihilation, while m_N and $M_{\rm Pl}$ the nucleon and Planck mass. The result is much smaller than the observed value in Eq. (15.7). Finally, there are excellent reasons, based on observed features of the cosmic microwave background radiation, to think that inflation took place during the history of our Universe. Any primordial baryon asymmetry would have been exponentially diluted away by inflation.

Therefore, it is widely assumed that the present huge asymmetry evolved from an initial state that was basically symmetric. The program of baryogenesis is to

[17] Remember Ref. [514] from 1933!

understand under which condition the baryon number of our Universe that vanishes at the initial time – which for all practical purposes is the Planck time – develops a non-zero value later on:

$$\Delta n_{\rm Bar}(t = t_{\rm Pl} \simeq 0) = 0 \stackrel{?}{\Longrightarrow} \Delta n_{\rm Bar}(t = \text{`today'}) \neq 0. \tag{15.9}$$

One could go further with this challenge: to explain how the observed baryon number is *dynamically* generated no matter[18] *what its initial value was*!

In a pioneering paper that appeared in 1967, Sakharov listed the three ingredients that are essential for the feasibility of such a program [570]:

Ingredient 1: Since the final and initial baryon numbers differ, there have to be baryon number violating transitions:

$$\mathcal{L}(\Delta n_{\rm Bar} \neq 0) \neq 0. \tag{15.10}$$

Ingredient 2: **CP** (and **C**) invariance has to be broken. Otherwise for every baryon number changing transition $N \to f$ there is its **CP** conjugate one $\overline{N} \to \overline{f}$ with $\Gamma(N \to f) = \Gamma(\bar{N} \to \bar{f})$. Therefore one needs

$$\Gamma(N \stackrel{\mathcal{L}(\Delta n_{\rm Bar} \neq 0)}{\longrightarrow} f) \neq \Gamma(\overline{N} \stackrel{\mathcal{L}(\Delta n_{\rm Bar} \neq 0)}{\longrightarrow} \overline{f}) \tag{15.11}$$

to produce non-zero net baryon number 'now'.

Ingredient 3: Unless one is willing to entertain thoughts of **CPT** violations, the baryon number and **CP** violating transitions have to proceed *out of thermal equilibrium*. For in thermal equilibrium time becomes irrelevant globally and **CPT** invariance reduces to **CP** symmetry, which has to be *avoided*:

$$\textbf{CPT} \text{ invariance} \stackrel{\text{thermal equilibrium}}{\Longrightarrow} \textbf{CP} \text{ invariance}. \tag{15.12}$$

These three conditions have to be satisfied *simultaneously*. The other side of the coin: once a baryon number has been generated through the concurrence of these three effects, it can be washed out again by these same effects in the evolution of the Universe. Sakharov himself came back to this challenge 17 years later, in another published paper with more refined tools and the title "Cosmological transitions with changes in the signature of the metric" [571].

15.4.2 *GUT baryogenesis*

Sakharov's paper was not noticed for several years (except from the author of Ref. [572]) until the concept of Grand Unified Theories (GUTs) emerged in 1974 [573]. The three necessary ingredients that Sakharov had identified started to be recognized in GUTs from the years 1978/79 [574, 575]:

Ingredient 1, see Eq. (15.10): Obviously baryon number changing transitions exist in GUTs. Placing quarks and leptons into common representations of the underlying gauge groups characterized the GUTs from the beginning – first and second examples of GUT models employed $SU(5)$ and $SO(10)$ groups. It means that

[18] It is a pun with the word 'matter'...

gauge interactions can change baryon and lepton numbers. The gauge bosons that are generically referred to as X bosons are characterized by at least two couplings to fermions that violate baryon and/or lepton number:

$$X \leftrightarrow qq, q\bar{\ell} \, . \tag{15.13}$$

Ingredient 2, see Eq. (15.11): X decays can be described in this simplified model by

$$\Gamma(X \to qq) = (1+\Delta_q)\Gamma_q \, , \;\; \Gamma(X \to q\bar{\ell}) = (1-\Delta_\ell)\Gamma_\ell \, , \tag{15.14}$$

$$\Gamma(\overline{X} \to \overline{qq}) = (1-\Delta_q)\Gamma_q \, , \;\; \Gamma(\overline{X} \to \bar{q}\ell) = (1+\Delta_\ell)\Gamma_\ell \, . \tag{15.15}$$

Thus one can probe discrete symmetries, namely **CPT**, **CP** and **C** invariance :

$$\textbf{CPT} \text{ invariance } \Longrightarrow \Delta_q\Gamma_q = \Delta_\ell\Gamma_\ell \tag{15.16}$$

$$\textbf{CP} \text{ invariance } \Rightarrow \Delta_q = 0 = \Delta_\ell \quad , \quad \textbf{C} \text{ invariance } \Rightarrow \Delta_q = 0 = \Delta_\ell \tag{15.17}$$

GUT models generally allow for several potential sources of **CP** violation.

Ingredient 3, see Eq. (15.12): GUTs of the SM $(SU(3)_C \times SU(2)_L \times U(1))$ are characterized by the mass (or energy) scale M_{GUT} where a phase transition takes place. For temperatures T well *above* M_{GUT} *all* quanta are relativistic with a number density

$$n(T) \propto T^3 \, . \tag{15.18}$$

This is expected on dimensional grounds, since the energy scale T is the only scale in such ultra-relativistic limit. For temperatures T *around* the phase transition M_{GUT}, X gauge bosons acquire a mass $M_X \sim \mathcal{O}(M_{\mathrm{GUT}})$ and their equilibrium number density becomes Boltzmann suppressed:

$$n_X(T) \propto (M_X T)^{\frac{3}{2}} \exp\left(-\frac{M_X}{T}\right) . \tag{15.19}$$

More X bosons will decay than they are regenerated from qq and $q\bar{l}$ collisions; ultimately the number of X bosons is brought down to the level described by Eq. (15.19). Yet that will take some time; the expansion in Big Bang cosmologies leads to a cooling rate that is so rapid that thermal equilibrium cannot be maintained through the phase transition. Then X bosons decay drop out of thermal equilibrium [576]. By ignoring the back production of X bosons in qq and $q\bar{\ell}$ collisions, one finds the estimate:

$$\frac{\Delta n_{\mathrm{Bar}}}{n_\gamma} \sim \frac{\frac{4}{3}\,\Delta_q\,\Gamma_q - \frac{2}{3}\,\Delta_\ell\,\Gamma_\ell}{\Gamma_{\mathrm{tot}}}\frac{n_X}{n_0} = \frac{\frac{2}{3}\,\Delta_q\,\Gamma_q}{\Gamma_{\mathrm{tot}}}\frac{n_X}{n_0} \, , \tag{15.20}$$

where n_X denotes the initial number density of X bosons and n_0 the number density of the light decay products.[19] The three essential conditions for baryogenesis are thus 'naturally' realized around the GUT scale in Big Bang cosmologies:

[19] Due to thermalization effects we can have $n_0 \gg 2n_X$.

- $\Gamma_q \neq 0$ represents baryon number violation;
- $\Delta_q \neq 0$ reflects **CP** violation;
- thermal equilibrium does not happen there.

This challenge had been formulated first in GUT models, and the answers were in the right ballpark. It was a highly attractive feature of GUTs – in particular, since this was *not* among the original motivations for constructing such theories.

However, let us mention that there are serious problems in any attempt to have baryogenesis occur at a GUT scale:

- A baryon number generated at huge temperatures is likely to be washed out in the subsequent evolution of the Universe. Thus GUT baryogenesis scenarios are unlikely to reproduce the observed baryon asymmetry of our Universe.
- Very little is known about the dynamical actors operating at GUT scales.

There have been attempts to overcome the first problem, creating a non-vanishing $B - L$ asymmetry by using specific matter representations, larger groups or Majorana neutrinos. At the very least, baryogenesis at GUT scales can be characterized as a proof of principle that the baryon number of our Universe can be understood as dynamically generated, That said, it would be premature to write-off GUT baryogenesis and there have been recent attempts to revive it, see for instance Ref. [577].

15.4.3 *Electroweak baryogenesis*

As its name suggests, electroweak baryogenesis refers to any mechanism that produces an asymmetry in the density of baryons during the electroweak phase transition in our Universe. It is a very active and attractive scenario at present [578], one reason being its high testability. The components of its dynamics are very well known; experimenters and theorists can find a common ground. There are common features of various realizations of electroweak baryogenesis:
(a) A well-studied phase transition, namely the spontaneous breaking

$$SU(2)_L \times U(1) \implies U(1)_{\rm QED} \tag{15.21}$$

takes place. At the electroweak phase transition, our very early Universe was out of thermal equilibrium, and one of the Sakharov's condition is satisfied.

(b) We know from the CKM analysis of the flavour sector that **CP** and **CP** violation operate in weak interactions at the EW scale; **CP** violation is well established in $\Delta S \neq 0 \neq \Delta B$ transitions and in March 2019 it was established in $\Delta C = 1$ ones as well [174]. This satisfies another of Sakharov conditions. Yet most authors agree that CKM theory fails to give the amount of **CP** violation necessary for baryogenesis by several orders of magnitude. On the other hand, ND scenarios of **CP** violation – in particular of the Higgs variety – can reasonably be called upon to perform this task.

What about baryon number violation? It is often not appreciated that the electroweak forces of the SM by themselves violate baryon number in a very subtle way. It follows by a "quantum anomaly": the baryon number current is conserved on the classical level, yet *not* on the quantum one:

$$\partial^\mu J_\mu^{\rm Bar} = \partial^\mu \sum_q (\overline{q}_L \gamma_\mu q_L) = \frac{g_2^2}{16\pi^2} {\rm Tr}\, G^{\mu\nu}\tilde{G}_{\mu\nu} \neq 0 \,; \tag{15.22}$$

g_2 denotes the $SU(2)_L$ gauge coupling, $G_{\mu\nu}$ the electroweak field strength tensor $G_{\mu\nu} = \tau_a \left(\partial_\mu A_\nu^a - \partial_\nu A_\mu^a + g\epsilon_{abc}A_\mu^b A_\nu^c\right)$ with τ_a being the $SU(2)$ generators and $\tilde{G}_{\mu\nu}$ its dual: $\tilde{G}_{\mu\nu} = \frac{1}{2}\epsilon_{\mu\nu\alpha\beta}G_{\alpha\beta}$. The right hand side of the Eq. (15.22) can be written as the divergence of a current:

$${\rm Tr}\, G^{\mu\nu}\tilde{G}_{\mu\nu} = \partial^\mu K_\mu \,, \quad K_\mu = 2\epsilon_{\mu\nu\alpha\beta}\, {\rm Tr}\left(A^\nu \partial^\alpha A^\beta - \frac{2}{3} i g A^\nu A^\alpha A^\beta\right). \tag{15.23}$$

We have encountered this situation in our discussion of the strong **CP** problem of QCD in **Sect. 10.2**: a conservation law is vitiated by a triangle anomaly on the quantum level. Although the offending term can be written as a *total divergence*, it still affects the dynamics of non-abelian gauge theories in a non-trivial way. There is an infinite numbers of inequivalent ground states that are differentiated by the value of their K charge, i.e. the space integral of K_0, the zeroth component of the K_μ current constructed from their gauge field configuration. This charge reflects differences in the gauge topology of the ground states; it is called the "topological" charge. The transition from one ground state to another, which represents a tunneling phenomenon, is accompanied by a change in baryon number. Elementary quantum mechanics tells us that this baryon number violation is described as a barrier penetration and exponentially suppressed at low temperatures (or energies) [213]:

$${\rm Prob}(\Delta n_{\rm Bar} \neq 0) \propto \exp(-16\pi^2/g^2) \sim \mathcal{O}(10^{-160}). \tag{15.24}$$

This suppression reflects the tiny size of the weak coupling.

There is a corresponding anomaly for the lepton number current: the lepton number is violated as well with this selection rule:

$$\Delta n_{Bar} = \Delta n_{Lept} \;; \tag{15.25}$$

$n_{Bar} - n_{Lept}$ – the difference between baryon and lepton numbers – is still conserved.

This is a case where the perturbative expansion is unable to describe all the dynamics of the theory. We need non-perturbative dynamics, where we can employ the following three classes of tools:

- "Instanton", which is a localized, **finite-action** solution of classical field equation for *imaginary time* τ. Instantons have often been used in the 'Dilute Instanton Gas Approximation' (DIGA) together with Lattice QCD [536, 579]. The analyses of only DIGA vs. only Lattice QCD give quite different results [580].
- "Topological soliton", which is a static, *stable*, **finite-energy** solution of the classical field equations for *real* time t.

- "Sphaleron",[20] which is a static, *unstable*, **finite-energy** solutions of the classical field equations for *real* time t.

Instantons and topological solitons are relevant for the equilibrium properties of the underlying theory, whereas sphalerons are relevant to the dynamics and thus for the origin of the cosmic matter-antimatter asymmetry [581].

Let us go back to $\Delta n_{\rm Bar}$: while its value is negligibly small at low temperatures, it can be significant at much higher temperatures (or energies). At sufficiently high energies the huge suppression of baryon number changing transition rates will evaporate, since the transition between different ground states can be achieved classically through a motion *over* the barrier. At which energy scale this will happen and how quickly baryon number violation will become operative? Some theoretical answers based on semi-quantitative observations had already been given without knowing the mass of 'the' Higgs boson, see for instance Refs. [576,582,583]. Of course nowadays the impact of the Higgs field on these analyses has to be included.[21]

The sphalerons carry the topological K charge. In the SM they induce effective multi-state interactions among left-handed fermions that change baryon and lepton number by three units each:

$$\Delta n_{\rm Bar} = \Delta n_{\rm Lept} = 3. \tag{15.26}$$

At high energies where the weak bosons W and Z are massless, the height of the transition barrier between different ground states vanishes likewise, and the change of baryon number can proceed in an unimpeded way and presumably faster than our Universe expands. Thermal equilibrium is then maintained, and any baryon asymmetry existing before this era is actually washed out.[22] Rather than generating a baryon number, sphalerons act to drive our Universe back to matter–antimatter symmetry at this point in its evolution.

At energies below the phase transition, i.e., in the broken phase of $SU(2)_L \times U(1)$ baryon number is conserved for all practical purposes, as pointed out above. The value of $\Delta n_{\rm Bar}$ as observed today can thus be generated only in the transition from the unbroken high energy to the broken low energy phase. With $\Delta n_{\rm Bar} \neq 0$ processes operating there, the issue now turns to the strength of the phase transition: is it relatively smooth like a second order phase transition or violent like a first order one? Only the latter scenario can support baryogenesis.

A large interesting theoretical work has been done on the thermodynamics of the SM in an expanding Universe. Employing perturbation theory and lattice studies, our community has arrived at the following conclusion: for light Higgs masses up

---

[20] Greek: $\sigma\phi\alpha\lambda\epsilon\rho o\zeta$ = 'slippery'.

[21] 'The' Higgs boson has been found with mass and features compatible with the SM. The Higgs sector plays an essential role to constrain electroweak baryogenesis and the discovery of the Higgs has narrowed down the possibilities of various models.

[22] To be more precise: only $n_{\rm Bar} + n_{\rm Lept}$ is erased within the SM, whereas $n_{\rm Bar} - n_{\rm Lept}$ remains unchanged.

to around 70 GeV, the phase transition is first order, while for larger masses it is second order [584]. Since 2012 we know that the neutral Higgs boson has a mass $\simeq$ 125 GeV [4], thus foreclosing baryogenesis from the (non-extended) SM.

15.4.4 *Leptogenesis driving baryogenesis*

An interesting approach, known under the name of baryogenesis through leptogenesis (or in short leptogenesis), is when the baryon number of the universe is seen as a reflection of a primary lepton asymmetry. If at some high energy scales a lepton number is generated, sphaleron processes can communicate this asymmetry to the baryon sector while preserving the difference $n_{\rm Bar} - n_{\rm Lept}$. This is usually referred to by saying that $B - L$, the difference between baryon and lepton number, is conserved. There are various ways in which such scenario can be realized. The simplest one requires hardly more than adding *heavy right handed Majorana* neutrinos to the SM. This assumption is highly attractive in any case, since it enables us to implement the see-saw mechanism, described in **Sect. 9.5**, for explaining why the observed neutrinos are practically massless. We list the main features of baryogenesis driven by leptogenesis, following Ref. [160], where it was conjectured for the first time:

- A primary lepton asymmetry is generated at high energies well above the electroweak phase transition:
  (a) Since a Majorana neutrino N is its own **CPT** mirror image, its dynamics necessarily violates lepton number. It will decay at least through the following channels:

$$N \to \ell \overline{H}\ , \ \ \overline{\ell} H \tag{15.27}$$

  with ℓ and $\overline{\ell}$ denoting a light lepton and anti-lepton and H and $\overline{H}$ a Higgs and anti-Higgs field.[23]
  (b) A **CP** asymmetry can arise

$$\Gamma(N \to \ell \overline{H}) \neq \Gamma(\overline{N} \to \overline{\ell} H) \tag{15.28}$$

  A mixing matrix naturally arises among neutrino generations, which can be expected to contain irreducible complex phases and thus induce **CP** violation in qualitative analogy to the CKM mechanism in the quark sector. The Majorana nature of the neutrinos introduces additional physics phases with respect to the case in which all fermions are Dirac particles. Moreover, the neutrino mass matrix is quite different from the mass matrix for charged leptons.
  (c) Depending on the smallness of the couplings, these neutrino decays can be sufficiently slow to occur *out* of thermal equilibrium at the time that the asymmetry is generated, that is around the energy scale where the Majorana masses emerge.

[23] It is somewhat similar to what we have discussed in **Sect. 15.4.2**.

- The resulting lepton asymmetry is transferred into a baryon number through sphaleron mediated processes in the *un*broken high energy phase of $SU(2)_L \times U(1)$:

$$\langle \Delta n_{\rm Lept} \rangle = \frac{1}{2}\langle \Delta n_{\rm Lept} + \Delta n_{\rm Bar} \rangle + \frac{1}{2}\langle \Delta n_{\rm Lept} - \Delta n_{\rm Bar} \rangle \implies \frac{1}{2}\langle \Delta n_{\rm Lept} - \Delta n_{\rm Bar} \rangle \, . \tag{15.29}$$

- $\Delta n_{\rm Bar}$ thus generated survives through the subsequent evolution of the Universe.

In the original scenario of leptogenesis, the heavy Majorana neutrino masses are typically close to the GUT scale, as suggested by natural GUT embedding of the seesaw mechanism. Such scenario is very difficult to test experimentally. There are other models that predict neutrinos of lower masses, which are within the reach of modern experiments. With the finally measured spin-zero H^0 meson, large-scale lattice simulations have been used to compute the sphaleron rate in the SM, together with the freeze-out temperature in our early Universe, where the Hubble rate wins over the baryon number violation rate [585]. These values are relevant for low scale leptogenesis scenarios.

15.4.5 *Wisdom – conventional and otherwise*

We understand how nuclei were formed in the universe, given protons and neutrons. Obviously it would be even more fascinating if we could also understand how these baryons were generated in the first place, but a quantitative successfully theory of baryogenesis has not been found yet. Yet we have learnt which dynamical ingredients are necessary:

- GUT scenarios for baryogenesis provide us with a proof of principle that such a program can be realized. In practical terms, however, they suffer from various shortcomings:
  (a) Since baryogenesis is generated at the GUT scales, very little is known about that dynamics and it is unlikely to learn much more about it.
  (b) It appears most likely that a baryon number produced at such high scales is subsequently washed out.
- The highly fascinating proposal of baryogenesis at the electroweak phase transition has attracted a large attention and deservedly so:
  (a) A baryon number emerging from this phase transition would be in no danger of being diluted substantially.
  (b) The dynamics involved here is known to a considerable degree and will be probed even more with ever increasing sensitivity over the coming years.
  However, in the SM the electroweak phase transition is of second order and thus not sufficiently violent for baryogenesis.
- The most intriguing tale – in our view – turns some of the vices of sphaleron dynamics into virtues, by attempting to understand the baryon number of our

Universe as a reflection of a *primary* lepton asymmetry. The required new dynamical entities – Majorana neutrinos and their decays – obviously would impact on our Universe in other ways as well.

The challenge to understand baryogenesis has already inspired our imagination, prompted the development of some very intriguing scenarios and thus has initiated many fruitful studies – and in the end we might even be successful in meeting it!

15.5 Multiverse – short history in the 'real world'?

By now a reader will have understood (we hope) why we titled this book: "New Era for **CP** Asymmetries" plus "Axions and Rare Decays of Hadrons and Leptons". The landscape of our Universe is quite 'complex'. As just seen in the previous **Sections**, our Universe is not close to being empty, since DM and Dark Energy are 'everywhere' – we have candidates for DM, although we have hardly candidates for Dark Energy. Yet we have not changed our mind: probing **CP** asymmetries and truly rare decays is determining to show the impact of ND in indirect ways. Yet we cannot ignore new concepts and schemes or even innovative ideas coming from the early 1980's. After all, this book is dedicated to Lev Okun; his record has shown he had broad horizons, and he was always very interested to new ideas.

An example is the complex 'string theory landscape':

- Classes of local SUSY (with gravity) are embedded in super-string theory (with 10 or 11 dimensions) to describe the dynamics including cosmologies.
- Novel challenges follow: the string theory landscape refers to a huge number of different vacua. All parameters of QFT, the masses and couplings of particles, depend on them, and one has to understand how to draw the fundamental ones from such large set of solutions.

An innovative 'travel' started in 1981 – 1985 with these important papers:

- "The Inflationary Universe: A Possible Solution to the Horizon and Flatness Problems" by Guth [586];
- "Creation of Universes from Nothing" by Vilenkin [587];
- "Wave Function of the Universe" by Hartle and Hawking [588];
- "The New Inflationary Universe Scenario" by Linde [589];
- "Natural Inflation" by Steinhardt [590].
- "Quantum Mechanics of Inflation" by Fischler, Ratra and Susskind [591].

It is amazing that some of these pioneers still quite disagree after a long time of discussions, thinking and back to discussions. In our view that is a good sign of vitality and progress.

"Cosmology" is the prime focus of these analyses. Very close to the previous century, our understanding of the Universe has changed crucially. The results of the Supernova Search Team Collaboration and Supernova Cosmology Project Collaboration become available in 1998 [592] and 1999 [593]. Three leaders of these collaborations (S. Perlmutter, A. Riess and B. Schmidt) were awarded the 2011 Nobel Prize for Physics for providing evidence that the expansion of the Universe is *accelerating.*[24] Our Universe is expanding in every direction since the Big Bang; it started with "time" at zero to exist. One question follows immediately: some time in the future (on the cosmology time scale) our Universe will come back to a 'singularity' in space and time, namely called 'Big Crunch'? Present data (and our understanding of those) tell us that 'Big Crunch' is unlikely to happen, favouring other theories for the Cosmic doomsday.

Another of the problems (or challenges) for present cosmology is the "horizon problem". It originates from the observed high degree of uniformity and thermal equilibrium of the cosmic microwave background. While it might at first seems quite natural for the hot gas of the early universe to be in good thermal equilibrium, this becomes awkward in the standard Big Bang picture because of the presence of a cosmological horizon. The horizon size of the universe is just how far a photon can have traveled since the Big Bang, since matter and energy can travel no faster than the vacuum speed of light. Two points in the universe separated by more than a horizon size have no way to reach thermal equilibrium, since they cannot have ever been in causal contact. This means that two regions in order to interact must be separated by no more than the distance that light can travel during the age of the Universe at the epoch in question. They must be within each other's "horizon". Since the universe was about 300,000 years old at the Era of Recombination the horizon size then was about 300,000 light years. Causal contact (and therefore thermal equilibrium) could occur within a radius of about 300,000 light years. Presently, our horizon is about 13 billion light-years, so two regions in opposite directions in the sky are separated by 26 billion light years and could not have been in contact with each other when matter and radiation went their separate ways.

In other terms, the problem is how to connect the cosmic light horizons just after the beginning of the Universe on one side and the other one at the opposite directions and our cosmic light horizon, namely for our observers. For example, one photon hits the North Pole, while the other one hits the South Pole. Can those two photons exchange any information from the time when they are released, namely just after the Big Bang? Those are *not* causally connected, since they are *outside* of each other's horizon in 'classical' physics. Cosmic inflation comes to rescue. In the inflationary scenario, our very early Universe has gone through an exponential expansion of space. To be a bit more explicit: the "inflationary" epoch started immediately from the Big Bang lasting a truly tiny fraction of a second. Before this period of inflation, the entire Universe has been flat, could have been in causal contact and equilibrate to a common temperature. Widely separated regions today

[24]The results of these collaborations have changed our understanding of our Universe. Therefore the award of this Nobel prize is obvious. On the other hand, it is based on the efforts of large teams; still, the rules are to give the Nobel prize to three people at the most.

were actually very close together in the early Universe, explaining why photons from these regions have (almost exactly) the same temperature.

In the Spring 2013, the European Space Agency (ESA) announced new results from a satellite called "Planck" [594, 595]. This spacecraft had mapped the cosmic microwave background radiation light emitted more than 13 billion years ago just after the Big Bang. It was said the new map confirmed the theory that the Universe began with a Big Bang followed by a tiny period of hyper-accelerated expansion–the "inflation". This expansion smoothed the Universe to such an extent that – billions of years later – it remains nearly uniform all over space and in every direction and 'flat' – as opposed to curved like a sphere, except tiny variations in the concentration of matter that account for the finely detailed hierarchy of starts, galaxies and galaxy clusters around our Universe.

We observe a Universe with a near-flat geometry, which is a very special case, where the amount of matter present is just sufficient to slow down its expansion, but insufficient to re-collapse it. This seems like a truly remarkable coincidence: if the Universe were not nearly flat, it would have either re-collapsed very early before making galaxies or expanded so rapidly that structures would not have formed. Yet avoiding a drastic growth of minor variations in the universe density of matter from the Big Bang era requires an extreme fine tuning of initial conditions. The inflation provide a solution to this 'flatness problem'. By a very rough analogy, inflation flattens the universe as a balloon smoothes out the wrinkles on its surface by expanding. Pedagogic introductions to these problems and possible solutions can be found in Ref. [596].

There are other ideas that have came up with "inflationary" cosmology – actually "eternal inflationary" cosmology, and different meanings of "multiverse".[25] In many inflationary models large quantum fluctuations produced during inflation may significantly increase the value of the energy density in some parts of the universe. This can activate a mechanism of production of new inflationary domains. Eternal inflationary refers to eternal process of self-reproduction of universes – once inflation starts, it continues indefinitely into the future.

Let us give three examples of pioneering ideas from our "avant-garde".

(1) In Ref. [597] titled "The Multiverse Interpretation of Quantum Mechanics" Bousso and Susskind argued: "... that the many-worlds of quantum mechanics and the many worlds of the multiverse are the same thing, and that the multiverse is necessary to give exact operational meaning to probabilistic predictions from quantum mechanics. Decoherence (the modern version of wave function collapse) is subjective in that it depends on the choice of a set of unmonitored degrees of freedom, the environment".

[25]The "multiverse" is also addressed in the context of super-string theories. This concept has attracted mostly theorists with philosophical tendency.

(2) In "A brief history of the multiverse" given by Linde [598], questions from theorists like "Why is the Cosmological Constant so Small?" are presented [599].[26] That question has stimulated an important paper of S. Weinberg about "Anthropic Bound on the Cosmological Constant" [600]. These are new ideas and suggestions.[27]

The main point of inflationary cosmology is to make our part of the Universe homogeneous by stretching any pre-existing inhomogeneities and by placing all possible 'defects' such as domain walls and monopoles far away from us, thus rendering them unobservable. If the Universe consists of different parts, each of these parts after inflation may become locally homogeneous and so large that its inhabitants will not see other parts of the Universe, so they may conclude, incorrectly, that the Universe looks the same everywhere. However, properties of different parts of the Universe may be dramatically different. In this sense, the Universe effectively becomes a multiverse consisting of different exponentially large locally homogeneous parts with different properties. In other terms: the Universe may be divided into many large parts with different laws of low-energy physics operating in each of them an inflationary multiverse. When the Universe inflated, these different domains separated, each with its own values for physical constants – again applying the anthropic principle.

Going back to the main topic of the this book, the obvious question is if we can maybe understand the huge asymmetry in matter vs. anti-matter in the light of these novel ideas. We just mention two references already quoted in **Sect. 15.4**. In "A New Mechanism for Baryogenesis" [601] Affleck and Dine propose that baryogenesis is controlled by a light scalar field which experienced quantum fluctuations during inflation. The second one is a 1984 paper from Sakharov with the title "Cosmological Transitions with a Change in Metric Signature" [571], written 17 years later after his famous one [570].

(3) The third example is quite surprising. In an article for the "Scientific American, August 6, 2014" Steinhardt 'disowns Inflation, the Theory he helped create'. As he said in "Scientific American, April 1, 2011" on the inflation debate: 'Is the theory at the heart of modern cosmology deeply flawed?' – he continued, together with A. Ijjas and A. Loeb, in the article 'Cosmology: Pop Goes the Universe', published on "Scientific American Feb. 2017" [602]. One can summarize their lessons quickly: 'The latest astrophysical measurements, combined with theoretical problems, cast doubt on the long-cherished inflationary theory of the early cosmos and suggest we need new ideas.'

Following the 2013 ESA announcements mentioned before [594,595], Ijjas, Loeb and Steinhardt discussed their 'ramifications'. Those led them to very different conclusions in a paper published in 2013 [603] and also in a paper of two of

[26] We asked ourselves in **Chapters 10** on strong **CP** asymmetry: why is $\bar{\theta}$ is so tiny? This question led to "axions" (**Chapters 11**).

[27] Most often arguments based on anthropic considerations imply the existence of a multiverse where a selection is operated.

them in 2017 [604]. Their results are already summarized in their 2013 title: 'Inflationary paradigm in trouble after Planck2013'. They claimed that results from the *Planck* satellite combined with earlier observations from WMAP, ACT, SPT and other experiments eliminate a wide spectrum of complex inflationary models and favor models with a single scalar field. More importantly, all the simplest inflaton models are disfavored statistically relative to those with plateau-like potentials. A restriction to plateau-like models has three independent serious drawbacks: it exacerbates both the initial conditions problem and the multiverse-unpredictability problem, and it creates a new difficulty that they call the inflationary 'unlikeliness problem'. They commented on problems reconciling inflation with a standard model spin-zero Higgs H^0 boson, established by two ATLAS and CMS. Thus they conclude that 'recent data disfavor the best motivated inflationary scenarios and introduces new, serious difficulties that cut to the core of the inflationary paradigm'. In Ref. [604] with the title 'Fully stable cosmological solutions with a non-singular classical bounds', Ijjas and Steinhardt went forward by 'constructing non-singular classical bouncing cosmological solutions that are non-pathological'.
Although several of these statements seem to use generic words as 'suggestions', 'disfavored *statistically*' and 'serious drawbacks', they did not make their case with one or two statements, but studying correlations both with the experimental and theoretical sides.

We will not give judgments on these three approaches; we cannot act as referees, since we have no record about working on Multiverse. However, it is wonderful to watch these discussions. It is at order a comment similar to others crossing the whole book: to make progress about the underlying dynamics in Universe(s), our community cannot go for a 'golden medal' process: we have to discuss many transitions and their connections.

Finally, let us remark that we have given *72 references* for this **Chapter** alone! Basically we have listed a large literature for modern Cosmology, leaving the reader to ponder them – as well as future analyses.

Chapter 16

Learn from the past and lessons for the future

Imagination creates reality.

R. Wagner

16.1 Looking back

'Mature' people often think that looking at 'history' helps to analyze what happens with a broader perspective. Indeed, the comprehensive study of kaon and hyperon physics has been instrumental in guiding our community to the SM. Let us go through some fundamental steps:

- Parity symmetry had been well tested in atomic and nuclear transitions up to 1954, when the $\theta - \tau$ puzzle appeared with the decays:

$$\theta^+ \to \pi^+\pi^0$$
$$\tau^+ \to \pi^+\pi^+\pi^- \ ;$$

  the FS of θ^+ carries positive parity, while for τ^+ it is negative, which is fine. However, with better data it was realized, with smaller and smaller uncertainties, that both θ^+ and τ^+ had the very same mass and width. There was a 'puzzle'!
- In 1956 T.D. Lee and C.N. Yang pointed out that **P** had *not* been tested in weak dynamics [24]; these gentlemen were true pioneers.[1]
  By the year 1957 data had shown that both **P** and **C** invariances were not only broken, but basically broken at 100% level in charged weak transitions [25]. Furthermore, data 'then' connected **P** and **C** asymmetries in a way that kept both **CP** and **T** invariant (and also **CPT**). The solution of the the $\theta - \tau$ puzzle was that we have a single QM state: $\theta^+ \equiv K^+ \equiv \tau^+$.
  The *production* rate of this state exceeded the decay rate by many orders of magnitude – this was the origin of the name 'strange particles'. It was explained by postulating a new quantum number – "strangeness" – conserved by the

[1] T.D. Lee and C.N. Yang got the 1957 Nobel Physics Prize – a record!

strong and electric forces, though not by the weak ones. This was the beginning of the second quark family, although that was realized only later. As long as **CP** symmetry is *un*broken, the name of "left" vs. "right" can be defined by "matter" vs. "anti-matter", or by charge conjugation **C**.

- At that point, we had two neutral strange mesons that can decay only weakly. One can describe their transitions with mass eigenstates, which are also **P** eigenstates: "parity even" $K_\oplus \to 2\pi$, and "parity odd" $K_\ominus \to 3\pi$. Ignoring **CP** violation, which is a good practical assumption, we have four observables for neutral kaons, namely the masses for $K_\oplus$ and $K_\ominus$ and their widths.
  With the measured values $m_\pi \sim 140$ MeV and $m_K \sim 500$ MeV, the decays of $K_\ominus \to 3\pi$ are very much suppressed in their kinetic spaces. Thus the widths of $K_\ominus \to 3\pi$ are very small compared to $K_\oplus \to 2\pi$; to describe the same situation with different words: $\tau(K_\ominus)/\tau(K_\oplus) \gg 1$.[2] If the pions were close to be massless on the scale of M_K, one could hardly differentiate $K_\ominus$ vs. $K_\oplus$: the FS of $K_\oplus$ and $K_\ominus$ would be given by 'many' pions – not just two vs. three pions; it would lead to $\tau(K_\ominus)/\tau(K_\oplus) \sim 1$.
  In 1970 Glashow, Illiopolous and Maiani claimed that the SM can produce the measured values of the neutral kaon mass difference $\Delta M_K \simeq 3.5 \cdot 10^{-12}$ MeV using one-loop Feynman diagrams, if the SM 'then' had two families of quarks: it was a *prediction* of charm quarks with $m_c \sim 1.5$ GeV [131]![3]
- Finding **CP** violation was a true revolution in 1964. One reason was that our community assumed **CP** conservation without thinking about – except Lev Okun: he pointed out in his book "Weak Interactions of Elementary Particles" in his 1963 Russian version [1] – i.e., before the 1964 experimental paper [2] – that it was crucial to continue to search for the long-lived neutral kaon decay into $\pi^+\pi^-$.
  From **CP** violation we learnt that the name of "left" vs. "right" are not merely labels, but that they are *dynamically distinct*, as can be easily ascertain by comparing the rates of the two **CP**-mirror image processes of the long-lived neutral kaon into $\pi^-\ell^+\nu_\ell$ and $\pi^+\ell^-\bar{\nu}_\ell$. Another implication, if **CPT** invariance holds, is that our Universe distinguishes between *past* and *future* even at the *microscopic* level, because of **T** violation.
  CP violation finally leads to the postulation of yet another family, the third one; but again, that was realized only later.

All these elements, which are now essential pillars of the SM, were ND at *that* time! Our community has learnt from fundamental dynamics, whenever ND appeared in different directions; we still need more information for the future, in particular about connections among different processes.

---

[2]The measured value is $\tau(K_\ominus)/\tau(K_\oplus) \sim 574$!

[3]The skepticism of the time may be best expressed by the quote 'Nature must be smarter than Shelly' (Glashow): the creator of this aphorism prefers to remain anonymous.

16.2 Present 'landscape'

Between 1964 and 2000 **CP** violation was established only in a single system: the transitions of K_L mesons, as a seemingly unobtrusive phenomenon. Yet even so we came to understand that it represents not only an intellectual insight, but has also many far-reaching consequences. In 1980 it had been *pre*dicted that **CP** violation should be sizable, even large in the SM – in particular in the transition of $B^0 \to J/\psi K_S$ [134]. The first observation of **CP** violation outside the kaon system was announced by BaBar and Belle collaborations in 2002 [605, 606].

The SM has described successfully so far the world of hadrons with quarks of three colors and three families. It has also described charged leptons and neutrinos in three families. In the lepton sector, though, neutrino oscillations have been established, which are beyond the SM. In the following we will summarize some main results in flavour physics of hadrons (and top quark) and in leptons dynamics.

16.2.1 *Two neutral kaons*

Assuming **CPT** invariance, the two neutral pseudoscalar mesons K^0 and $\bar{K}^0$ have the same mass and width. As discussed in **Chapter 6**, in (linear) QM K^0 - $\bar{K}^0$ oscillations lead to

$$|K_S\rangle = p|K^0\rangle + q|\bar{K}^0\rangle \quad , \quad |K_L\rangle = p|K^0\rangle - q|\bar{K}^0\rangle \tag{16.1}$$

$$M(K_L) - M(K_S) > 0 \quad , \quad \Gamma(K_S) - \Gamma(K_L) > 0\,; \tag{16.2}$$

Eq. (16.1) defines the QM states K_L and K_S, where L means both *larger* mass and *longer* lifetime, and S indicates *smaller* mass and *shorter* lifetime; it is a nice mnemonic. In terms of **CP** eigenstates, **CP** symmetry leads to the non-leptonic transitions: $K_\ominus \to 3\pi$, and $K_\oplus \to 2\pi$. In presence of **CP** violation, K_L is mostly $K_\ominus$, but its amplitude has a small $K_\oplus$ component.[4] One can observe $K_L \to \pi^+\pi^-/\pi^0\pi^0$ (and $K_L \to l^\pm\nu\pi^\mp$) decays and quantify the amount of **CP** violation with the ratios of non-leptonic amplitudes for $K_L \to \pi\pi$ vs. $K_S \to \pi\pi$: $|\eta_{00}|, |\eta_{+-}| \neq 0$. *Indirect* **CP** violation is described by $|\epsilon_K| = (2|\eta_{+-}| + |\eta_{00}|)/3$, while *direct* one by $\mathrm{Re}(\epsilon'/\epsilon_K) \simeq (1 - |\eta_{00}/\eta_{+-}|)/3$. Two papers [24, 128] had a revolutionizing impact on the perception of our Universe, and how we analyze the elements of its Grand Design. Present data give:

$$|\epsilon_K| = (2.228 \pm 0.011) \cdot 10^{-3} \tag{16.3}$$

$$\mathrm{Re}(\epsilon'/\epsilon_K) = (1.66 \pm 0.23) \cdot 10^{-3}\,. \tag{16.4}$$

These values are the world averages from NA48 [241] and KTeV [242, 607] collaborations. The SM with three families of quarks produces the measured value of $|\epsilon_K|$. The situation for *direct* **CP** asymmetry is much less clear, as discussed in details in **Sect. 7.2**.

[4] Analogously, K_S has a very tiny $K_\ominus$ component.

16.2.2 *Truly rare decays of kaons*

The SM predictions for $K^+ \to \pi^+\bar{\nu}\nu$ and $K_L \to \pi^0\bar{\nu}\nu$ are well controlled [165, 190], as discussed in **Sect. 6.5.10**. Let us compare with present data

$$\begin{aligned}
\mathrm{BR}(K^+ \to \pi^+\bar{\nu}\nu)|_{\mathrm{SM}} &= (8.39 \pm 0.30) \cdot 10^{-11} \left[\frac{|V_{cb}|}{40.7 \cdot 10^{-3}}\right]^{2.8} \left[\frac{\phi_3/\gamma}{73.2^0}\right]^{0.74} \\
\mathrm{BR}(K_L \to \pi^0\bar{\nu}\nu)|_{\mathrm{SM}} &= (3.36 \pm 0.05) \cdot 10^{-11} \left[\frac{|V_{cb}|}{40.7 \cdot 10^{-3}}\right]^{2} \left[\frac{\phi_3/\gamma}{73.2^0}\right]^{2} . \\
\mathrm{BR}(K^+ \to \pi^+\bar{\nu}\nu)|_{\mathrm{PDG2020}} &= (17 \pm 11) \cdot 10^{-11} \qquad (16.5) \\
\mathrm{BR}(K_L \to \pi^0\bar{\nu}\nu)|_{\mathrm{PDG2020}} &< \ \ 300 \cdot 10^{-11} .
\end{aligned}$$

A search for the $K^+ \to \pi^+X^0$ decay, where X^0 is a long-lived feebly interacting particle, is performed through an interpretation of the $K^+ \to \pi^+\bar{\nu}\nu$ analysis of data collected in 2017 by the NA62 experiment at CERN [608]. Further progress is expected in studies of the $K^+ \to \pi^+\bar{\nu}\nu$ decays from NA62, since it will resume data-taking in 2021, after modifications of the NA62 beam line, and the installation of an additional beam spectrometer station and a veto counter.

The present limit on $\mathrm{BR}(K_L \to \pi^0\bar{\nu}\nu)$ came from the J-PARC KOTO experiment [194]. The KOTO experiment will continue its analysis with data gathered in 2016-2018, and 2019.[5] By $\sim$ 2024 the KOTO collaboration might have the sensitivity to reach the SM prediction. Furthermore they also provide a limit given by PDG2020 [8]: $\mathrm{BR}(K_L \to \pi^0X^0) < 2.4 \cdot 10^{-9}$, where X^0 is an *invisible* boson recoiling against the π^0.

16.2.3 *Transitions of beauty mesons*

Let us summarize the 'landscape' of oscillations and **CP** asymmetries in beauty mesons B^0, $B^\pm$ and B^0_s. For neutral B mesons, in contrast with the neutral K system, the lifetime difference between the two mass eigenstates is small compared with the mixing frequency shaped by the difference in masses. This difference in behavior is due to the larger mass of the B meson implying a greater phase space for flavor-specific decays in the B system. It dominates the partial width (in contrast to the K system), giving equivalent contributions (by **CPT** symmetry) to the width of both neutral B eigenstates. The resulting lack of decay suppression of either eigenstates implies nearly equivalent lifetimes.

A large $B^0 - \bar{B}^0$ oscillations had been predicted due to very heavy top quarks in the SM; that has been confirmed in 1987 by ARGUS [610] and also by CLEO in 1989 [611]. Data from PDG2020 give [8]

$$\frac{\Delta m(B^0)}{\Gamma(B^0)} = 0.769 \pm 0.004 \qquad \frac{\Delta\Gamma(B^0)}{\Gamma(B^0_s)} = 0.001 \pm 0.010 \qquad (16.6)$$

$$\frac{\Delta m(B^0_s)}{\Gamma(B^0_s)} = 26.89 \pm 0.07 \qquad \frac{\Delta\Gamma(B^0_s)}{\Gamma(B^0_s)} = 0.129 \pm 0.006 \qquad (16.7)$$

[5]At the ICHEP 2020 conference it was shown the upper limit of $\mathrm{BR}(K_L \to \pi^0\nu\bar{\nu}) < 71 \cdot 10^{-11}$ from 2016-2018 data set.

These values are quite consistent with expectations from the SM.

One can probe *indirect* **CP** violation by considering the **CP**-violating asymmetry in rates as a function of time:

$$a_{\mathbf{CP}}(t) = \frac{\Gamma(B^0(t) \to f) - \Gamma(\bar{B}^0(t) \to f)}{\Gamma(B^0(t) \to f) - \Gamma(\bar{B}^0(t) \to f)} \tag{16.8}$$

with flavour-not-specific FS. Resolving the time-dependent rates $\Gamma(t)$, we obtain

$$a_{\mathbf{CP}}(t) = C_{\mathbf{CP}} \cos \Delta mt - S_{\mathbf{CP}} \sin \Delta mt \; . \tag{16.9}$$

In the absence of **CP** violation, $S_{\mathbf{CP}}$ and $C_{\mathbf{CP}}$ must both go to zero. $C_{\mathbf{CP}}$ is only nonzero when the ratio of the amplitude norms differs from unity, which is the signature of direct **CP** violation. A non-zero $S_{\mathbf{CP}}$ is the signature of indirect **CP** violation (or mixing induced one), and it may be different from zero even when there is no direct **CP** violation or $|p/q| = 1$. **CP** violation has been observed in the tree-dominated $b \to c\bar{c}s$ transitions, and PDG2020 gives

$$S_{\mathbf{CP}}(B^0 \to J/\psi K^0) = 0.701 \pm 0.017 \; . \tag{16.10}$$

It is consistent with SM predictions, yet an impact of ND could still hide there. No non-zero value has been found yet for $S_{\mathbf{CP}}(B_s^0 \to J/\psi\phi)$. This decay is the analogue of the decay $B^0 \to J/\psi K^0$, with the spectator d-quark replaced by an s-quark, but the different CKM couplings induce a smaller predicted **CP** asymmetry. Indirect **CP** violation has also been observed in modes governed by the tree-dominated $b \to c\bar{u}d$ and $b \to c\bar{c}d$ transitions, in the mode $B^0 \to \pi^+\pi^-$ and in various modes related to the $b \to q\bar{q}s$ (penguin) transitions.

Decays to the final states $K^\mp\pi^\pm$ provided the first observations of *direct* **CP** violation in both B^0 and B_s^0 systems. Sizable or large direct **CP** asymmetries have been found in the two-body FS due to interference between tree and penguin diagrams:

$$A_{\mathbf{CP}}(B^0 \to K^+\pi^-) = -0.082 \pm 0.006 \;\; , \;\; A_{\mathbf{CP}}(B_s^0 \to \pi^+K^-) = +0.263 \pm 0.035 \; .$$

It is surprising that $A_{\mathbf{CP}}(B_s^0 \to \pi^+K^-)$ is larger than $A_{\mathbf{CP}}(B^0 \to K^+\pi^-)$, since in the SM contribution to the $b \Rightarrow d$ penguin diagram should be smaller than for the $b \Rightarrow s$ one, because of the smaller CKM factor. Lipkin [124] had used (broken) U-spin symmetry to describe this situation; he had suggested the following equality as a test of the validity of the SM[6]:

$$\Delta_{\text{U-spin}} = \frac{A_{\mathbf{CP}}(B^0 \to K^+\pi^-)}{A_{\mathbf{CP}}(B_s^0 \to \pi^+K^-)} + \frac{\Gamma(B_s^0 \to \pi^+K^-)}{\Gamma(\bar{B}^0 \to K^-\pi^+)} = 0 \; .$$

This equality has been checked by LHCb with data from run-1 [125]:

$$\Delta_{\text{LHCb}} = -0.11 \pm 0.04 \pm 0.03 \; .$$

With the present experimental precision one cannot conclude that there is evidence for deviation from zero.

The analyses of many-body FS give crucial lessons on the impact of non-perturbative QCD. Examples of **CP** violation in three body decays have been discussed in **Sect. 7.4.2**:

[6]See **Sect. 4.3**.

- LHCb data of run-1 have shown both averaged and 'regional' **CP** asymmetries of $B^+ \to K^+\pi^+\pi^-$ and $B^+ \to K^+K^+K^-$: positive asymmetry at low $m_{\pi^+\pi^-}$ just below m_{ρ^0} and negative asymmetry both at low and high $m_{K^+K^-}$ values [244]

$$\begin{aligned}
\langle A_{\mathbf{CP}}(B^+ \to K^+\pi^+\pi^-)\rangle &= +0.032 \pm 0.008|_{\text{stat}} \pm 0.004|_{\text{syst}} \pm 0.007|_{\psi K^\pm} \\
A_{\mathbf{CP}}(B^\pm \to K^\pm\pi^+\pi^-)|_{\text{'region'}} &= +0.678 \pm 0.078|_{\text{stat}} \pm 0.032|_{\text{syst}} \pm 0.007|_{\psi K^\pm} \\
\langle A_{\mathbf{CP}}(B^+ \to K^+K^+K^-)\rangle &= -0.043 \pm 0.009|_{\text{stat}} \pm 0.003|_{\text{syst}} \pm 0.007|_{\psi K^\pm} \\
A_{\mathbf{CP}}(B^\pm \to K^\pm K^+K^-)|_{\text{'region'}} &= -0.226 \pm 0.020|_{\text{stat}} \pm 0.004|_{\text{syst}} \pm 0.007|_{\psi K^\pm}
\end{aligned}$$

- Data from more CKM suppressed B^+ decays data have shown larger averaged and large regional **CP** asymmetries: large *positive* asymmetries at low and high $m^2_{\pi^+\pi^-}$ phase-space regions, and large *negative* asymmetry at low $m^2_{K^+K^-}$ ones [8, 244]:

$$\begin{aligned}
\langle A_{\mathbf{CP}}(B^\pm \to \pi^\pm\pi^+\pi^-)\rangle &= +0.117 \pm 0.021|_{\text{stat}} \pm 0.009|_{\text{syst}} \pm 0.007|_{\psi K^\pm} \\
A_{\mathbf{CP}}(B^\pm \to \pi^\pm\pi^+\pi^-)|_{\text{'region'}} &= +0.584 \pm 0.082|_{\text{stat}} \pm 0.027|_{\text{syst}} \pm 0.007|_{\psi K^\pm} \\
\langle A_{\mathbf{CP}}(B^\pm \to \pi^\pm K^+K^-)\rangle &= -0.141 \pm 0.040|_{\text{stat}} \pm 0.018|_{\text{syst}} \pm 0.007|_{\psi K^\pm} \\
A_{\mathbf{CP}}(B^\pm \to \pi^\pm K^+K^-)|_{\text{'region'}} &= -0.648 \pm 0.070|_{\text{stat}} \pm 0.013|_{\text{syst}} \pm 0.007|_{\psi K^\pm}
\end{aligned}$$

- In all previous decays both averaged and regional **CP** asymmetries have the same sign, in different cases, which is not trivial at all. It is interesting to observe that this intriguing pattern of **CP** violation occurs in phase space regions not associated to any known resonant structure, as stated by Ref. [244], since variation of the **CP** asymmetry across the Dalitz plot is expected to be related to the changes in strong phase associated with hadronic resonances. That leaves way to new mechanisms for **CP** asymmetries, to be incorporated in models for amplitude analyses of charmless three-body B decays.
- The LHCb collaboration has continued to measure **CP** asymmetries in $B^+ \to \pi^+\pi^+\pi^-$ with run-1 [609], reporting results on the amplitude structure obtained by employing three decay models, indicated as the isobar model, the K matrix formalism and the 'quasi-model-independent' (QMI) approach. Good agreement is found between all three models and the data. Ref. [609] has shown progress in analyzing Dalitz plots, in particular with the focus on the impact of S waves. One has to continue with $B^\pm \to \pi^\pm K^+K^-$. Although the experimental situations are different, it would be interesting to compare the results due to the re-scattering $\pi\pi \Leftrightarrow \bar{K}K$ to appreciated the impact of non-perturbative QCD. Likewise for **CP** asymmetries in $B^\pm \to K^\pm\pi^+\pi^-/K^\pm K^+K^-$.

So far, the LHCb collaboration has shown the analyses mostly from run-1. A much larger statistics is available in run-2 for three- four-body FS, hence in the future we will likely see more refined analyses, as discussed in **Sect. 7.4.3.**

16.2.4 *A new era for charmed mesons*

$D^0 - \bar{D}^0$ oscillation has been established and the mixing parameters measured: from PDG2020 we have

$$x_D = (5.0^{+1.3}_{-1.4}) \cdot 10^{-3} , \qquad y_D = (6.2 \pm 0.7) \cdot 10^{-3} . \tag{16.11}$$

Until 2018 no **CP** asymmetries had been found for SCS (singly Cabibbo suppressed) transitions; a new gate opened in 2019: the LHCb collaboration established *direct* **CP** asymmetries in $\Delta A_{\mathbf{CP}} = A_{\mathbf{CP}}(D^0 \to K^+K^-) - A_{\mathbf{CP}}(D^0 \to \pi^+\pi^-)$ [174]. PDG2020 gives the value:

$$\Delta_{\mathbf{CP}} = (-1.61 \pm 0.29) \cdot 10^{-3} . \tag{16.12}$$

The next step is to probe *indirect* **CP** violation in the same $D^0 \to K^+K^-$ and $D^0 \to \pi^+\pi^-$ channels. However (as said for beauty mesons) it is also crucial to probe *many-body* FS of D^0, D^+ and D_s^+ in SCS ones in averaged and "regional" phase space.

Probing **CP** asymmetries in DCS (doubly Cabibbo suppressed) decays is another, quite delicate, challenge that is awaiting. We will need much more data and use even more refined tools to deal with backgrounds on the experimental side, but there is a prize in store: since the SM gives hardly any violation there, it is a hunting region for ND.

16.3 Better understanding of flavor forces

Despite its phenomenological successes, the CKM ansatz does not provide us with a a complete understanding of **CP** violation. Since **CP** asymmetries enter through the quark mass matrices, their sources are related to three central mysteries of fundamental dynamics:

- How are fermion masses generated (including neutrinos)? Indeed, the Higgs technology has been very successful. It 'works' with the Yukawa couplings for quarks, but is there a deeper reason – like SUSY?
- Are there true family structures for quarks and leptons? Compare the quarks vs. leptons, in particular neutrinos! Not only their masses are very different, but also their mixing patterns.
- Why are there three families rather than one? Are there more families – for instance do we need sterile neutrinos?

As discussed in **Sect. 15.4**, there is an additional challenge, also an exciting one: the huge predominance of matter over anti-matter observed in our Universe. It requires **CP** violation, if it has to be understood as *dynamically generated* rather than merely reflecting the initial conditions. There is no chance that the SM can provide the right conditions. It is possible, but unlikely, that the *underlying* source of ND comes from flavor quark dynamics. Cosmology observations bring about other questions. What about an asymmetry in DM? What about a connection with the asymmetry in the known matter? What about a connection with Dark Energy? We have theory and particles candidates – yet more thinking is not enough: we need connections among observables!

CP studies are particularly suited to be employed as a high sensitivity way to probe ND *indirectly*; i.e., when new dynamical degrees of freedom enter only through quantum corrections.[7] Yet one cannot focus on one or two golden goals, even with high precision; it is crucial to measure **CP** asymmetries looking at different directions, in particular at beauty and charm hadrons transitions with many-body FS.

16.3.1 CP *asymmetries in beauty, charm and strange baryons*

CP asymmetries have been established in strange (1964), beauty (2001) and charm mesons (2019). One expects to find them also in baryon decays. We briefly summarize what the status is:

(a) All **CP** violation searches in baryon decays prior to LHC era (HyperCP, FOCUS, CLEO, CDF experiments) have given results consistent with **CP** symmetry.

(b) As discussed in **Sect. 7.6**, the LHCb Collaboration has analyzed the $\Lambda_b^0 \to p\pi^-\pi^+\pi^-$ decay, which is particularly well suited for **CP** violation searches due to a rich resonant structure in the decay. Exploiting the full LHC run-1 data set, it has found the first evidence of **CP** violation in a baryon decay with a significance of 3.3σ, including systematic uncertainties [613].

(c) A more recent analysis of the $\Lambda_b^0 \to p\pi^-\pi^+\pi^-$ decay from LHCb has included data from run-2, precisely data collected with an integrated luminosity of 6.6 fb^{-1} from 2011 to 2017 at centre-of-mass energies of 7, 8 and 13 TeV, superseding the previous results [255]. LHCb has observed **P** violation at a significance of 5.5σ when considering both local and integrated luminosity. A remarkable experimental achievement. New data do not clarify the situation on **CP** asymmetry: the highest significance of **CP** asymmetry is only 2.9σ. These new results are marginally compatible with the no **CP**-violation hypothesis. We have to wait for run-3, that will start in the spring of 2022.

(e) If hyperons are polarized, direct tests on **CP** symmetry can be conducted by simultaneously measuring the angular distributions of the hyperon and anti-hyperon decay products. Since a **CP**-violating effect is expected to be small, high precision is required. Precise **CP** tests on hyperon-anti-hyperon pairs can be performed f.e. in the process $e^+e^- \to J/\psi \to \bar{\Lambda}\Lambda/\Sigma^+\Sigma^-$ with large data samples. The BES III experiment has collected $10^{10} J/\psi$, the world's largest data sample produced directly from electron-positron annihilation, which allows for several stringent precision tests

[7] Yet 'direct' results tend to be more rewarding. On April 10, 2019, 'we' have seen an amazing example to establish direct impact, besides indirect one, of fundamental dynamics: it was unveiled the first direct visual evidence of a super-massive black hole and its shadow by 'The Event Horizon Telescope Collaboration' [612]. Most members of 'our' community are 'sure' that Black Hole do exist, actually that they have large impact on known matter. However, it is important to have it in pictures, after years of hard working and comparing observations to extensive computer models.

of **CP** symmetry. BES III has measured $A_{CP,\bar{\Lambda}\Lambda} = -0.006 \pm 0.012 \pm 0.007$, which is actually the most sensitive test for Λ **CP**-violation, and preliminary results for $A_{CP,\sigma^+\sigma^-} = -0.015 \pm 0.037 \pm 0.008$, which is consistent with **CP** conservation [614].

(f) In the future the LHCb collaboration could measure $J/\psi(3100) \to \bar{\Lambda}\Lambda \to [\bar{p}\pi^+][p\pi^-]$ from 13 - 14 TeV pp collisions and probe **CP** asymmetry below 10^{-4}. This and other hadron decays have also been analyzed by BES III, but it is a real challenge to find it in huge backgrounds. Measuring semi-regional asymmetry could give us novel lessons about non-perturbative QCD at least – or the impact of ND.

(g) One can test for baryons the same transitions $c \to u\bar{d}d(\bar{s}s)$ that led to the first observation of **CP** violation in charmed meson decays. Such SCS decays can include significant contributions from loop-level amplitudes, within which ND can enter. PDG2020 gives for SCS and DCS decays:

$$\begin{aligned} \mathrm{BR}(\Lambda_c^+ \to pK^+K^-) &= (1.06 \pm 0.06) \cdot 10^{-3} \\ \mathrm{BR}(\Lambda_c^+ \to p\pi^+\pi^-) &= (4.61 \pm 0.28) \cdot 10^{-3} \\ \mathrm{BR}(\Lambda_c^+ \to pK^+\pi^-) &= (0.111 \pm 0.018) \cdot 10^{-3} . \end{aligned} \tag{16.13}$$

(h) LHCb has tested direct **CP** asymmetry in SCS baryon decays [615]; the result is:

$$\Delta_{\mathbf{CP}} = A_{\mathbf{CP}}(\Lambda_c^+ \to pK^+K^-) - A_{\mathbf{CP}}(\Lambda_c^+ \to p\pi^+\pi^-) = (3 \pm 11) \cdot 10^{-3} .$$

which is compatible with no **CP** violation. The SM predicts basically zero **CP** asymmetry in DCS decays, and non-zero value of $A_{\mathbf{CP}}(\Lambda_c^+ \to pK^+\pi^-)$ would show the impact of ND.

16.3.2 *Dynamics of top quarks in a complex 'landscape'*

As discussed in **Chapter 8** top quarks decay *before* they can produce top hadrons [51]. They carry the unbroken quantum number "color" and evolve in connection with other states to produce hadrons without "color" in the end. One can directly measure top properties from decays. Top quark are predominantly produced through QCD interactions and dominant decays through weak interactions–besides, their are interesting for their large coupling to the Higgs.

It is worth underlining that:

(a) Both the Tevatron and LHC have measured the $\bar{t}t$ production cross-section and, thanks to the large available event samples, also differential cross-sections, which allow even more stringent tests of perturbative QCD as description of the production mechanism. The information from the data with a pair of $\bar{t}t$ quarks depends basically on the connections between productions and decays.

(b) Top physics represents an interesting windows on ND: effective field theory can connect contributions from different types of measurements [616]; flavor changing neutral currents in top sector can arise in ND scenarios [617], and so on.

(c) There is a special case, namely to measure $pp \to \bar{t}\, H^0\, t + X$ with the color connection with X in the finished run-2 of LHC and the future run-3. The associated production of $\bar{t}t$ pairs with Higgs boson has been already done both by ATLAS [618] and CMS [620] at more than 5σ. **CP** structure have been also investigated [619,620]. More results are expected in run-3.

(d) In future pp collisions one can discuss even more challenging FS, namely $pp \to t\,H^0 + \bar{X}$ and $pp \to H^0\,\bar{t} + X$[8]; i.e., single production of t and $\bar{t}$. Once one has learnt how to deal with this huge background with*out* connection with Higgs plus top quarks, we have a good chance to find the impact of ND – in particular with **CP** asymmetries.

ND scenarios, suggested for various reasons (mainly theoretical ones), like the gauge hierarchy problem, can 'naturally' get many more sources for **CP** asymmetries. It is realistic to start with simple versions of such models; however, we should continue with truly motivated models.

16.3.3 *Status of neutrino dynamics*

In the SM one gets three families of quarks and leptons; each has two charged quarks and one charged lepton and one neutrino, which is massless; thus there is no source for **CP** violation in lepton dynamics. However, in this sector we have gone beyond the SM and learnt that the three neutrinos do oscillate; there are two set of eigenstates that are truly different, namely non-zero mass eigenstates ν_i with $i = 1, 2, 3$ and flavor eigenstates ν_e, ν_μ and ν_τ. The observations still leave room for two possible scenarios[9]:

$$m(\nu_1) < m(\nu_2) \ll m(\nu_3) \quad \text{'Normal Order'} \tag{16.14}$$

$$m(\nu_3) \ll m(\nu_1) < m(\nu_2) \quad \text{'Inverted Order'} \tag{16.15}$$

The quantum mixing of neutrino flavour eigenstates as neutrinos travel over large distances provides a way to probe another potential source of **CP** violation: a complex phase δ_{CP} in the neutrino mixing matrix.

CP violation has not been established yet outside the quark sector, but very recently the T2K experiment in Japan has reported the strongest hint of lepton **CP** violation, based on an analysis of nine years of neutrino-oscillation data [319].

For the "normal" order (favoured by T2K and other experiments), and averaged over all other oscillation parameters, a 3σ confidence-level interval for δ_{CP} is between -3.41 and -0.03, while for the "inverted" one the interval between -2.54 and -0.32 and was found. Averaged over all oscillation parameters, $\delta_{\text{CP}} = 0$ is disfavoured at 3σ confidence (although for some allowed values of the mixing angle θ_{23} it is still within the 3σ bound). Further data are required to confirm this results, but this recent finding strengthens previous observations and offers hope for a future discovery of leptonic **CP** violation.

[8] In principle, $\bar{X}$ carries color $\bar{3}$ and X color 3; in the real world it does not matter.

[9] Most of the literature says there are only two "players" in this competition: $m_1 < m_2 \ll m_3$ or $m_3 \ll m_1 < m_2$. However, the two authors of Ref. [621] claim that quasi-degenerate spectrum $m_1 \sim m_2 \sim m_3$ is still possible.

Another compelling question in the neutrino landscape is: are neutrinos Dirac fermions like quarks or Majorana particles? The solution to this problem could be crucial for understanding the origin of the smallness of neutrino masses. Moreover, three Dirac neutrinos would have only one source for **CP** violation like quarks, namely *one observable phase* δ_{CP}; instead, Majorana ones could have *three observable phases* (see **Sect. 9.5** and in particular in **Sect. 9.5.1**).

16.3.4 *Future neutrino oscillations*

There are several experiments in the neutrino sector; let us mention two future experiments with broad programs (and very different detectors):

(a) The new DUNE experiment at Fermilab will start in 2029 [622]. Its main science goals are: (1) search for the origin of matter; (2) shed light on the unification of the forces in our Universe; (3) learn more about 'neutron stars' and 'black holes'.

(b) The new Hyper-Kamiokande (Hyper-K) experiment in Japan, whose construction has been approved in early 2020, and the beginning of its data-taking is expected around 2027 [623]. Hyper-K is the next generation underground water Cherenkov detector that builds on the highly successful Super-Kamiokande experiment. The detector has an 8.4 times larger effective volume than its predecessor, and it will be located along the T2K neutrino beam line. It will utilize an upgraded J-PARC beam with 2.6 times beam power and it is expected to operate for 20 years, providing some quite stringent tests of the SM. Its main goals are: (1) to establish **CP** asymmetries in neutrino oscillations (2) determining the ordering of the neutrino masses; (3) cosmic neutrino observation in 'our' sun, 'our' galactic center and/or a supernova explosion; (4) search for proton decays.

The programs just mentioned are quite broad; let us make a short summary of the significant points.[10]

(1) One wants to establish **CP** asymmetries also in leptonic dynamics. In our Universe we have at present a huge asymmetry in matter vs. anti-matter: we expect that it is *dynamically* generated. As discussed in **Sect. 15.4**, and in particular in **Sect. 15.4.4**, the knowledge of leptonic dynamics is essential: leptogenesis could drive baryogenesis!
(2) A heavy neutrino N would be a Majorana one; its dynamics has to violate lepton number with at least these classes of decays: $N \to \ell\bar{H}$, $\bar{\ell}H$ where ℓ, $\bar{\ell}$ are light leptons and anti-leptons, and H, $\bar{H}$ are Higgs and anti-Higgs fields, respectively.
CP asymmetry leads to $\Gamma(N \to l\bar{H}) \neq \Gamma(\bar{N} \to \bar{l}H))$. In the leptogenesis scenario, the lepton asymmetry, arising from the out of equilibrium and **CP**-violating decays, is transferred into a baryon asymmetry that survives through the subsequent evolution of our Universe.

[10] For a good review, see for instance Ref. [621].

(3) Oscillations between three neutrinos give $\Delta m^2_{32} \sim +[-]2.5 \cdot 10^{-3}$ (eV)^2 for normal [inverted] hierarchy [8]. To decide which is correct gives a better understanding of neutrino dynamics.
(4) The SM describes electromagnetic, weak and strong interactions based on the gauge group $SU(3)_C \times SU(2)_L \times U(1)$. Fans of symmetries 'think' that Grand Unified Theories (GUT) have to exist to combine these interactions into a single force.[11] How can one test this idea? Maybe the proton is not truly stable? PDG2020 gives the actual value of the proton mean life $3.6 \cdot 10^{29}$ years in invisible modes (or $10^{31} - 10^{33}$ years for 'golden' channels).[12]
(5) A sensitive probe to whether neutrinos are Dirac or Majorana states is the neutrino-less double β decay: $[A, Z] \to [A, Z+2] + e^- + e^-$, where A/Z are the numbers of atoms/protons. In terms of elementary particles, one can 'paint' this decays as $nn \to p\, e^- e^-\, p$. If the neutrino is a Dirac ν state, $\nu \neq \bar{\nu}$ are different; thus no mixing (and no neutrino-less double β decay) can happen. On the other hand, if the neutrino is a Majorana one, mixing happens between ν and $\bar{\nu}$; i.e., they are described by the same field.
If one assumes that the Majorana neutrino mass is the only source of lepton number violation ($\Delta L = \pm 2$) at low energies, the half-lives of neutrino-less double β decays are determined by direct detection through the relation (see Ref. [159])

$$(T^{0\nu}_{1/2})^{-1} = G_{0\nu}|M_{0\nu}|^2 \langle m_{\beta\beta}\rangle^2 \tag{16.16}$$

$G_{0\nu}$ is the phase space integral for the final atomic state, $|M_{0\nu}|$ is a nuclear matrix element of the transition, and $\langle m_{\beta\beta}\rangle$ is the effective Majorana mass of ν_e. The goal is to find non-zero value for $\langle m_{\beta\beta}\rangle$; so far we have only limits.
Actually, the strongest bounds come from the decay of ^{136}Xe measured by KamLAND-Zen 400 and from the decay of ^{76}Ge measured by GERDA.
From ^{136}Xe [624]:

$$T^{0\nu}_{1/2} > 1.07 \cdot 10^{26}\,\text{years} \qquad [^{136}\text{Xe}] \tag{16.17}$$
$$\langle m_{\beta\beta}\rangle < (0.061 - 0.165)\,\text{eV}\;; \tag{16.18}$$

The upper limit mass range in $\langle m_{\beta\beta}\rangle$ shows the uncertainty in the calculation in the nuclear matrix element.
From ^{76}Ge [625] the lifetime limit is slightly better:

$$T^{0\nu}_{1/2} > 1.8 \cdot 10^{26}\,\text{years} \qquad [^{76}\text{Ge}] \tag{16.19}$$

but the upper limit mass range is higher because of the less favorable nuclear matrix element and phase space with respect to ^{136}Xe.
It is instructive to add that, although worse in life-time sensitivity, the CUORE experiment [626] yields a competitive upper limit mass range because of the

[11]GUTs ignore gravity; the "final" theory should include it.
[12]For a very special case and a somewhat remarkable coincidence: $n-\bar{n}$ oscillation time $> 8.6 \cdot 10^7$ s for free n or $> 2.7 \cdot 10^8$ s for bound n [8].

more favorable nuclear matrix element and phase space of the ^{130}Te target nucleus:

$$T^{0\nu}_{1/2} > 3.2 \cdot 10^{25}\,\text{years} \qquad [^{130}\text{Te}] \tag{16.20}$$
$$\langle m_{\beta\beta}\rangle < (0.075 - 0.350)\,\text{eV}\;; \tag{16.21}$$

The situation might change in the future, taking into account that a rich and aggressive program is foreseen in all underground laboratories around the world. For a recent experimental review see the agenda at Neutrino 2020 conference.[13]

It is crucial that these five items are deeply connected, not independent!

16.4 Search for ND in EDMs

As discussed in **Sect. 12.2** and **Chapter 13**, a non-zero value of a EDM indicates **T** and **CP** (via **CPT** invariance) asymmetries. The SM provides two sources of EDMs, the **CP** violating phase in the CKM matrix of the weak sector and the QCD $\bar{\theta}$ term in the strong sector. However, three-loops amplitudes are needed to generate CKM-induced EDMs of quarks in the SM, and CKM-induced EDMs of charged leptons are even more suppressed, arising at four-loop order. The resulting values can hardly be reached. Furthermore, the contributions from the strong sector depend on the unknown parameters $\bar{\theta}$; in turn, its value is strongly constrained by the experimental searches for EDMs.[14] EDMs give us another hunting region for ND with a decent chance to find its existence and maybe also its features. Thus a quest for EDMs has to continue with renewed vigor.

To date, there is no experimental observation of an EDM of an elementary particle or non-degenerate bound quantum system. The most stringent limits have been obtained for the EDMs of the electron, neutron, thallium and mercury atoms. In particular, the best limit for the neutron EDM is reported in Ref. [442] and discussed in **Sect. 13.1**:

$$|d_N| < 1.8 \cdot 10^{-26}\, e\,\text{cm} \tag{16.22}$$

EDMs in different systems constrain **CP**-violating interactions differently, then it is important to look for non-zero EDMs in multiple systems. During this search our community can apply not ‘only’ to atomic and nuclear physics and HEP, but also to solid physics and its technologies. Here we do *not* go for accuracy – we go for sensitivity to fundamental dynamics. The amount of **CP** violation in the fundamental interactions has always been an important measurable. Finding a statistically significant EDM would help us to better understand the sources of **CP** violation.[15]

[13] Neutrino 2020, https://indico.fnal.gov/event/43209/timetable/.

[14] As discussed in **Chapter 11**, some version of axions may solve that problem.

[15] Even if physicists have not found yet a non-zero value for any EDM and got only a limit after all their hard work, they have still got a training place, and learnt good lessons to apply to different regions of dynamics. Once again, the load of future work will go mostly on the shoulders of the experimental colleagues.

16.5 The cathedral builder's paradigm leading to 'Novel SM'

CP violation is both a fundamental phenomenon and a mysterious one. It exists in our Universe, but contains a message that has not been decoded yet. In our judgement it would be unrealistic to expect progress through 'pure thinking'. We strongly believe that we have to appeal to our Universe through experimental efforts to provide us with the pieces still missing from the puzzles. **CP** studies are essential for obtaining complete information on the dynamics driving the generations and differentiation of fermions, gauge and scalar bosons and others – and there is a wide realm open for them.

In our quest for ND we have to probe the dynamical ingredients of numerous and multi-layered manifestations. Accordingly, we search for them in many phenomena:

- During the last five years there have been claims [117] that the SM might not reproduce the measured value of direct **CP** asymmetries in $K_L \to \pi\pi$. Another challenge is to investigate the strongly suppressed $K^+ \to \pi^+ \bar{\nu}\nu$ decay and later measure the BR$(K_L \to \pi^0 \bar{\nu}\nu)$. The predictions from the SM at the level of several$\cdot 10^{-11}$ are well controlled, having small uncertainties.
- The LHCb collaboration has found **CP** violation for the first time in charm hadrons, namely direct **CP** asymmetry in SCS decays $D^0 \to K^+K^-$ vs. $D^0 \to \pi^+\pi^-$ [174]; it is only the beginning. It is interesting to analyze DCS channels (driven by the $c \to d\bar{s}u$ decays), where the SM gives hardly non-zero values.
- It is crucial to probe *regional* asymmetries with three- and four-body FS in the decays of beauty and charm hadrons. Measuring Dalitz plots with accuracy should *not* been seen as back-up information – their analysis is crucial to understand the underlying dynamics.
- We have to measure **CP** asymmetries in baryons, namely Λ_b^0, Ξ_b, Λ_c, Ξ_c and hyperons through different 'roads'. One possibility is to probe **CP** asymmetries in J/ψ decays into baryon and anti-baryon– for example, **CP** asymmetries in $J/\psi \to \bar{\Lambda}\Lambda \to [\bar{p}\pi^+][p\pi^-]$ can be probed by the LHCb collaboration with future run-3/4 below the level of 10^{-4}.
- It seems we have to wait for Belle II to probe **CP** asymmetries in the decays of τ leptons beyond $\bar{K}^0 - K^0$ oscillations [627].
- The landscape is very different for the transitions of top quarks. Since they do not hadronize, one cannot follow the beaten paths to probe **CP** asymmetries. The observation of Higgs boson production in association with top quarks at the LHC [618, 628] provides an opportunity to probe the **CP** properties of the Yukawa coupling of the Higgs boson to the top quark. The possibility of the presence of a pseudoscalar admixture which introduces a second coupling to the top quark has not yet been excluded by data. Any measured **CP**-odd contribution would be a sign of physics beyond the SM.
- Some **CP** asymmetries are predicted with high *parametric* reliability. New theoretical technologies will allow us to translate such parametric reliability into *quantitative* accuracy.

- Our community has searched for EDMs in atoms, neutrons (and combinations with protons) and charged leptons in several laboratories of Europe and the North America. There are excellent reasons to continue this search. The point is not go for accuracy, but for sensitivity to ND. The SM cannot produce 'background', when one probes EDMs. Many models of **CP** violation based on ND are eagerly awaiting their turn in the wings. Even when the impact of ND is insignificant in other processes, it can generate observable effects for EDMs.

This list, although incomplete, makes it clear that front-line researches are pursued at many laboratories and locations all over our world. Techniques from several different branches of physics (atomic-, nuclear-, high energy- and astro-physics) are harnessed in this endeavor, together with a wide range of set-ups. Experiments are performed at the lowest temperatures that can be realized on our Earth (ultra-cold neutrons) and at the highest ones in collisions produced at CERN with LHC (and future ones) and astrophysics and cosmology in our Universe. All of that is dedicated to one profound goal.

Now we can explain what we mean by the 'cathedral builders paradigm'.[16] The building of cathedrals required inter-regional collaborations, front-line technology (for the period) from many different fields, and long commitment; it had to be based on solid foundations and it took time.[17] The analogy to the ways and needs of high energy physics is obvious, but it goes deeper than that. At first sight a cathedral looks like a very complicated and confusing structure with something here and something there. Yet further scrutiny reveals that a cathedral is more appropriately characterized as a complex rather than a complicated structure, one that is multi-faceted and multi-layered with a *coherent* theme! We cannot (at least for first rate cathedrals) remove any of its elements without diluting its architectural soundness and intellectual message.

16.6 Somewhat final words

What **CP** and **T** invariance and their limitations can teach us about Nature's Grand Design? We seem closer to the beginning than the end of the 'travel' we have outlined. Such a program of inquiry is exciting, but neither easy nor quick! Yet we have to keep the following in mind: insights into Nature's Grand Design, which can be obtained from a comprehensive and detailed program of fundamental dynamics, led by studies of **CP** asymmetries and rare decays, are essential, cannot be obtained any other way and cannot become obsolete!

[16]The cathedral builders paradigm' has already been used in Ref. [12]; however, it has been updated here.

[17]To be honest: some cathedrals were not built on solid foundation and/or good technology; those did not make it outside the Middle Ages. A special case is the leaning Tower of Pisa; some readers might remember the connection with physics due to Galilei's experiment – actually, it might have been a thought experiment, prompt by his view inside the Pisa Cathedral.

Going back to history: the Greek philosopher Aristotle had published in a well-known book "On the Heavens" around 350 BCE the following statement: "...a man, being just as hungry as thirsty, and placed in between food and drink, must necessarily remain where he is and starve to death." The 14th century French philosopher Jean Buridan gave an illustration of the paradox in the conception of free will, that became known as 'Buridan's ass'. That shows the difference between the thinking of philosophers and physicists – in particular if we consider in quantum mechanics the EPR (Einstein-Podolsky-Rosen) correlations – not a paradox! – or 'quantum entanglement'.[18]

According to the standard cosmological model, our Universe consists of three parts: Known Matter with $\simeq$ 4.5%, Dark Matter (DM) $\sim$ 26.5% and Dark Energy (DE) $\sim$ 69%.[19] The latter two have been established only due to gravity so far, and bring new challenges:

- We have candidates for DM like SUSY, axions etc., as discussed above. We have the tools to find them somewhere in astrophysics and cosmology. There is also another 'road', namely to search for their impact on **CP** asymmetries and rare decays. What we need is more data, more time – and novel ideas.
- Let us be honest: our community does not seem to have good candidates for DE and no decent 'roads' to probe them beyond gravity.

There are also significant challenges to deal with known matter:

(1) There are still unfinished 'jobs' in our analysis of weak decays of strange hadrons (including strange *baryons*).
(2) A determined effort has to be mounted for studying the **CP** properties of τ leptons and top quarks; in the latter case to describe the connection of top quarks with Higgs bosons.
(3) A comprehensive, detailed and high-statistics analysis of the decays of beauty and charm hadrons is bound to provide us with essential information on fundamental dynamics. It represents a unique opportunity where data can be compared with predictions that are or will be numerically accurate.
(4) The quest for EDMs has to continue with vigor both in leptons and baryons.
(5) To understand mass generation, we should endeavor to acquire all information concerning it, and that includes the observable content of the non-diagonal mass matrices. No direct observation of new fields – like SUSY partners – can supersede that information. On the contrary, it would be of great help by providing us with essential input for our predictions.

Some readers might share the notion that 'fables' can give intriguing lessons about the real world. Here is an example taken from the legend about the Cumaean Sybil: 'She had offered nine books with all her prophecies to the last Roman king

[18]It was crucial to establish large **CP** violation in $e^+e^- \to \Upsilon(4S) \to \bar{B}^0 B^0 \to [J/\psi K_S][l^\pm \nu X^\mp]$.
[19]One would have expected that one part gets close to 100%, with the other two closer to 0%.

Fig. 16.1 On the left: Renaissance and contemporary art (picture taken by IIB); on the right, La Loge (Theatre box) by Pierre Auguste Renoir (1874), The Courtauld Gallery (London) [Courtauld Institute].

Tarquinius Superbus (Tarquin the Proud) for sale. Considering the asking price too stiff, he declined. She then threw three of the books into a fire to burn them, and asked the same price for the remaining six books. He still refused, whereupon she burnt three more books. Then he relented and bought the leftover three books for the original asking price'.[20] The experimentalists among us will recognize that Tarquinius Superbus acted like the typical funding agency that asks for 'de-scoping' your project only to end up to pay the same price for less. The theorists will claim that if we had nine flavors to study, we would have already figured out the dynamics underlying the flavor enigma and even more. Actually we have six 'books', namely three families of quarks and three families of leptons. We have candidates for the missing three books: Higgs dynamics and electroweak processes; SUSY; EDMs from leptons and baryons; connection with DM and others.

A concise summary is provided by two examples from art: (a) The connection from very different regions: renaissance and contemporary art, see **Fig. 16.1** on the left. It is inside a building just south of the center place in Krakow.[21] (b) A painting from Renoir describing the situation in a balcony of a theater, see **Fig. 16.1** on the right. One occupant, the gentleman, seems to look at the 'data', while the other one, the lady, appears as if she wants to put the spotlight on the beauty of her attire – the beauty of the 'theory'.

[20]The legend also tells us that the purchase did not help the last king of Roma – he was thrown out.

[21]It is not easy to find it.

Chapter 17

Epilogue

Discussions are crucial to make decisions: let us mention two very famous examples from the history:
(a) In the summer of 1519 in Germany there was a 'disputation' (= discussion) between J. Eck, A. Karlstadt, and M. Luther in person.[1] In the series of events that led to the Diet of Worms, marking the definitive breach of Luther from Catholicism, few were more significant than this public theological debate.
(b) In 1632 Galileo had written the "Dialogue Concerning the Two Chief World Systems". He compared the Copernican system with the Ptolemaic one with discussions between three 'actors' Salviati, Sagredo and Simplicio. The first two names were connected with real persons: Salviati was a scientist and Sagredo a mathematician. Both were close friends of Galileo; on the other hand, they had passed away by 1621. The third one – Simplicio – who defends the Ptolemy position, was not a real person, as the name shows[2]; it is obvious that he was not an intelligent scientist – actually worse. In this dialogue new interpretative paradigms of reality (the scientific method) emerge, which are at the basis of modern science.

At this point it seems appropriate to comment on a debate involving L.B. Okun, published on an article with the title: "Trialogue on the number of fundamental constants" [629].[3] It is very unusual – and therefore it is very interesting, never mind whether a reader agrees with the statements inside or not. The choice of the fundamental constants are a reflection of our vision of the universe, the actual and the future (theory of everything) one. The three authors are true experts: L.B. Okun, G. Veneziano and M.J. Duff. The article is atypically organized already from the 'Abstract': "... Okun develops the traditional[4] approach with three constants, Veneziano argues in favor of at most two (within superstring theory), while Duff advocates zero".

---

[1] Officially Eck was called the 'winner' then.
[2] The Italian for "simple" (as in "simple minded") is "semplice".
[3] Strictly spaeking, trialogue is misnamed, since the Greek root of dia in dialogue means through and not two, while "logos" means words, thoughts.
[4] Actually, we disagree that Okun's work can be described as 'traditional': Lev Okun was not at all traditional, as we have shown throughout this book.

In this book we have followed the community of HEP theorists and used $c = 1 = \hbar$, to make the equations shorter. However, this convention does not help the reader to understand the underlying dynamics of special relativity and quantum mechanics. The goal of this Epilogue is to go deeper.

The "Trialogue" [629] gave three 'Statements', each of these written by *one* author:

- 'Statement I': "Fundamental constants: parameters and units" – *L.B. Okun*;
- 'Statement II': "Fundamental units in physics: how many, if any?" – *G. Veneziano*;
- 'Statement III': "A party political broadcast on behalf of the Zero Constants Party" – *M.J. Duff*.

We summarize these Statements with a few comments below. We add here that on the article the three authors have organized their statements differently, following a less methodical way than usual, which is not surprising at all. It is the right of 'mature' people *not* to follow 'fashions'.

17.1 Statement I

Okun said there are two kinds of fundamental constants of our Nature: dimensionless (like $\alpha = e^2/\hbar c \simeq 1/137$) and dimensionful ones, which are c = velocity of light, $\hbar$ = quantum of action (and angular momentum), and G = Newton gravitational constant. He referred to the former as fundamental parameters and the latter as basic units.

Physics consists of measurements, formulae – and 'words'. In his part, which contains no new formula, Okun dealt mainly with 'words'. He believes that an adequate language is crucial in physics; the absence of accurately defined terms or the use of ill defined terms leads to confusion and proliferation of wrong statements.

Okun claimed that it is necessary and sufficient to have three basic units in order to reproduce in an experimentally meaningful way the dimensions of all physical quantities:

- "c" is not only the velocity of light, but, more significantly, is the maximal velocity of any object in Nature. This fundamental character would not be diminished in a world without photons. At a first sight α may look superior to "c" because its value does not depend on the choice of units, whereas the numerical value of "c" depends explicitly on the units of length and time and hence on conventions. However, "c" is more fundamental than α, because it is the basis of relativity theory unifying space and time, as well as energy, momentum and mass.
- Likewise for "$\hbar$": it is the quantum of the angular momentum and a natural unit of the actions. When the angular momentum or the action are close to "$\hbar$", the whole realm of quantum mechanics appears. Fermions, with half-integer

angular momentum, obey the Pauli exclusion principle, which is so basic for the structure of atoms, nuclei and neutron stars.

- The status of "G" (and its derivatives, the Planck mass m_P, length l_P and time t_P) is very different from that of "c" and "$\hbar$" and its position not quite firm as theirs: quantum theory of gravity is still under construction. The majority of experts connects their hopes with extra spatial dimensions and superstrings. However, at present, the bridge between superstrings and experimental physics exists only as wishful thinking.

There is a hope that the values of dimensionless fundamental parameters might be ultimately calculated in the framework of the Theory of Everything (TOE). The universal character of the three basic units (and hence Planck derivatives) makes natural their use in General Relativity with futuristic quantum gravity and in the TOE. In practice the use of these units is realized with what Okun calls 'The art of putting $c = 1$, $\hbar = 1$, $G = 1$'.

17.2 Statement II

Veneziano wrote a 1986 letter on the number of dimensional fundamental constants in string theory, where he came to the surprising conclusions that *two* constants – with dimensions of space and time – were both necessary and sufficient. Later he became aware of S. Weinberg's 1983 paper [630], whose views he incorporated in his work of the following 10 years. S. Weinberg defined constants to be fundamental, if one cannot calculate their values in terms of more fundamental constants. In the paper we are discussing, Veneziano still subscribes to the contents of his old papers, even if he believes he might have expressed some specific points differently. He has the impression that, in the end, the disagreement with Okun and Duff is more in the words than in physics.

He started with two statements on which the three authors seemed to agree:

(1) Physics deals with dimensionless quantities like $\alpha = e^2/\hbar c$ and the ratio of the electron and proton mass m_e/m_p.
(2) One introduces 'units' for a physical quantity q as a fixed quantity u_q *of the same kind*: $q = (q/u_q)u_q$; thus q/u_q is a number.

Then he posed three questions:

Q_1 : Are units arbitrary?
Q_2 : Are there units that are more fundamental than others?
Q_3 : How many units (fundamental or not) are necessary?

Veneziano thought that the three authors agree on 'yes' for Q_1, since only q_i/q_j matter and these ratios do not depend on the choice of units. However the answers for Q_2 and Q_3 depend on the framework under scrutiny. Therefore he analyzed Q_2 and Q_3 within three distinct frameworks and provided, for each case, answers A_2 and A_3. Here we discuss only two distinct frameworks.

Fundamental units in QFT + General Relativity (GR): On the QFT (and QM) level "c" and "$\hbar$" are more fundamental units than any other: A_2 is a qualified 'yes'. Indeed, one can apply S. Weinberg's criterion [630] and ask: can we compute "c" and "$\hbar$" in terms of more fundamental units? In this framework the answer A_3 is the most probable one.

A_3 is most probably three: in QFT + GR 'we' cannot compute everything that is observable in terms of "c" and "$\hbar$" plus dimensionless constants *without* also introducing some mass scale. Unlike the case of "c" and "$\hbar$" it is much less obvious what to choose as the mass scale: on the basis of the mentioned Weinberg's criterion, the Planck mass M_P does not look like a good choice, since it is very hard – even conceptually – to compute, for instance, m_e or m_p in terms of M_P. What about two units? Veneziano said there is a deeper reason why QFT and GR needs a separate unit for mass. QFT is affected by UV divergences that need to be renormalized. It forces us to introduce a cut-off which has nothing to do "c", "$\hbar$" and M_P, and has to be 'removed' in the end. However, remnants of the cut-off remain in the renormalized theory. Perhaps one day string theory (or other unified theory) will connect the hadronic mass scales, as the proton mass m_p, with the Planck scale M_P, but QFT and GR cannot.

Fundamental units in modern quantum string theory (QST)/M-theory: Veneziano pointed out one can give A_2 and A_3 together in the context of first-quantized string theory in the presence of background fields. The beautiful feature of this formulation is that all possible parameters of string theory – dimensional and dimensionless alike – are replaced by background fields whose vacuum expectation value one hopes to be able to determine *dynamically*. He stands by the statements of his previous works: the fundamental constants of Nature are the constants of the vacuum in QST. He believed that all known consistent string theories correspond to perturbations around different vacua of a single, yet unknown, 'M-theory'. Although the number of physically not equivalent non-perturbative vacua of the 'M-theory' is not known, he expected 'we' need only one unit of length and one of time, namely two fundamental units.

17.3 Statement III

In his abstract Duff synthetically summarizes the three positions in the historical order: the Manifesto of Okun's Three Constants Party – Planck's constant $\hbar$, the velocity of light c and Newton's constant G – was followed by Veneziano's Two Constant Party – string length $\lambda_{\rm str}$ and "c" – and his platform of the Zero Constant Party, stating that none of these units is fundamental.

The false propaganda of the Three Constants Party: Duff had to confess that for a long time he was a card-carrying member of the Three Constants Party. However, the faith in this 'dogma' was shaken by papers of Veneziano with the TOE. Based on what he refers as the M-theory revolution, he said that $\hbar$, c and G are nothing but conversion factors – e.g. energy to frequency $E = \hbar\omega$ or energy to mass $E = mc^2$. Thus none of these units or conversion factors is fundamental – namely the Zero Constants Party. Put in different words, one can have any number of "fundamental" constants, depending on how different are the units one employs. The attitude of the Zero Constants Party is that the most economical choice is to use natural units where there are no conversion factors at all. In the natural units favored by the Zero Constants Party there are no dimensions at all, and $\hbar = c = G = ... = 1$ may be imposed literally and without contradiction.

What about theories with time-varying constants? Many notable physicists have entertained the notion that G or c are changing in time (for some reason, time-varying $\hbar$ is not as popular). Duff stated that the time-variation in the physical laws is best described in terms of time-varying dimensionless parameters of the SM, rather than dimensionful constants. In a paper published in 2001 [631], astrophysical data were presented suggesting a time-varying fine structure constant.[5] In the context of M-theory which starts out with no parameters at all, these SM parameters would appear as vacuum expectation values of scalar fields. Duff affirmed that replacing parameters by scalar fields is the only sensible way to implement time varying constants of our Universe.

17.4 Debate

Here we enlist some points of discussions among the three authors.

Okun's comments about Veneziano's Statement II: (a) He noted with satisfaction that over years the discussion was making progress, as some arguments of Veneziano's previous works were not repeated (b) He advocated for consistency. It seemed to him inconsistent to keep two units (c, λ_s) explicitly in the equations, while substituting by unity the third one ($\hbar$). (c) He could not agree that the electron mass or G_{Fermi} are as good for the role of fundamental units as the Planck mass or G.

Okun's comments about Duff's Statement III: Mike introduced a definition of fundamental constants with the help of an alien with whom one can exchange only dimensionless numbers. The fundamental constants are those whose values can be communicated to the alien: thus no fundamental units. Okun replied that the use of the natural units $c = \hbar = G = 1$, which cannot be considered literally, corresponds to using the same three fundamental units. Okun also said that the 'alien definition' is misleading. Dimensionless variables have impact in a theory of physics; however, it does not mean that every problem should be explicitly presented in dimensionless form. Sometimes one can use dimensional units and compare their ratios with ratios of other dimensional units. If "c" is changed, while the dimensions of atoms are not changed, electromagnetic and optical properties of the atoms would change drastically because of the change of α, which is the ratio of electron velocity

[5] This paper has been cited about 600 times!

in hydrogen atom to that of light. Using proper language (terms and semantics) three fundamental units are the only possible basis for self-consistent description of fundamental physics. Okun did not understand why Duff disagreed with him in these considerations.

In turn, Veneziano conceded to Okun that, given the fact that momentum and energy are logically distinct from lengths and time for ordinary objects, insisting on the use of the same (or of reciprocal) units for both sets can be pedagogically confusing; he agreed that the set of "c", "$\hbar$" and $\lambda_{\rm str}$ define the most practical set of fundamental units at present – within QST.

Veneziano disagreed with Duff on two items:
(a) *The alien story*: Can one communicate to an alien our values for "c" and $\lambda_{\rm str}$ and check whether they agree with ours? Yes, since one can tell her/him our definition of "cm" and "s (=second)" in terms of a physical system like the H atom. If the alien comes up with the same numbers for the fundamental units–for example, $c \simeq 3 \times 10^{10}$ cm/s – the 'alien story' supports rather the fact that 'we' *have* fundamental units of length and time. Duff seemed to agree with the alien's reply – however he concluded that "c" is *not* a fundamental unit: a completely rescaled world, in which both "c" and the velocity of the electron in the H atom are twice as large, is indistinguishable from ours. (b) *Reducing fundamental units to conversion factors*: Duff said that fundamental units are just conversion factors in order to convert quantities into others, and, eventually, everything into pure numbers. However, this is not the important point for Veneziano. It is that among arbitrary units there are a few which are fundamental, in the sense that dramatic new phenomena occur when a quantity became $\mathcal{O}(1)$ in the latter units.

Veneziano agreed with Okun that relativity has a fundamental unit of speed (its maximal value), QM has a fundamental unit of action (a minimal uncertainty) and string theory has a fundamental unit of length (the characteristic size of strings). QST appears to provide the missing third fundamental unit of the three-constants system. These three units form a very convenient system except that, in string theory, one can identify the Planck constant with the string length, eliminating the necessity of a third unit besides those needed for lengths and time intervals. At the same time Veneziano agrees with Duff that all that matters are pure numbers, but added the comment: observation showed that relativity and QM provide in string theory units of length and time which look, at present, more fundamental than any other. The number of distinct units is a matter of choice and convenience, and also depends on our understanding of the underlying physical laws. Within QFT and GR it looks mostly useful to reduce this number to three, but there is no obvious candidate for the third unit after c and $\hbar$. Within QST the third unit naturally emerges as being the string length $\lambda_{\rm str}$. However, Veneziano thought one can keep only two units "c" and $\lambda_{\rm str}$, which allows to *express* space and time intervals *in terms* of pure numbers. Space and time are distinguished and "c" is introduced as a fundamental unit of speed and not as a trivial conversion factor.

Duff agreed with Okun that the finiteness (minimum angular momentum, maximum velocity) of the conversion factors is important, but no significance should be attached to their values, and one can have as many or as few of them as she/he likes. He claimed that the issue of what is fundamental would continue to go round and around until they would all agree on an operational definition of "fundamental constants". He did not believed this could be the already mentioned definition of Weinberg [630], according to which constants are fundamental, if one cannot calculate their values in terms of more fundamental constants, not just because the calculation is too hard, but because we do not know of anything more fundamental. Duff coined his own, based on experiments which would tell us whether the alien's universe has the same or different constants of nature as ours. His conclusions are that, according to his definition, the dimensionless parameters, such as the fine structure constant, are fundamental, whereas all dimensional constants, including $\hbar$, c and G, are not.

Time variation of fundamental units? This was another point of reflection for Veneziano. He conceded that discussions indicated that time variation of a fundamental unit as "c" has no meaning, unless one can specify what else, having the same units, is kept fixed. Only the time variation of dimensionless constants, such as α for an atom, has an intrinsic physical meaning. However, he claimed with an example that, in principle, the time variation of dimensionless constants has a physical meaning for a specific system because it represents the time variation of some physical dimensional quantity with respect to the fundamental unit provided by string theory.

17.5 Artistic presentation

One has to give credit to these Gentlemen for their patience in discussing their points of view. We have given only a brief summary of what they said more than 15 years ago, and reported in Ref. [629]. We chose to describe our 'understanding' through an artistic presentation, a 1514 print from Dürer,[6] see **Fig. 17.1**. It combines several possible interpretations: a person is using a tool (the loosely held compass which forms an 'angle'), is thinking, needs time (see 'sandglass' or 'hourglass' just above her head), and is inspired by the 'sun' and 'rainbow', etc. etc. There are also two puzzles: (1) Above her head and east of the 'sandglasses' there is the 'magic square 4×4'[7]; a reader might see a reference to 'algebra'. (2) Below the 'rainbow' and west of her face one can see the 'truncated rhombohedron'; again, a reference to 'geometry'.

[6]Does Dürer's print give the reader the feeling that physicists get when they go after fundamental dynamics? That is our 'job' – we enjoy it, even if we know that there will be no final result.

[7]On the fourth line of the square one sees in the middle '15' and '14' = 1514. One also see '4' and '1', describing the family name 'D(ürer)' and the first name 'A(lbrecht)' of the artist, respectively.

Fig. 17.1 Melancholy (Melencolia I), engraving by Albrecht Dürer (1514) [Metropolitan Museum of Art, New York: online Art Collection].

Acknowledgments: This work was supported by the NSF under the grant numbers PHY-1520966 & PHY-1820860 and by the INFN, Istituto Nazionale di Fisica Nucleare (Italy). We thank S. Bianco, Alberto Correa dos Reis, I. Bediaga, M.D. Pennington for useful discussions and support. On a personal side, G.R. thanks her family, especially aunt Iole and Eleonora, and all her friends, for their great support after the loss of her mother Gioia, a joy in name and in fact, during the last stage of the manuscript preparation.

Bibliography

[1] L.B. Okun: *Weak Interactions of Elementary Particles*, Pergamon, 1965; [the Russian original appeared in 1963, i.e. *before* the discovery of CP violation.]

[2] J.H. Christenson, J.W. Cronin, V.L. Fitch, R. Turlay: *Evidence for the 2π Decay of the K_2^0 Meson*, Phys. Rev. Lett. 13 (1964) 132.

[3] J.J. Sakurai: *Invariance Principles and Elementary Particles*, Princeton, NJ: Princeton University Press, 1964; L.I. Schiff: *Quantum Mechanics*, New York, McGraw-Hill Inc., 1968; T.-P. Cheng, L.-F. Lee: *Gauge Theory of Elementary Particles*, Oxford, Oxford University Press, 1982; T.D. Lee: *Particle Physics and Introduction to Field Theory*, New York: Harwood Academic Publishing GmbH, 1988; M. Kaku: *Quantum Field Theory – A Modern Introduction*, Oxford: Oxford University Press, 1993; M.E. Peskin, D.V. Schroeder: *An Introduction to Quantum Field Theory*, Reading, MA, Addison-Wesley, 1995; C.M. Becchi, G. Ridolfi: *An introduction to relativistic processes and the standard model of electroweak interactions*, Springer, 2014; M. D. Schwartz: *Quantum Field Theory and the Standard Model*, Cambridge University Press, 2014.

[4] G. Aad et al. [ATLAS]: *Observation of a new particle in the search for the Standard Model Higgs boson with the ATLAS detector at the LHC* Phys. Lett. B 716 (2012) 1; S. Chatrchyan et al. [CMS]: *Observation of a New Boson at a Mass of 125 GeV with the CMS Experiment at the LHC* Phys. Lett. B 716 (2012) 30.

[5] H. Georgi, S. Glashow: *Unity of All Elementary-Particle Forces*, Phys. Rev. Lett. 32 (1974) 438.

[6] G. Aad et al. [ATLAS]: *Evidence for the spin-0 nature of the Higgs boson using ATLAS data*, Phys. Lett. B 726 (2013) 120.

[7] V. Khachatryan et al. [CMS]: *Constraints on the spin-parity and anomalous HVV couplings of the Higgs boson in proton collisions at 7 and 8 TeV*, Phys. Rev. D 92 (2015) 012004.

[8] P.A. Zyla et al. (Particle Data Group): *Review of Particle Physics*, Prog. Theor. Exp. Phys. 2020, (2020) 083C01, online version only 2020.

[9] E. Schrödinger: *An undulatory theory of the mechanics of atoms and molecules*, Phys. Rev. 28 (1926) 1049.

[10] E. Noether, *Invariante Variationsprobleme* Nachr. v. d. Ges. d. Wiss. zu Göttingen (1918) 235-257; for an English version see: arXiv:[physics]/0503066 [physics.hist-ph].

[11] D. E. Neuenschwander: *Emmy Noether's Wonderful Theorem*, (2011) The Johns Hopkins University Press.
[12] I.I. Bigi, A.I. Sanda: *CP Violation, Second Edition*, Cambridge Monographs on Particle Physics, Nuclear Physics and Cosmology, Cambridge Univ. Press, 2009.
[13] T.D. Lee: *Particle Physics and Introduction to Field Theory*, New York, Harwood Academic Publishers GmbH, 1988.
[14] C. Jarlskog (ed.): *CP Violation*, Advanced Series on Directions in High Energy Physics - Vol. 3, World Scientific, 1988, ISBN 9971-50-560-6. [It is a very good book, not only historically. There are many remarks still valid today.]
[15] A. Angelopoulos et al. [CPLEAR]: *First direct observation of time reversal non-invariance in the neutral kaon system*, Phys. Lett. B 444 (1998) 43.
[16] J. P. Lees et al. [BaBar]: *Observation of Time Reversal Violation in the B^0 Meson System*, Phys. Rev. Lett. 109 (2012) 211801.
[17] E.P. Wigner: *Uber die Operation der Zeitumkehr in der Quantenmechanik*, Nachr. Ges. Wiss. Göttingen, Math.-Physik. Kl. 32 (1932) 546.
[18] H.A. Kramers: *Théorie générale de la rotation paramagnétique dans les cristaux*, Proc. Acad. Sci. Amsterdam 33 (1930) 959.
[19] A. Abragam, B. Bleaney: *Electron Paramagnetic Resonance of Transition Ions*, Oxford: Clarendon Press, 1970.
[20] A. Einstein, B. Podolsky, N. Rosen: *Can Quantum-Mechanical Description of Physical Reality Be Considered Complete?* Phys. Rev. 47 (1935) 777.
[21] N.F. Ramsey: *Earliest Criticisms of Assumed P and T Symmetries*, M. Skalsey et al. (eds.), *Time Reversal – The Arthur Rich Memorial Symposium*, AIP Conf. Proc. 270, New York, American Institute of Physics, 1993.
[22] Physics Today, January 2013.
[23] E.M. Purcell, N.F. Ramsey: *On the Possibility of Electric Dipole Moments for Elementary Particles and Nuclei*, Phys. Rev. 78 (1950) 807.
[24] T.D. Lee, C.N. Yang: *Question of Parity Conservation in Weak Interactions*, Phys. Rev. 104 (1956) 254.
[25] C.S. Wu et al.: *Experimental Test of Parity Conservation in Beta Decay*, Phys. Rev. 105 (1957) 1413; R.L. Garwin, L.M. Lederman, M. Weinrich, *Observations of the Failure of Conservation of Parity and Charge Conjugation in Meson Decays: The Magnetic Moment of the Free Muon*, Phys. Rev. 105 (1957) 1415.
[26] J.H. Smith, E.M. Purcell, N.F. Ramsey: *Experimental limit to the electric dipole moment of the neutron*, Phys. Rev. 108 (1957) 120.
[27] N.F. Ramsey: *Electric Dipole Moment of the Neutron*, Annu. Rev. Nucl. Part. Sci. 40 (1990) 1-14; K.F. Smith et al., *A Search for the Electric Dipole Moment of the Neutron*, Phys. Lett. B 234 (1990) 191.
[28] L.I. Schiff: *Quantum Mechanics*, New York, McGraw-Hill Inc., 1968.
[29] G. Lüders: *Proof of the TCP theorem*, Ann. Phys. 2 (1957) 1.
[30] R.F. Streater, A.S. Wightmann: *PCT, Spin and Statistics, and All That*, New York, Benjamin, 1964.
[31] L. Laudau: *On The Conservation Laws for Weak Interactions*, Nucl. Phys. 3 (1957) 127-131. [He said there: "I would like to express my deep appreciation to *L. Okun*,

B. Ioffe and A. Rudik for discussions from which the idea of this part of the present paper emerged."]

[32] L. B. Okun, *Mirror particles and mirror matter: 50 years of speculations and search*, Phys. Usp. 50 (2007) 380.

[33] A. Pais: *CP violation: the first 25 years*, CP Violation in Particle and Astrophysics, J. Tran Than Van (ed.), Edition Frontières, Gif-sur-Yvette, France, 1989 [An exciting historical account of the progress of our field during this period].

[34] B. Laurent, M. Roos: Phys. Rev. Lett. 13 (1964) 269; *ibid.* 15 (1964) 104.

[35] J.W. Cronin: *CP Symmetry Violation: The Search for Its Origin*, Rev. Mod. Phys. 53 (1981) 373; Science 212 (1981) 1221.

[36] J.W. Cronin: *The discovery of CP violation*, Eur. Phys. J. H 36 (2011) 487.

[37] M. Gell-Mann, A. Pais: *Behavior of Neutral Particles under Charge Conjugation* Phys. Rev. 97 (1955) 1387.

[38] J.W. Cronin: *A Life in High-Energy Physics: Success Beyond Expectations*, Ann. Rev. Nucl. Part. Science, Vol. 64 (2014) 1.

[39] J.W. Cronin et al.: *Production of Hadrons with Large Transverse Momentum at 200-GeV and 300-GeV*, Phys. Rev. Lett. 31 (1973) 1426.

[40] J.W. Cronin et al.: *Production of hadrons with large transverse momentum at 200, 300, and 400 GeV*, Phys. Rev. D 11 (1975) 3105.

[41] J.W. Cronin: *Recent results from the Pierre Auger Observatory*, arXiv:astro-ph/0911.4714

[42] J.W. Cronin: *The 1953 Cosmic Ray Conference at Bagneres de Bigorre: the Birth of Sub-Atomic Physics*, Eur. Phys. J. H 36 (2011) 183.

[43] K. M. Watson: *Some General Relations between the Photoproduction and Scattering of π Mesons*, Phys. Rev. D 95 (1954) 228.

[44] I.I. Bigi, V.A. Khoze, N.G. Uraltsev, A.I. Sanda: *CP Violation*, C. Jarlskog (Editor), p. 175-248, World Scientific (1988); N.G. Uraltsev, Proceedings of DPF-92, arXiv:[hep-ph]/9212233.

[45] I.I. Bigi: *Could Charm and τ Transitions be the 'Poor Princesses' of Deeper Understanding of Fundamental Dynamics? – or – Finding Novel Forces?*, Front. Phys. 10 (2015) 101203.

[46] L. Wolfenstein: *Final-state interactions and CP violation in weak decays*, Phys. Rev. D 43 (1991) 151.

[47] M. Gell-Mann: *Nonleptonic Weak Decays and the Eightfold Way*, Phys. Rev. Lett. 12 (1964) 155; *A schematic model of baryons and mesons*, Phys. Lett. 8 (1964) 214.

[48] C. Zweig: *Developments in the Quark Theory of Hadrons*, CERN Rep. 8419/TH 412.

[49] H.J. Lipkin: *Lie Groups for Pedestrians*, North-Holland Publ. Co., 1965.

[50] B. Hall, *Lie Groups, Lie Algebras, and Representations: An Elementary Introduction*, Springer (2003).

[51] I.I. Bigi, Y. Dokshitzer, V. Khoze, J. Kühn, P. Zerwas: *Production and decay properties of ultra-heavy quarks*, Phys. Lett. B 181 (1986) 157.

[52] W. Pauli to A. Pais, Letter 1682 in W. Pauli, in *Wissenschaftlicher Briefwechsel*, Vol. IV, Part II (1999) Springer-Verlag, edited by K. V. Meyenn.

[53] I.I. Bigi, M. Shifman, N.G. Uraltsev, A. Vainshtein: *Pole mass of the heavy quark: Perturbation theory and beyond*, Phys. Rev. D 50 (1994) 2234; M. Beneke, V. Braun:

Heavy Quark Effective Theory beyond Perturbation Theory: Renormalons, the Pole Mass and the Residual Mass Term, Nucl. Phys. B 426 (1994) 301.

[54] I. Bigi, M. Shifman, N. Uraltsev: *Aspects of Heavy Quark Theory*, Ann. Rev. Nucl. Part. Sci. 47 (1997) 591.

[55] M. Shifman: *New and Old About Renormalons*, arXiv:[hep-ph]/310.1966 [hep-th]; contribution to the book: *QCD and Heavy Quarks – In Memoriam Nikolai Uraltsev*, I.I. Bigi, P. Gambino, Th. Mannel (editors), World Scientific Publishing Co., 2015.

[56] N. Gray, D.J. Broadhurst, W. Grafe, K. Schilcher: *Three-loop relation of quark $\overline{MS}$ and pole masses*, Z. Phys. C 48 (1990) 673.

[57] D. J. Broadhurst, N. Gray, K. Schilcher, *Gauge invariant on-shell Z(2) in QED, QCD and the effective field theory of a static quark*, Z. Phys. C 52 (1991) 111.

[58] K. G. Chetyrkin, M. Steinhauser, *Short distance mass of a heavy quark at order α_s^3*, Phys. Rev. Lett. 83 (1999) 4001.

[59] K. Melnikov, T. v. Ritbergen, *The Three loop relation between the MS-bar and the pole quark masses*, Phys. Lett. B 482 (2000) 99.

[60] P. Marquard, A. V. Smirnov, V. A. Smirnov, M. Steinhauser, *Quark Mass Relations to Four-Loop Order in Perturbative QCD*, Phys. Rev. Lett. 114 (2015) 142002.

[61] I. I. Bigi, M. A. Shifman, N. G. Uraltsev, A. I. Vainshtein: *The Pole mass of the heavy quark. Perturbation theory and beyond*, Phys. Rev. D 50 (1994) 2234.

[62] M. Beneke: *More on ambiguities in the pole mass*, Phys. Lett. B 344 (1995) 341.

[63] I. I. Bigi, M. A. Shifman, N. Uraltsev, A. I. Vainshtein: *High power n of m(b) in beauty widths and n=5 $\to$ infinity limit*, Phys. Rev. D 56 (1997) 4017; I. I. Bigi, M. A. Shifman, N. G. Uraltsev, A. I. Vainshtein: *Sum rules for heavy flavor transitions in the SV limit*, Phys. Rev. D 52 (1995) 196.

[64] M. Voloshin: *"Optical" sum rule for form factors of heavy mesons*, Phys. Rev. D 46 (1992) 3062.

[65] S. Bianco, F.L. Fabbri, D. Benson, I. Bigi: *A Cicerone for the physics of charm*, Riv. Nuovo Cim. 26N7 (2003) 1.

[66] M. Beneke: *A Quark mass definition adequate for threshold problems*, Phys. Lett. B 434 (1998) 115.

[67] A.H. Hoang, Z. Ligeti, A.V. Manohar: *B decay and the Upsilon mass*, Phys. Rev. Lett. 82 (1999) 277; A.H. Hoang, Z. Ligeti, A.V. Manohar: *B decays in the upsilon expansion*, Phys. Rev. D 59 (1999) 074017.

[68] A. Pineda: *Determination of the bottom quark mass from the $\Upsilon(1S)$ system*, JHEP 0106 (2001) 022.

[69] N. Uraltsev: *Heavy quark expansion in beauty: recent successes and problems*, talk at Continuous Advances in QCD 2004, arXiv:[hep-ph]/0409125.

[70] M. Bona et al. [UT fit]: *Model-independent constraints on $\Delta F = 2$ operators and the scale of New Physics*, JHEP 0803 (2008) 049.

[71] M. Goldberger, S. Treiman: *Conserved Currents in the Theory of Fermi Interactions* Phys. Rev. 110 (1958) 1178.

[72] E. Shuryak: *Hadrons Containing a Heavy Quark and QCD Sum Rules*, Nucl. Phys. B 198 (1982) 83.

[73] S. Nussinov, W. Wetzel: *Comparison of Exclusive Decay Rates for $b \to u$ and $b \to c$ Transitions*, Phys. Rev. D 36 (1987) 130; M. Voloshin, M. Shifman: *On Production of D and D^* Mesons in B Meson Decays*, Sov. J. Nucl. Phys. 47 (1988) 511; N. Isgur,

M. Wise: *Weak Decays of Heavy Mesons in the Static Quark Approximation*, Phys. Lett. B 232 (1989) 113; *Weak transition form-factors between heavy Mesons*, Phys. Lett. B 237 (1990) 527.

[74] E. Eichten, B. Hill: *An Effective Field Theory for the Calculation of Matrix Elements Involving Heavy Quarks*, Phys. Lett. B 234 (1990) 511; H. Georgi, *An Effective Field Theory for Heavy Quarks at Low-energies*, Phys. Lett. B 240 (1990) 447.

[75] I. Bigi et al.: *Sum rules for heavy flavor transitions in the small velocity limit*, Phys. Rev. D 52 (1995) 196.

[76] L.L. Foldy, S.A. Wouthuysen: *On the Dirac theory of spin 1/2 particle and its non relativistic limit*, Phys. Rev. 78 (1950) 29.

[77] M. Luke: *Effects of subleading operators in the heavy quark effective theory*, Phys. Lett. B 252 (1990) 447.

[78] K. Wilson: *Non-Lagrangian models of current algebra*, Phys. Rev. 179 (1969) 1499. [The title might confuse readers now.]

[79] M. Voloshin, M. Shifman: *Pre-asymptotic Effects in Inclusive Weak Decays of Charmed Particles*, Sov. J. Nucl. Phys. 41 (1985) 120.

[80] J. Chay, H. Georgi, B. Grinstein: *Lepton energy distributions in heavy meson decays from QCD*, Phys. Lett. B 247 (1990) 399.

[81] I.I. Bigi, N.G. Uraltsev, A. Vainshtein: *Non-perturbative corrections to inclusive beauty and charm decays: QCD versus phenomenological models*, Phys. Lett. B 293 (1992) 430.

[82] E. Bagan, P. Ball, V. M. Braun, P. Gosdzinsky: *Theoretical update of the semileptonic branching ratio of B mesons*, Phys. Lett. B 342 (1995) 362; Erratum: [*Phys. Lett.* B 374 (1996) 363].

[83] I.I. Bigi, M. Shifman, N.G. Uraltsev, A. Vainshtein: *QCD predictions for lepton spectra in inclusive heavy flavor decays* Phys. Rev. Lett. 71 (1993) 496.

[84] P. Gambino: *Inclusive Semileptonic B Decays and* $|V_{cb}|$, arXiv:[hep-ph]/1501.00314, contribution to the book: *QCD and Heavy Quarks – In Memoriam Nikolai Uraltsev*, I.I. Bigi, P. Gambino, Th. Mannel (editors), World Scientific Publishing Co., 2015.

[85] F.J. Dyson: *Divergence of perturbation theory in quantum electrodynamics*, Phys. Rev. 85 (1952) 631.

[86] W. E. Thirring: *On the divergence of perturbation theory for quantized fields*, Helv. Phys. Acta 26 (1953) 33.

[87] L. N. Lipatov: *Divergence of the Perturbation Theory Series and the Quasiclassical Theory*, Sov. Phys. JETP 45 (1977) 216 [Zh. Eksp. Teor. Fiz. 72 (1977) 411].

[88] E. B. Bogomolny: *Calculation Of Instanton - Anti-instanton Contributions In Quantum Mechanics*, Phys. Lett. B 91 431 (1980); J. Zinn-Justin: *Multi - Instanton Contributions in Quantum Mechanics*, Nucl. Phys. B 192 (1981) 125.

[89] G. 't Hooft: *Can We Make Sense Out of Quantum Chromodynamics?*, Subnucl. Ser. 15 (1979) 943.

[90] N. Uraltsev: *Heavy quark expansion in beauty and its decays*, Proc. Int. Sch. Phys. Fermi 137 (1998) 329.

[91] P. Argyres, M. Unsal: *A semiclassical realization of infrared renormalons*, Phys. Rev. Lett. 109 (2012) 121601; G. V. Dunne, M. Üoensal: *Continuity and Resurgence: towards a continuum definition of the* $\mathbb{CP}(N\text{-}1)$ *model*, Phys. Rev. D 87 (2013) 025015.

[92] M. Shifman: *Resurgence, operator product expansion, and remarks on renormalons in supersymmetric Yang-Mills theory*, J. Exp. Theor. Phys. 120 (2015) 386.

[93] M. A. Shifman: *Quark-hadron duality*, Boris Ioffe Festschrift, At the Frontier of Particle Physics / Handbook of QCD, ed. M. Shifman (World Scientific, Singapore, 2001).

[94] I. Bigi, N.G. Uraltsev: *A vademecum on Quark-Hadron Duality*, Int. J. Mod. Phys. A 16 (2001) 5201.

[95] G. 't Hooft: *A Two-Dimensional Model for Mesons*, Nucl. Phys. B 75 (1974) 461.

[96] J. Steinberger: *On the Use of subtraction fields and the lifetimes of some types of meson decay*, Phys. Rev. 76 (1949) 1180. [Actually the situation is somewhat easier to connect the charges of leptons e and ν_e and protons and neutrons.]

[97] D.J. Gross, F. Wilczek: *Asymptotically Free Gauge Theories - I* Phys. Rev. D 8 (1973) 3497.

[98] H.D. Politzer: *Reliable Perturbative Results for Strong Interactions?*, Phys. Rev. Lett. 26 (1973) 1346.

[99] G. 't Hooft: *Conference on Lagrangian Field Theory*, Marseille, 1972.

[100] M. Gell-Mann, R. J. Oakes, B. Renner: *Behavior of current divergences under* $SU(3) \times SU(3)$, *Phys. Rev.* 175 (1968) 2195.

[101] M. Gell-Mann, M.L. Goldberger, W.E. Thirring: *Use of Causality Conditions in Quantum Theory*, Phys. Rev. 95 (1954) 1612.

[102] R. de L. Kronig: *On the Theory of Dispersion of X-Rays*, J. Opt. Soc. Am. 12 (1926) 547.

[103] H. A. Kramers: *La diffusion de la lumière par les atomes*, Atti Cong. Intern. Fisica, Como 2 (1927) 545.

[104] F. Niecknig, B. Kubis: *Dispersion-theoretical analysis of the* $D^+ \to K^-\pi^+\pi^+$ *Dalitz plot*, JHEP 1510 (2015) 142; C. Hanhart: *Proceedings of CHARM 2013, Modelling low-mass resonances in multi-body decays*, arXiv:[hep-ph]/1311.6627; B. Kubis: *The role of final-state interactions in Dalitz plot studies*, arXiv:[hep-ph]/1108.5866; S. Gardner, U.-G. Meissner: *Rescattering and chiral dynamics in* $B \to \rho\pi$ *decay*, Phys. Rev. D 65 (2002) 094004.

[105] J. Donoghue: lecture given at the International School on Effective Field Theory, arXiv:[hep-ph]/9607351.

[106] M. Kaku: *Quantum Field Theory – A Modern Introduction*, Oxford: Oxford University Press, 1993, Sect. 6.10 "Dispersion Relations".

[107] G. 't Hooft: *A Two-Dimensional Model for Mesons*, Nucl. Phys. B 75 (1974) 461; C. Callan, N. Coote, D. Gross: *Two-Dimensional Yang-Mills Theory: A Model of Quark Confinement*, Phys. Rev. D 13 (1976) 1649; M. Einhorn, S. Nussinov, E. Rabinovici: *Meson Scattering in Quantum Chromodynamics in Two-Dimensions*, Phys. Rev. D 15 (1977) 2282.

[108] S. Okubo: *Phi meson and unitary symmetry model*, Phys. Lett. 5 (1963) 1975; G. Zweig, *An SU(3) model for strong interaction symmetry and its breaking. Version 1*, CERN Report No. 8419/TH412 (1964); J. Iizuka: *Systematics and phenomenology of meson family*, Prog. Theor. Phys. Suppl. 37 (1966) 21.

[109] I.I. Bigi, in: *Proceedings of Charm Physics*, edited by Ming-Han Ye and Tao Huang, QCD161:S12:1987A, pp. 339-425 [in particular pp. 370-389]; SLAC-PUB-4349.

[110] M.A. Shifman, A.I. Vainshtein, V.I. Zakharov: *QCD and resonance physics: Theoretical foundation*, *Nucl.Phys.* B 147 (1979) 385; 448; V.A. Novikov, M.A. Shifman, A.I. Vainshtein, V.I. Zakharov, *Wilson's Operator Expansion: Can It Fail?*, Nucl. Phys. B 249 (1985) 445.

[111] J.D. Bjorken: *New symmetries in heavy flavor physics*, Proc. 4th Rencontres de Physique de la Vallee d'Aoste, La Thuille, Italy, Conf. Proc. C 900318 (1990) 583; I.I. Bigi et al.: *On measuring the kinetic energy of the heavy quark inside B mesons*, Phys. Lett. B 339 (1994) 160.

[112] N. Uraltsev: *New exact heavy quark sum rules*, Phys. Lett. B 501 (2001) 86.

[113] G.P. Lepage: *Lattice QCD for Novices*, arXiv:hep-lat/0506036.

[114] K. Symanzik: *Continuum Limit and Improved Action in Lattice Theories. 1. Principles and ϕ^4 Theory*, Nucl. Phys. B 226 (1983) 187.

[115] H. B. Nielsen and M. Ninomiya: *No Go Theorem for Regularizing Chiral Fermions*, Phys. Lett. B 105 (1981) 219.

[116] R. Abbott et al. [RBC and UKQCD]: *Direct CP violation and the $\Delta I = 1/2$ rule in $K \to \pi\pi$ decay from the Standard Model*, Phys. Rev. D 102 (2020) 5, 054509.

[117] A.J. Buras, J.-M. Gerard: *Final state interactions in $K \to \pi\pi$ decays: $\Delta I = 1/2$ rule vs ϵ'/ϵ*, Eur. Phys. J. C 77 (2017); *Upper bound on ϵ'/ϵ parameters $B_6^{(1/2)}$ and $B_8^{(3/2)}$ from large N QCD and other news*, JHEP 12 (2015) 008.

[118] G. Colangelo, J. Gasser and H. Leutwyler: *$\pi\pi$ scattering*, Nucl. Phys. B 603 (2001) 125.

[119] A. J. Buras and J. M. Gérard: *Isospin-breaking in ε'/ε: impact of η_0 at the dawn of the 2020s*, Eur. Phys. J. C 80 (2020) 701.

[120] L.H. Karsten, J. Smith: *Lattice fermions: Species doubling, chiral invariance and the triangle anomaly*, Nucl. Phys. B 183 (1981) 103.

[121] M. Golterman: *Lattice Chiral Gauge Theories*, Nucl. Phys. B, Proc. Suppl 94 (2001) 189.

[122] D.B. Kaplan: *A Method for Simulating Chiral Fermions on the Lattice*, Phys. Lett. B 288 (1992) 342.

[123] D. Grabowska, D.B. Kaplan: *Nonperturbative Regulator for Chiral Gauge Theories?*, Phys. Rev. Lett 116 (2016) 211602.

[124] H. Lipkin: *Is observed direct CP violation in $B_d \to K^+\pi^-$ due to new physics? Check standard model prediction of equal violation in $B_s \to K^-\pi^+$*, Phys. Lett. B 621 (2005) 126.

[125] R. Aaij et al. [LHCb]: *Measurement of CP asymmetries in two-body B_s^0-meson decays to charged pions and kaons*, Phys. Rev. D 98 (2018) 032004.

[126] I.I. Bigi: *CP Asymmetries in Many-Body Final States in Beauty and Charm Transitions*, arXiv:[hep-ph]/1509.03899, pg. 16 in the beginning of "Sect. 3 $\Delta B \neq 0$ forces" there.

[127] Nobel award 1979, based on papers from 1968: S. Weinberg, *A Model of Leptons*, Phys. Rev. Lett. 19 (1967) 1264; A. Salam (1968): *Elementary Particles Theory: Relativistic Groups and Analyticity; Eighth Nobel Symposium*, ed. N. Svartholm; S.L. Glashow: *Partial-symmetries of weak interactions*, Nucl. Phys. 22 (1961) 579.

[128] M. Kobayashi, T. Maskawa: *CP Violation in the Renormalizable Theory of Weak Interaction*, Prog. Theor. Phys. 49 (1973) 652.

[129] R. Mohapatra: *Renormalizable model of weak and electromagnetic interactions with CP violation* Phys. Rev. D 6 (1972) 203.
[130] R.A. Horn, C.R. Johnson: *Matrix Analysis*, Cambridge University Press, 1990.
[131] S. Glashow, J. Illiopolous, L. Maiani: *Weak Interactions with Lepton-Hadron Symmetry*, Phys. Rev. D 2 (1970) 1285.
[132] C. Jarlskog: *Commutator of the Quark Mass Matrices in the Standard Electroweak Model and a Measure of Maximal CP Violation*, Phys. Rev. Lett. 55, 1039 (1985).
[133] L. Wolfenstein: *Parametrization of the Kobayashi-Maskawa Matrix*, Phys. Rev. Lett. 51 (1983) 1945.
[134] I.I. Bigi, A.I. Sanda: *Notes on the Observability of of CP Violation in B Decays*, Nucl. Phys. B 193 (1981) 85; $B^0 - \bar{B}^0$ *mixing and violations of CP symmetry*, Phys. Rev. D 29 (1984) 1393; *CP Violation in Heavy Flavor Decays: Predictions and Search Strategies*, Nucl. Phys. B 281 (1987) 41.
[135] M. Bona et al. [UTfit]: *The Unitarity Triangle Fit in the Standard Model and Hadronic Parameters from Lattice QCD: A Reappraisal after the Measurements of* $\Delta\ m(s)$ *and* $BR(B \longrightarrow \tau\nu(\tau))$ JHEP 10 (2006) 081.
[136] Y.H. Ahn, H-Y. Cheng, S. Oh: *Wolfenstein Parametrization at Higher Order: Seeming Discrepancies and Their Resolution* Phys. Lett. B 703 (2011) 571.
[137] I.I. Bigi: *3- and 4-Body Final States in B, D and* τ *Decays about Features of New Dynamics with CPT Invariance or "Achaeans outside Troy"*, arXiv:[hep-ph]/1306.6014, talk given at FPCP2013 "Flavor Physics and CP Violation 2013".
[138] J. Chadwick, *The Existence of a Neutron*, Proc. R. Soc. Lond. A 136 (1932).
[139] E. Fermi: *Tentativo di una Teoria Dei Raggi* β, Nuovo Cim. 11 (1934) 1.
[140] E. Fermi: *Tentativo di una teoria dell'emissione dei raggi* β, La Ric. Scientifica, Anno IV, Vol. 2 N. 12 (1934); for a German version: *Versuch einer Theorie der* β*-Strahlen. I* [An attempt of a theory of beta radiation], Z. Phys. 88 (1934) 161.
[141] H. Bethe, R. Peierls: *The 'neutrino'*, Nature 133 (1934) 532.
[142] B. Pontecorvo: *Inverse beta process*, Chalk River Laboratory Report PD-205 (1946) reprinted in Camb. Monogr. Part. Phys. Nucl. Phys. Cosmol. 1 (1991) 25.
[143] C.L. Cowan et al.: *Large liquid scintillation detectors*, Phys. Rev. 90 (1953) 493.
[144] F. Reines and C.L. Cowan: *Detection of the free neutrino*, Phys. Rev. 92 (1953) 830.
[145] B. Pontecorvo: *Mesonium and anti-mesonium*, *Sov.Phys.JETP* 6 (1957) 429 [Zh. Eksp. Teor. Fiz. 33 (1957) 549].
[146] M. Goldhaber, L. Grodzins, A. W. Sunyar: *Helicity of Neutrinos*, Phys. Rev. 109 (1958) 1015.
[147] Z. Maki, M. Nakagawa, S. Sakata: *Remarks on the unified model of elementary particles*, Prog. Theor. Phys. 28 (1962) 870.
[148] B. Pontecorvo: *Neutrino Experiments and the Problem of Conservation of Leptonic Charge*, Sov. Phys. JETP 26 (1968) 984 [Zh. Eksp. Teor. Fiz. 53 (1967) 1717].
[149] V. N. Gribov, B. Pontecorvo: *Neutrino astronomy and lepton charge*, Phys. Lett. B 28 (1969) 493.
[150] R. Davies, *Solar neutrinos. II: Experimental*, Phys. Rev. Lett. 12 (1964) 303.
[151] B. Aharmim et al. [SNO]: *Independent Measurement of the Total Active* ^8B *Solar Neutrino Flux Using an Array of* ^3He *Proportional Counters at the Sudbury Neutrino Observatory*, Phys. Rev. Lett. 101 (2008) 111301.

[152] A.B. McDonald: *Nobel Lecture: The Sudbury Neutrino Observatory: Observation of flavor change for solar neutrinos*, Rev. Mod. Phys. 88 (2016) 030502.

[153] T. Kajita: *Nobel Lecture: Discovery of atmospheric neutrino oscillations*, Rev. Mod. Phys. 88 (2016) 030501.

[154] S. Abe et al. [KamLAND]: *Precision Measurement of Neutrino Oscillation Parameters with KamLAND* Phys. Rev. Lett. 100 (2008) 221803.

[155] P. Adamson; et al [MINOS]: *Measurement of the neutrino mass splitting and flavor mixing by MINOS*, Phys. Rev. Lett. 106 (2011) 181801.

[156] F. P. An et al. [Daya Bay]: *Observation of electron-antineutrino disappearance at Daya Bay*, Phys. Rev. Lett. 108 (2012) 171803.

[157] E. Majorana, *Teoria simmetrica dell'elettrone e del positrone*, Nuovo Cimento 14 (1937) 171.

[158] G. Segre, B. Hoerlin: *The Pope of Physics, Enrico Fermi and the Birth of the Atomic Age*, Henry Holt & Company, 2016. [There are several reason to read it, well beyond Majorana.]

[159] F.T. Avignone III, St.R. Elliott, J. Engel: *Double Beta Decay, Majorana Neutrinos, and Neutrino Mass*, Rev. Mod. Phys. 80 (2008) 481.

[160] M. Fukugita, T. Yanagida: *Physics and Astrophysics of Neutrinos*, Springer-Verlag, Tokyo, 1994.

[161] St.R. Elliott, M. Franz: *Majorana fermions in nuclear, particle, and solid-state physics*, Rev. Mod. Phys. 87 (2015) 137.

[162] S. Gardner, X. Yan: *CPT, CP, and C transformations of fermions, and their consequences, in theories with B-L violation*, Phys. Rev. D 93 (2016) 096008.

[163] P. Higgs: *Broken symmetries, massless particles and gauge fields*, Phys. Lett. 12 (1964) 132, *Broken Symmetries and the Masses of Gauge Bosons*, Phys. Rev. Lett. 13 (1964) 508; F. Englert, R. Brout: *Broken Symmetry and the Mass of Gauge Vector Mesons* Phys. Rev. Lett. 13 (1964) 321; G. Guralnik, C. Hagen, T. Kibble: *Global Conservation Laws and Massless Particles* Phys. Rev. Lett. 13 (1964) 585.

[164] K. Inoue et al.: *Aspects of Grand Unified Models with Softly Broken Supersymmetry*, Prog. Theor. Phys. C 68 (1982) 927; *Renormalization of Supersymmetry Breaking Parameters Revisited*, ibid. C 71 (1984) 413; I.E. Ibanez, G.G. Ross: $SU_L(2) \times U(1)$ *Symmetry Breaking as a Radiative Effect of Supersymmetry Breaking in GUTS*, Phys. Lett. B 110 (1982) 215.

[165] A.J. Buras: *The Return of Kaon Flavor Physics*, 24th Cracow Epiphany Conference on Advances in Heavy Flavour Physics, arXiv:[hep-ph]/1805.11096. [It makes it clearly the connection with the impact of LQCD.]

[166] E. Cortina Gil et al. [NA62], *First search for $K^+ \to \pi^+ \nu\bar{\nu}$ using the decay-in-flight technique*, Phys. Lett. B 791 (2019) 156.

[167] E. Cortina Gil et al. [NA62]: *An investigation of the very rare $K \to \pi\nu\nu$ decay*, JHEP 11 (2020) 042.

[168] K. Shiomi [KOTO], *$K_L \to \pi^0 \nu\bar{\nu}$ at KOTO*, CKM 2014, arXiv:[hep-ph]/1411.4250.

[169] A. Paul, I.I. Bigi, St. Recksiegel: *$D^0 \to \gamma\gamma$ and $D^0 \to \mu^+\mu^-$ Rates on an Unlikely Impact of the Littlest Higgs Model with T-Parity*, Phys. Rev. D 82 (2010) 094006; *On $D \to X_u l^+ l^-$ within the Standard Model and Frameworks like the Littlest Higgs Model with T Parity*, Phys. Rev. D 83 (2011) 114006; A. Paul, A. de La Puente,

I.I. Bigi: *Manifestations of warped extra dimension in rare charm decays and asymmetries* Phys. Rev. D 90 (2014) 014035.

[170] A. Baldini et al., *MEG Upgrade proposal*, arXiv:[hep-ph]/1301.7225.

[171] W. H. Bertl et al. [SINDRUM II], *A Search for muon to electron conversion in muonic gold*, Eur. Phys. J. C 47 (2006) 337-346.

[172] L. Bartoszek et al. [Mu2e]: *Technical Design Report*, arXiv:[hep-ph]/1501.05241 TDR 2014 FERMILAB-TM-2594, FERMILAB-DESIGN-2014-01.

[173] J.C. Hardy, I.S. Towner: *Super-allowed $0^+ \to 0^+$ nuclear β decay: 2014 critical survey, with precise results of V_{ud} and CKM unitarity*, Phys. Rev. C 91 (2015) 025501.

[174] R. Aaij et al. [LHCb]: *Observation of CP violation in charm decays, Phys. Rev. Lett.* 122 (2019) 211803.

[175] T.D. Lee, C.S. Wu: *Weak Interactions: Decays of neutral K mesons*, Ann. Rev. Nucl. Sci. 16 (1966) 471.

[176] L. Wolfenstein, *Violation of CP Invariance and the Possibility of Very Weak Interactions*, Phys. Rev. Lett. 13 (1964) 562.

[177] A. Pais: *Some Remarks on the V-Particles*, Phys. Rev. 86 (1952) 663.

[178] T. Inami, C.S. Lim: *Effects of Superheavy Quarks and Leptons in Low-Energy Weak Processes $K_L \to \mu^+\mu^-$, $K^+ \to \pi^+\nu\bar{\nu}$ and $K^0 \leftrightarrow \bar{K}^0$*, Prog. Theor. Phys. 65 (1981) 297; Erratum: Prog. Theor. Phys. 65 (1981) 172.

[179] G. Buchalla, A. J. Buras, M. E. Lautenbacher: *Weak decays beyond leading logarithms*, Rev. Mod. Phys. 68 (1996) 1125.

[180] J. Brod and M. Gorbahn: *ϵ_K at Next-to-Next-to-Leading Order: The Charm-Top-Quark Contribution*, Phys. Rev. D 82 (2010) 094026.

[181] S. Aoki et al. [Flavour Lattice Averaging Group], *FLAG Review 2019*, arXiv:[hep-ph]/1902.08191.

[182] J.S. Bell, J. Steinberger: *Weak interactions of kaons*, Proc. 1965 Oxford Int. Conf. on Elementary Particles, Rutherford High Energy Laboratory, Didcot (1966) 195.

[183] P.K. Kabir: *What is not invariant under time reversal?*, Phys. Rev. D 2 (1970) 540.

[184] I. I. Y. Bigi and A. I. Sanda: *On limitations of T invariance in K decays*, Phys. Lett. B 466 (1999) 33.

[185] L.M. Sehgal, M. Wanninger: *CP violation in the decay $K_L \to \pi^+\pi^-e^+e^-$*, Phys. Rev. D 46 (1992) 1035; L.M. Sehgal: *Rare K decays*, arXiv:[hep-ph]/9411374; *CP and T violation in the decay $K_L \to \pi^+\pi^-e^+e^-$ and related processes*, arXiv:[hep-ph]/9908338.

[186] J.K. Elwood, M.B. Wise, M.J. Savage: $K_L \to \pi^+\pi^-e^+e^-$, Phys. Rev. D 52 (1995) 5095; J.K. Elwood et al.: *Final state interactions and CP violation in $K_L \to \pi^+\pi^-e^+e^-$*, Phys. Rev. D 53 (1996) 4078.

[187] J. Adams et al. [kTeV]: *Measurement of branching fraction of the decay $K_L \to \pi^+\pi^-e^+e^-$*, Phys. Rev. Lett. 80 (1998) 4123; A. Lai et al. [Na48]: *Observation of the decay $K_S \to \pi^+\pi^-e^+e^-$*, Phys. Lett. B 599 (2004) 197.

[188] I.I. Bigi, A.I. Sanda: *On limitations of T invariance in K decays*, Phys. Lett. B 466 (1999) 33.

[189] G. Isidori et al.: *Light-quark loops in* $K \to \pi\nu\bar{\nu}$, Nucl. Phys. B 718 (2005) 319; A. Buras et al., *Charm quark contribution to* $K^+ \to \pi^+\nu\bar{\nu}$ *at next-to-next-to-leading order*, JHEP 11 (2006) 002.

[190] A.J. Buras: *Kaon Theory News*, Talk presented at HEP-EPS in Vienna, July 2015, arXiv:[hep-ph]/1510.00128.

[191] Y. Grossman, Y. Nir: $K_L \to \pi^0$ *neutrino anti-neutrino beyond the standard model*, Phys. Lett. B 398 (1997) 163.

[192] A. V. Artamonov et al. [BNL-E949]: *Study of the decay* $K^+ \to \pi^+\nu\bar{\nu}$ *in the momentum region* $140 < P_\pi < 199\ MeV/c$, Phys. Rev. D 79 (2009) 092004.

[193] G. Ruggiero [NA62]: *Status of the CERN NA62 Experiment*, J. Phys. Conf. Ser. 800 (2017) 012023.

[194] J.K. Ahn et al. [KOTO]: *Search for the* $K_L \to \pi_0\nu\bar{\nu}$ *and* $K_L \to \pi_0 X$ *decays at the J-PARC KOTO experiment*, Phys. Rev. Lett. 122 (2019) 021802.

[195] J.K. Ahn et al. [KOTO Collab.]: *Study of the* $K_L \to \pi^0\nu\bar{\nu}$ *decay at the J-PARC KOTO experiment*, Phys. Rev. Lett. 126 (2021) 121801.

[196] E. Cortina Gil et al. [NA62]: *Measurement of the very rare* $K^+\pi^+\nu\bar{\nu}$ *decay*, arXiv:[hep-ex]2103.15389.

[197] F. Ambrosino et al.: *KLEVER Project*, arXiv:[hep-ph]/1901.03099.

[198] N. Christ et al.: *Lattice QCD study of the rare kaon decay* $K^+ \to \pi^+\bar{\nu}\nu$ *at a near-physical pion mass*, Phys. Rev. D 100 (2019) 114506; Ziyuan Bai et al.: $K^+ \to \pi^+\bar{\nu}\nu$ *decay amplitude from lattice QCD*, Phys. Rev. D 98 (2018) 074509.

[199] S. W. Herb et al.: *Observation of a Dimuon Resonance at 9.5-GeV in 400-GeV Proton-Nucleus Collisions*, Phys. Rev. Lett. 39 (1977) 252.

[200] *Culture and History, Revisiting the b revolution*, CERN Courier, 19 May 2017.

[201] S. Behrends et al. [CLEO]: *Observation of Exclusive Decay Modes of B Flavored Mesons*, Phys. Rev. Lett. 50 (1983) 881.

[202] E. Fernandez et al., *Lifetime of Particles Containing B Quarks*, Phys. Rev. Lett. 51 (1983) 1022.

[203] N. Lockyer et al., *Measurement of the Lifetime of Bottom Hadrons*, Phys. Rev. Lett. 51 (1983) 1316.

[204] H. Albrecht et al. [ARGUS], *Observation of* B^0*-*$\bar{B}^0$ *Mixing*, Phys. Lett. B 192 (1987) 245.

[205] M. Staric et al. [BELLE]: *Evidence for* D^0*-*$\bar{D}^0$ *Mixing*, Phys. Rev. Lett. 98 (2007) 211803.

[206] B. Aubert et al. [BaBar]: *Evidence for* D^0*-*$\bar{D}^0$ *Mixing*, Phys. Rev. Lett. 98 (2007) 211802.

[207] T. Aaltonen et al. [CDF]: *Evidence for* D^0*-*$\bar{D}^0$ *mixing using the CDF II Detector*, Phys. Rev. Lett. 100 (2008) 121802.

[208] R. Aaij et al. [LHCb]: *Measurement of* D^0*-*$\bar{D}^0$ *Mixing Parameters and Search for CP Violation Using* $D^0 \to K^+\pi^-$ *Decays*, Phys. Rev. Lett. 111 (2013) 251801.

[209] I.I. Bigi, N.G. Uraltsev: D^0*-*$\bar{D}^0$ *oscillations as a probe of quark hadron duality*, Nucl. Phys. B 592 (2001) 92.

[210] R. Aaij et al. [LHCb]: *Updated determination of* D^0*-*$\bar{D}^0$ *mixing and CP violation parameters with* $D^0 \to K^+\pi^-$ *decays*, Phys. Rev. D 97 (2018) 031101; R. Aaij et al. [LHCb]: *Observation of the mass difference between neutral charm-meson eigenstates*, arXiv:[hep-ex]/2106.03744.

[211] D.M. Asner, A.J. Schwartz: *Review of Particle Physics 2020 - 69.* $D^0 - \bar{D}^0$ *Mixing*

[212] R. Aaij et al. [LHCb]: *Measurement of the CP asymmetry in* $B^+ \to K^+\mu^+\mu^-$ *decays*, Phys. Rev. Lett. 111 (2013) 251801.

[213] G. t' Hooft: *Symmetry Breaking Through Bell-Jackiw Anomalies*, Phys. Rev. Lett. 37 (1976) 8; *Computation of the Quantum Effects Due to a Four-Dimensional Pseudoparticle*, Phys. Rev. D 14 (1976) 3432.

[214] V. A. Kuzmin, V. A. Rubakov, M. E. Shaposhnikov: *On the Anomalous Electroweak Baryon Number Nonconservation in the Early Universe*, Phys. Lett. B 155, 36 (1985).

[215] M. Baldo-Ceolin et al.: *A New experimental limit on neutron-anti-neutron oscillations*, Z. Phys. C 63 (1994) 409.

[216] B. Misra, E.C.G. Sudarshan: *The Zeno's paradox in quantum theory*, Journ. Math. Physics 18 (1977) 756.

[217] E. Friedman, A. Gal: *Realistic calculations of nuclear disappearance lifetimes induced by n anti-n oscillations*, Phys. Rev. D 78 (2008) 016002.

[218] V.A. Kuzmin: *CP non-invariance and charge asymmetry of the universe*, Izv. Akad. Nauk Arm.SSR Fiz. 10 (1971) 2088; R.N. Mohapatra, R.E. Marshak: *Local B-L Symmetry of Electroweak Interactions, Majorana Neutrinos and Neutron Oscillations*, Phys. Rev. Lett. 44 (1980) 1316.

[219] K.S. Babu, Mohapathra: *Determining Majorana Nature of Neutrino from Nucleon Decays and* $n - \bar{n}$ *oscillations*, Phys. Rev. D 91 (2015) 1, 013008; J.M. Arnold, B. Fornal, M.B. Wise: *Simplified models with baryon number violation but no proton decay*, Phys. Rev. D 87 (2013) 075004; K.S. Babu, R.N. Mohapatra, S. Nasri: *Post-Sphaleron Baryogenesis*, Phys. Rev. Lett. 97 (2006) 131301.

[220] D.G. Phillips II et al.: *Neutron-antineutron oscillations: Theoretical status and experimental prospects*, Physics Reports 612 (2016) 1.

[221] K. Abe et al. [SK]: *Search for di-nucleon decay into pions at Super-Kamiokande*, Phys. Rev. D 91 (2015) 072006.

[222] I. I. Bigi: *Matter-antimatter oscillations and CP violation as manifested through quantum mysteries*, Rept. Prog. Phys. 70 (2007) 1869.

[223] A. Gal: *Limits on n anti-n oscillations from nuclear stability*, Phys. Rev. C 61 (2000) 028201.

[224] E. Rinaldi et al.: *Neutron-Antineutron Oscillations from Lattice QCD*, Phys. Rev. Lett. 122 (2019) 162001; *Lattice QCD determination of neutron-antineutron matrix elements with physical quark masses*, Phys. Rev. D 99 (2019) 074510.

[225] C. Theroine: *A neutron-antineutron oscillation experiment at the European Spallation Source*, Nuclear and Particle Physics Proceeding Vol. 273-275 (2016) 156-161.

[226] R. Garcia-Martin et al.: *The Pion-pion scattering amplitude: Improved analysis with once subtracted Roy-like equations up to 1100 MeV*, Phys. Rev. D 83 (2011) 074004; R. Garcia-Martin, J.R. Pelacz, F.J. Yndurain: *Experimental status of the* $\pi\pi$ *isoscalar S wave at low energy:* $f_0(600)$ *pole and scattering length*, Phys. Rev. D 76 (2007) 074034.

[227] M.R. Pennington: *Translating Quark Dynamics into Hadron Physics (and back again)* Proceed. of MESON2002, arXiv:[hep-ph]/0207220; J.R. Pelaez et al.: *The nature of the lightest scalar meson, its* N_c *behaviour and semi-local duality*, Proceed. of the Hadron 2011, arXiv:[hep-ph]/1109.2392; J.R. Pelaez: *Recent progress*

on light scalars: from confusion to precision using dispersion theory, arXiv:[hep-ph]/1301.4431.

[228] R.H. Dalitz: *CXII. On the analysis of tau-meson data and the nature of the tau-meson*, Philosophical Magazine and Journal of Science, 44 (1953) 1068.

[229] G. Valencia: *Angular Correlations in the Decay $B \to VV$ and CP Violation*, Phys. Rev. D 39 (1989) 3339.

[230] I.I. Bigi: *Heavy flavor physics: On its more than 50 years of history, its future and the Rio manifesto*, Frascati Physics Series, Vol. XX, Proc. of the "5th Heavy Quarks at Fixed Target", Editors I. Bediaga, J. Miranda, A. Reis, pgs. 549 - 586 [arXiv:[hep-ph]/0012161]; I.I. Bigi: *Charm physics: Like Botticelli in the Sistine Chapel*, Proc. of KAON2001, Pisa, 2001, pp. 417-429 [arXiv:[hep-ph]/0107102].

[231] I.I. Bigi: *CP Violation in τ decays at SuperB and Super-Belle II Experiments - like Finding Signs of Dark Matter*, TAU 2012 WS, Nuclear Physics B Proceedings Supplement (2013) 1 - 4, arXiv:[hep-ph]/1210.2968; I.I. Bigi: *Heavy flavor physics: On its more than 50 years of history, its future and the Rio manifesto*, Rio de Janeiro 2000, Heavy quarks at fixed target, arXiv:[hep-ph]/0012161.

[232] G. Durieux, Y. Grossman: *Probing CP violation systematically in differential distributions*, Phys. Rev. D 92 (2015) 076013.

[233] Z. Bai et al. [RBC and UKQCD]: *Standard Model Prediction for Direct CP Violation in $K \to \pi\pi$ Decay*, Phys. Rev. Lett. 115 (2015) 212001.

[234] H. Gisbert, A. Pich: *Direct CP violation in $K^0 \to \pi\pi$: Standard Model Status*, arXiv:[hep-ph]/1712.06147.

[235] R. Abbott et al.: [RBC and UKQCD]: *Direct CP violation and the $\Delta I = 1/2$ rule in $K \to \pi\pi$ decay from the Standard Model*, arXiv:[hep-ph]2004.09440.

[236] A.J. Buras, J.-M. Gerard: *Isospin-Breaking of ϵ'/ϵ: Impact of η_0 at the Dawn of the 2020s*, [arXiv:2005.08976 [hep-ph]]; J. Aebischer, Ch. Bobeth, A.J. Buras: *ϵ'/ϵ in the Standard Model at the Dawn of the 2020s*, arXiv:[hep-ph]/2005.05978.

[237] V. Cirigliano *et. al.*: *Theoretical status of ϵ'/ϵ*, arXiv:[hep-ph]/1912.04736.

[238] H. Burkhardt et al. [NA31]: *First Evidence for Direct CP Violation*, Phys. Lett. B 206 (1988) 169; G. D. Barr et al. [NA31]: *A New measurement of direct CP violation in the neutral kaon system*, Phys. Lett. B 317 (1993) 233.

[239] V. Fanti et al. [NA48]: *A New measurement of direct CP violation in two pion decays of the neutral kaon*, Phys. Lett. B 465 (1999) 335.

[240] V. Fanti et al. [NA48]: *Performance of an electromagnetic liquid krypton calorimeter*, Nucl. Instrum. Meth. A 344 (1994) 507; G. D. Barr et al. [NA48]: *Performance of an electromagnetic liquid krypton calorimeter based on a ribbon electrode tower structure*, Nucl. Instrum. Meth. A 370 (1996) 413.

[241] J. R. Batley et al. [NA48]: *A Precision measurement of direct CP violation in the decay of neutral kaons into two pions*, Phys. Lett. B 544 (2002) 97.

[242] E. Abouzaid et al. [KTeV]: *Precise Measurements of Direct CP Violation, CPT Symmetry, and Other Parameters in the Neutral Kaon System*, Phys. Rev. D 83 (2011) 092001.

[243] I.I. Bigi: *The Dynamics of Beauty & Charm Hadrons and top quarks in the Era of the LHCb and Belle II and ATLAS/CMS*, Motto: Non-perturbative QCD and Many-body Final States, Acta Physica Polonica B 49 (2018) 1021.

[244] R. Aaij et al. [LHCb]: *Measurement of CP violation in the phase space of $B^\pm \to K^\pm\pi^-\pi^+$ and $B^\pm \to K^+K^-K^\pm$ decays*, Phys. Rev. Lett. 111 (2013) 101801; R. Aaij et al. [LHCb]: *Measurement of CP violation in the phase space of $B^\pm \to K^+K^-\pi^\pm$ and $B^\pm \to K^+K^-\pi^\pm$ decays*, Phys. Rev. Lett. 112 (2014) 011801.

[245] I. Bediaga, I.I. Bigi, J. Miranda, A. Reis: *CP asymmetries in three-body final states in charged D decays and CPT invariance*, Phys. Rev. D 89 (2014) 074024.

[246] M. Williams: *Observing CP Violation in Many-Body Decays*, Phys. Rev. D 84 (2011) 054015.

[247] I.I. Bigi, H. Yamamoto: *Interference between Cabibbo allowed and doubly forbidden transitions in $D \to K_S K_L + \pi$'s decays*, Phys. Lett. B 349 (1995) 363.

[248] Y.Q. Chen et al. (Belle Collaboration) *Dalitz analysis of $D^0 \to K^-\pi^+\eta$ decays at Belle*, Phys. Rev. D 102 (2020) 1, 012002.

[249] S. Bianco, I.I. Bigi: *2019/20 Lessons from $\tau(\Omega_c^0)$ and $\tau(\Xi_c^0)$ and CP asymmetry in charm decays*, Int. J. Mod. Phys. A 35 (2020) 24, 2030013.

[250] I.I. Bigi: *Bridge between Hadrodynamics and HEP: Regional CP Violation in Beauty & Charm Decays*, PoS (ICHEP2016) 531, arXiv:[hep-ph]/1608.06528.

[251] R. Aaij et al. [LHCb]: *Measurement of matter-antimatter differences in beauty baryon decays*, Nature Physics 13 (2017) 391.

[252] *LHCb sees first hints of CP violation in baryons*, CERN Courier, Vol. 57, Num. 2, March 2017, 10.

[253] G. Durieux: *CP violation in multibody decays of beauty baryons*, JHEP 1610 (2016) 005.

[254] R. Aaij et al. [LHCb]: *Search for CP violation using triple product asymmetries in $\Lambda_b^0 \to pK^-\pi^+\pi^-$, $\Lambda_b^0 \to pK^-K^+K^-$ and $\Xi_b^0 \to pK^-K^-\pi^+$ decays*, JHEP 08 (2018) 039.

[255] R. Aaji et al. [LHCb]: *Search for CP violation and observation of P violation in $\Lambda_b^0 \to p\pi^-\pi^+\pi^-$ decays*, Phys. Rev. D 102 (2020) 051101.

[256] A. Meri: *Searches for CP violation in multi-body baryon decays at LHCb*, CERN seminar 22nd Oct. 2019.

[257] I.I. Bigi, X.-W. Kang, H.-B. Li: *CP Asymmetries in Strange Baryon Decays*, Chin. Phys. C 42 (2018) 013101.

[258] M. Jacob, G. C. Wick: *On the general theory of collisions for particles with spin*, Annals Phys. 7 (1959) 404; [Annals Phys. 281 (2000) 774].

[259] H. Chen, R. G. Ping: *Helicity amplitude analysis of hyperon nonleptonic decays in J/ψ or $\psi(2S)$ decays*, Phys. Rev. D 76 (2007) 036005.

[260] I.I. Bigi: *Probing CP Asymmetries in Charm Baryons Decays*, arXiv:[hep-ph]/1206.4554.

[261] CDF and D0: *Forward-backward asymmetries in top-antitop quark production at the Tevatron*, FERMILAB-CONF-16-386-PPD, CDF Note 11206 and D0 Note 6492.

[262] R. Hawkings: *Inclusive $t\bar{t}$ cross-section measurements at LHC*, arXiv:[hep-ph]/1711.07400; R. Lysak: *Top quark properties at Tevatron*, arXiv:[hep-ph]/1711.09686.

[263] 2017 PDG: "72. The Top Quark".

[264] P. Nason: *Renormalons and the Top Quark Mass Measurement*, arXiv:[hep-ph]/1901.04737.

[265] A. Abada et al.: *The Lepton Collider*, Eur. Phys. J. Special Topics 228 (2019) 261.

[266] M. Aicheler et al. [CERN]: *The Compact Linear Collider (CLIC) - Project Implementation Plan* (2019), arXiv:[hep-ph]/1903.08655.

[267] H. Aihara et al.: *The International Linear Collider. A Global Project* (2019), arXiv:[hep-ph]/1901.09829.

[268] J. P. Delahaye et al.: *Muon Colliders* (2019), arXiv:[hep-ph]/1901.06150.

[269] C.R. Schmidt, M. Peskin: *A Probe of CP violation in top quark pair production at hadron supercolliders*, Phys. Rev. Lett. 69 (1992) 410.

[270] A. Giammanco: *TOP2017: Experimental Summary*, arXiv:[hep-ph]/1712.02177.

[271] M. Owen: *Top2018: Experimental Summary*, arXiv:[hep-ph]/1901.11516.

[272] R. Aaij et al. [LHCb]: *Measurement of forward top pair production in the dilepton channel in pp collisions at* $\sqrt{s} = 13$ *TeV*, JHEP 08 (2018) 174.

[273] A. Jung: *Experimental summary of the 12th International Workshop on Top Quark Physics*, Beijing, 22-27 September 2019.

[274] L. Reina: *Theory Summary of the 12th International Workshop on Top-Quark Physics*, Beijing, 22-27 September 2019.

[275] M. Aaboud et al. [ATLAS]: *Measurement of colour flow using jet-pull observables in* $t\bar{t}$ *events with the ATLAS experiment at* $\sqrt{s} = 13$ *TeV*, Eur. Phys. J. 78 (2018) 847.

[276] M.R. Buckley, D. Goncalves: *Boosting the Direct CP Measurement of the Higgs-Top Coupling*, Phys. Rev. Lett.116 (2016) 091801.

[277] S. Bar-Shalom, D. Atwood, G. Eilam, A. Soni: *CP nonconservation in* $e^+e^- \to t\bar{t}g$, Z. Phys. C 72 (1996) 79.

[278] The Review of Particle Physics (2020): "Summary Tables"; 2020 Review of Particle Physics, "11. Status of Higgs Boson Physics", revised 2019 by M. Carena et al.

[279] [ATLAS]: *Evidence for the associated of the Higgs boson and a top quark pair the with the ATLAS detector*, [arXiv:[hep-ph]/1712.08891.

[280] Priyanka et al. [CMS]: *First measurement of tW production cross-section at* $\sqrt{s} = 13$ *TeV with CMS*, arXiv:[hep-ph]/1901.04179.

[281] [ATLAS and CMS]: *Combination of inclusive and differential* $t\bar{t}$ *charge asymmetry measurements using ATLAS and CMS data at* $\sqrt{s} = 7$ *and 8 TeV*, arXiv:[hep-ph]/1709.05327.

[282] M. Aaboud et al. [ATLAS]: *Measurements of* $t\bar{t}$ *differential cross-sections of highly boosted top quarks decaying to all-hadronic final states in pp collisions at* $\sqrt{s} = 13$ *TeV using the ATLAS detector*, JHEP 04 (2018) 033.

[283] A. M. Sirunyan et al. [CMS]: *Measurements of the top quark mass using single top quark events in proton-proton collisions at* $\sqrt{s} = 8$ *TeV*, Eur. Phys. J. C 77 (2017) 354.

[284] M. Aaboud et al. [ATLAS]: *Measurements of differential cross-sections of a single top quark produced in association with a W boson at* $\sqrt{s} = 13$ *TeV with ATLAS*, Eur. Phys. J. C 78 (2018) 186.

[285] V. Khachatryan et al. [CMS] *Search for CP violation in top-pair events with lepton+jets channel at* $\sqrt{s} = 8$ *TeV*, Report number: CMS-PAS-TOP-16-001.

[286] M. Aaboud et al. [ATLAS]:*CP violation in b-hadron decays using top-pair events in 8 TeV ATLAS data*, PoS EPS-HEP2017 (2018) 754.

[287] R.G.H. Robertson and D.A. Knapp: *Direct measurements of neutrino mass*, Ann. Rev. Nucl. Part. Sc. 38 (1988) 185.

[288] M. Fertl: *Review of absolute neutrino mass measurements*, Hyperfine Interact 239 (2018) 52.

[289] M. Agostini et al. [Borexino]: *Experimental evidence of neutrinos produced by CNO fusion cycle in the Sun*, Nature 587 (2020) 577-582.

[290] P. Anselmann et al.: *Final results of the Cr-51 neutrino source experiments in GALLEX*, Phys. Lett. B 420 (1998) 114.

[291] J. Abdurashitov et al.: *Results from SAGE*, Phys. Lett. B 328 (1994) 234.

[292] L. Wolfenstein: *Neutrino Oscillations in Matter*, Phys. Rev. D 17 (1978) 2369; S.P.Mikheyev, A.Yu. Smirnov: *Resonance Amplification of Oscillations in Matter and Spectroscopy of Solar Neutrinos*, Sov.J.Nucl.Phys. 42 (1985) 913, Yad.Fiz. 42 (1985) 1441.

[293] A. Smirnov: *The Mikheyev-Smirnov-Wolfenstein (MSW) Effect*, arXiv:[hep-ph]/ 1901.11473.

[294] R. Davis: *A review of the Homestake solar neutrino experiment*, Prog. Part. Nucl. Phys. 32 (1994) 13.

[295] W. Hampel et al. [GALLEX]: *GALLEX solar neutrino observations: Results for GALLEX IV*, Phys. Lett. B 447 (1999) 127.

[296] K.S. Hirata et al. [Kamiokande-II]: *Observation of B-8 Solar Neutrinos in the Kamiokande-II Detector*, Phys. Rev. Lett. 63 (1989) 16.

[297] J. Bahcall, *Solar neutrinos. I: Theoretical*, Phys. Rev. Lett. 12 (1964) 300; for a recent calculation: N. Vinyoles et al.: *A new Generation of Standard Solar Models*, Astrophys.J. 835 (2017) 2, 202.

[298] G. Aardsma et al. [SNO]: *A Heavy Water Detector to Resolve the Solar Neutrino Problem*, Phys. Lett. B 194 (1987) 321; A. Bellerive et al. [SNO]: *The Sudbury Neutrino Observatory*, Nucl. Phys. B 908 (2016) 30.

[299] Y. Oyama et al. [KAMIOKANDE]: *Experimental Study of Upward Going Muons in Kamiokande*, Phys. Rev. D 39 (1989) 1481.

[300] C.B. Bratton et al.:[IMB] *IMB: Nucleon Decay Search Status Report*, Proceedings, Grand Unification*, 84-97 (1981); J. Lo Secco, *History of "Anomalous" Atmospheric Neutrino Events: A First Person Account*, arXiv:[hep-ph]/1606.00665.

[301] K. Abe et al. [SK]: *Atmospheric neutrino oscillation analysis with external constraints in Super-Kamiokande I-IV*, Phys. Rev. D 97 (2018) 7, 072001.

[302] Z. Li et al. [SK]: *A Measurement of the Tau Neutrino Cross Section in Atmospheric Neutrino Oscillations with Super-Kamiokande*, Phys. Rev. D 98 (2018) 052006.

[303] M.G. Aartsen et al.[IceCube] *Evidence for High-Energy Extraterrestrial Neutrinos at the IceCube Detector*, Science 342 (2013) 1242856.

[304] M.G. Aartsen et al. [IceCube]: *Measurement of Atmospheric Neutrino Oscillations at 6 - 56 GeV with IceCube DeepCore*, Phys. Rev. Lett. 120 (2018) 071801.

[305] K.Abe et al. [T2K]: *Search for CP Violation in Neutrino and Antineutrino Oscillations by the T2K Experiment with* $2.2 \cdot 10^{21}$ *Protons on Target*, Phys. Rev. Lett. 121 (2018) 171802.

[306] K.Abe et al. [T2K]: *T2K measurements of muon neutrino and antineutrino disappearance using* 3.13×10^{21} *protons on target*, arXiv:[hep-ex]/2008.07921.

[307] N. Agafonova et al. [OPERA]: *Final results of the search for* $\nu_\mu \to \nu_e$ *oscillations with the OPERA detector in the CNGS beam*, JHEP 1806 (2018) 151.

[308] N. Agafonova et al. [OPERA]: *Final results of the OPERA experiment on ν_τ appearance in the CNGS beam*, Phys. Rev. Lett. 120 (2018) 211801.

[309] P. Adamson et al. [MINOS]: *An improved measurement of muon antineutrino disappearance in MINOS*, Phys. Rev. Lett. 108 (2012) 191801.

[310] P. Adamson et al. [MINOS]: *Precision constraints for three-flavor neutrino oscillations from the full MINOS+ and MINOS data set*, Phys. Rev. Lett. 125 (2020) 131802.

[311] M.A. Acero et al. [NOvA]: *New constraints on oscillation parameters from ν_e appearance and ν_μ disappearance in the NOvA experiment*, Phys. Rev. D 98 (2018) 032012.

[312] M.A. Acero et al. [NOvA]: *First measurement of neutrino oscillation parameters using neutrinos and antineutrinos by NOvA*, Phys. Rev. Lett. 123 (2019) 151803.

[313] M. Agostini et al. [BOREXino]: *Comprehensive measurement of pp-chain solar neutrinos*, Nature 562 (2018) 505.

[314] M. Agostini et al. [BOREXino]: *Simultaneous precision spectroscopy of pp, ^7Be, and pep solar neutrinos with Borexino Phase-II*, Phys. Rev. D 100 (2019) 082004.

[315] D. Adey et al. [Daya Bay]: *Measurement of the Electron Antineutrino Oscillation with 1958 Days of Operation at Daya Bay*, Phys. Rev. Lett. 121 (2018) 241805.

[316] G. Bak et al. [RENO]: *Measurement of reactor antineutrino oscillation amplitude and frequency at RENO*, Phys. Rev. Lett. 121 (2018) 201801.

[317] H. de Kerret et al. [Double Chooz]: *Double Chooz θ_{13} measurement via total neutron capture detection*, Nature Phys. 16 (2020) 558-564.

[318] K. Abe et al. [T2K]: *Observation of Electron Neutrino Appearance in a Muon Neutrino Beam*, Phys. Rev. Lett. 112 (2014) 061802.

[319] K.Abe et al. [T2K]: *Constraint on the matter–antimatter symmetry-violating phase in neutrino oscillations*, Nature 580 (2020) 339.

[320] K. Nakamura, S.T. Petcov: *Review of Neutrino masses, mixing and oscillations*, PDG2018.

[321] F. Capozzi et al.: *Global constraints on absolute neutrino masses and their ordering*, Phys. Rev. D 95 (2017) 096014.

[322] M. Fukugita, T. Yanagida: *Baryogenesis Without Grand Unification*, Phys. Lett. B 174 (1986) 45.

[323] M. Goeppert-Meyer: *Double beta-disintegration*, Phys. Rev. 48 (1935) 512.

[324] S.R. Elliott, A. Hahn, M.K. Moe: *Direct Evidence for Two Neutrino Double Beta Decay in ^{82}Se*, Phys. Rev. Lett. 59 (1987) 2020.

[325] *KATRIN: Karlsruhe Tritium Neutrino Experiment*, see https://www.katrin.kit.edu.

[326] I. C. Barnett et al.: *An apparatus for the measurement of the transverse polarization of positrons from the decay of polarized muons*, Nucl. Instrum. Meth. A 455 (2000) 329.

[327] I. I. Bigi and A. I. Sanda: *A 'Known' CP asymmetry in tau decays*, Phys. Lett. B 625 (2005) 47.

[328] Y. Grossman, Y. Nir: *CP Violation in $\tau \to \nu\pi K_S$ and $D \to \pi K_S$: The Importance of $K_S - K_L$ Interference*, JHEP 1204 (2012) 002.

[329] J. P. Lees et al. [BaBar]: *Search for CP Violation in the Decay $\tau^- -> \pi^- K^0_S (>= 0\, \pi^0 s)\nu_\tau$*, Phys. Rev. D 85 (2012) 031102; [Erratum: Phys. Rev. D 85 (2012) 099904].

[330] E. P. Shabalin: *U(1) problem, θ term and CP violation*, Sov. J. Nucl. Phys. 36 (1982) 575.

[331] J. R. Ellis, M. K. Gaillard: *Strong and Weak CP Violation*, Nucl. Phys. B 150 (1979) 141.

[332] I. B. Khriplovich, A. I. Vainshtein: *Infinite renormalization of Theta term and Jarlskog invariant for CP violation*, Nucl. Phys. B 414 (1994) 27.

[333] R. Peccei, H. Quinn: *CP Conservation in the Presence of Instantons*, Phys. Rev. Lett. 38 (1977) 1440; *Constraints Imposed by CP Conservation in the Presence of Instantons*, Phys. Rev. D 16 (1977) 1791.

[334] S. Weinberg: *A New Light Boson*, Phys. Rev. Lett. 40 (1978) 223; F. Wilczek: *Problem of Strong P and T Invariance in the Presence of Instantons*, Phys. Rev. Lett. 40 (1978) 279.

[335] C.G. Callan, R. Dashen, D. Gross: *The Structure of the Gauge Theory Vacuum*, Phys. Lett. B 63 (1976) 334.

[336] R. Jackiw, C. Rebbi: *Vacuum Periodicity in a Yang-Mills Quantum Theory*, Phys. Rev. Lett. 37 (1976) 172.

[337] R.D. Peccei: *The Strong CP problem and axions*, Lect. Notes Phys. 741 (2008) 3 [arXiv:[hep-ph]/0607268].

[338] D. G. Sutherland: *Current algebra and some non-strong mesonic decays*, Nucl. Phys. B 2 (1967) 433.

[339] R.D. Peccei: *Why PQ?*, invited talk at the Axion 2010 Conference arXiv:[hep-ph]/1005.0643.

[340] S. Weinberg: *The U(1) Problem*, Phys. Rev. D 11 (1975) 3583.

[341] S.L. Adler: *Axial vector vertex in spinor electrodynamics*, Phys. Rev. 177 (1969) 2426; J.S. Bell, R. Jackiw: *A PCAC puzzle: $\pi^0 \to \gamma\gamma$ in the σ model*, Nuovo Cimento 60 (1969) 47.

[342] W.A. Bardeen: *Anomalous Ward identities in spinor field theories*, Phys. Rev. 184 (1969) 1848.

[343] R. Jackiw: *Current algebra and some non-strong mesonic decays*, Diverse topics in theoretical and mathematical physics, World Scientific 1995, p. 95.

[344] V. Baluni: *CP Violating Effects in QCD*, Phys. Rev. D 19 (1979) 2227.

[345] R.J. Crewther et al.: *Chiral Estimate of the Electric Dipole Moment of the Neutron in Quantum Chromodynamics*, Phys. Lett. B 88 (1979) 123, Erratum B 91 (1980) 487.

[346] R.D. Peccei: in *CP Violation*, C. Jarlskog (ed.), World Scientitic, 1988, pp. 503-551.

[347] J. Gasser, H. Leutwyler: *Quark Masses*, Phys. Report 87 (1982) 77.

[348] S. Khlebnikov, M. Shaposhnikov: *Brane-worlds and theta-vacua*, Phys. Rev. D 71 (2005) 104024; *Extra Space-time Dimensions: Towards a Solution to the Strong CP Problem*, Phys. Lett. B 203 (1988) 121.

[349] R.N. Mohapatra, G. Senjanovic: *Exact Left-Right Symmetry and Spontaneous Violation of Parity*, Phys. Rev. D 12 (1975) 1502; R.N. Mohapatra in: *CP Violation*, Singapore: World Scientific, 1988; G. Ecker, W. Grimus, H. Neufeld: *Spontaneous CP Violation in Left-right Symmetric Gauge Theories*, Nucl. Phys. B 247 (1984) 70; P. Langacker, S.U. Sankar: *Bounds on the Mass of W(R) and the W(L)-W(R) Mixing Angle ξ in General $SU(2)_L \times SU(2)_R \times U(1)$ Models*, Phys. Rev. D 40 (1989) 1569.

[350] T.D. Lee: *A Theory of Spontaneous T Violation*, Phys. Rev. D 8 (1973) 1226; *Phys. Rep.* 96 (1979); S. Weinberg: *Gauge Theory of CP Violation*, Phys. Rev. Lett. 37 (1976) 657; L. Wolfenstein, Y.L. Wu: *Sources of CP violation in the two Higgs doublet model*, Phys. Rev. Lett. 73 (1994) 1762; S. Kraml et al.: *Report of the CPNSH Workshop, May 2004 - Dec. 2005* [arXiv:[hep-ph]/0608079]; D. Chang, W.-Y. Keung, T.C. Yuan: *Chromoelectric dipole moment of light quarks through two loop mechanism* Phys. Lett. 251 (1990) 608; I.I. Bigi, N.G. Uraltsev: *Induced Multi - Gluon Couplings and the Neutron Electric Dipole Moment*, Nucl. Phys. B 353 (1991) 321.

[351] Y.B. Zeldovich, I.B. Kobzarev, L.B. Okun: *Realistic Quark Models and Astrophysics*, Sov. Phys. JETP 40 (1975) 1. [On the list of very important articles of L. Okun it is item 23.]

[352] S. Barr, A. Zee: *Solution of the Strong CP Problem by Color Exchange*, Phys. Rev. Lett. 55 (1985) 2253.

[353] A. Nelson: *Naturally Weak CP Violation*, Phys. Lett. B 136 (1983) 387.

[354] S. Barr: *Solving the Strong CP Problem Without the Peccei-Quinn Symmetry*, Phys. Rev. Lett. 53 (1984) 329; *A Natural Class of non Peccei-Quinn Models*, Phys. Rev. D 30 (1984) 1805.

[355] Th. Kaluza: *Zum Unitätsproblem der Physik*, Sitz. Preuss. Akad. Wiss. Kl (1921) 966; O. Klein: *Quantum Theory and Five-Dimensional Theory of Relativity*, Z. Phys. 37 (1926) 895.

[356] P. Sikivie: *Experimental Tests of the 'Invisible' Axion*, Phys. Rev. Lett. 51 (1983) 1415.

[357] N.A. Bardeen, R.D. Peccei, T. Yanagita: *Constraints on variant axion models*, Nucl. Phys. B 279 (1987) 401.

[358] N.A. Bardeen, S.H. Tye: *Current Algebra Applied to Properties of the Light Higgs Boson*, Phys. Lett. B 74 (1978) 229.

[359] S. Yamada: *Proc. 1983 Int. Symp. on Lepton and Photon Interactions at High Energies*, Cornell, 1983.

[360] S. Adler et al. [E687]: *Evidence for the decay* $K^+ \to \pi^+ \nu\bar{\nu}$, Phys. Rev. Lett. 79 (1997) 2204; E949, A.V. Artamonov et al.: *Upper Limit on the Branching Ratio for the Decay* $\pi^0 \to \nu\bar{\nu}$, Phys. Rev. D 72 (2005) 091102.

[361] J.K. Ahn et al. [KOTO]: *A new search for the* $K_L \to \pi^0 \nu\bar{\nu}$ *and* $K_L \to \pi^0 X$ *decays*, PTEP 2017 (2017) 2, 021C01.

[362] A. Ringwald, L.J. Rosenberg, G. Rybka: *112. Axions and Other Similar Particles*, PDG2019.

[363] J. Kim: *Weak Interaction Singlet and Strong CP Invariance*, Phys. Rev. Lett. 43 (1979) 103; M.A. Shifman, A.I. Vainshtein, V.I. Zakharov: *Can Confinement Ensure Natural CP Invariance of Strong Interactions?*, Nucl. Phys. 166 (1980) 493.

[364] M. Dine, W. Fischler, M. Srednicki: *A Simple Solution to the Strong CP Problem with a Harmless Axion*, Phys. Lett. 104 (1981) 199; A.P. Zhitnitskii: *On Possible Suppression of the Axion Hadron Interactions*, Sov. J. Nucl. Phys. 31 (1980) 260.

[365] G. Ruoso et al.: *Limits on light scalar and pseudoscalar particles from a photon regeneration experiment*, Z. Phys. C56 505 (1992); R. Cameron et al.: *Search for nearly massless, weakly coupled particles by optical techniques*, Phys. Rev. D 47 (1993) 3707.

[366] P. Pugnat et al. [OSQAR]: *Search for weakly interacting sub-eV particles with the OSQAR laser-based experiment: results and perspectives*, Eur. Phys. J. C 74 (2014) 3027 [arXiv:[hep-ph]/1306.0443]; R. Ballou et al. [OSQAR]: Phys. Rev. D 92 (2015) 092002 [arXiv:[hep-ph]/1506.08082].

[367] E. Zavattini et al. [PVLAS]: *Experimental observation of optical rotation generated in vacuum by a magnetic field*, Phys. Rev. Lett. 96 (2006) 110406 [Erratum: Phys. Rev. Lett. 99 (2007) 129901].

[368] F. Della Valle et al. [PVLAS]: *The PVLAS experiment: measuring vacuum magnetic birefringence and dichroism with a birefringent Fabry–Perot cavity*, Eur. Phys. J. C 76 (2016) 24.

[369] H. Primakoff: *Photoproduction of neutral mesons in nuclear electric fields and the mean life of the neutral meson*, Phys. Rev. 81 (1951) 899.

[370] V. Anastassopoulos et al. [CAST]: *New CAST Limit on the Axion-Photon Interaction*, Nature Phys. 13 (2017) 584 [arXiv:[hep-ph]/1705.02290]; E. Arik et al. [CAST]: JCAP 0902 (2009) 008 [arXiv:0810.4482]; S. Aune et al. [CAST]: Phys. Rev. Lett. 107 (2011) 261302 [arXiv:[hep-ph]/1106.3919]; M. Arik et al. [CAST]: Phys. Rev. Lett. 112 (2014) 091302 [arXiv:[hep-ph]/1307.1985]; M. Arik et al. [CAST]: Phys. Rev. D 92 (2015) 021101 [arXiv:[hep-ph]/1503.00610].

[371] E. Aprile et al. [XENON]: *Excess Electronic Recoil Events in XENON1T*, Phys. Rev. D 102 (2020) 072004.

[372] G.G. Raffelt: *Lect. Notes Phys.741* (2008) 51-71 [arXiv:[hep-ph]/0611350]; *Axions: Motivation, limits and searches*, J. Phys. A 40 (2007) 6607.

[373] P. Sikivie: *Dark matter axions and caustic rings*, arXiv:[hep-ph]/9709477; *Caustic rings of dark matter*, Phys. Lett. B 432 (1998) 139.

[374] F. Wilczek: *Two Applications of Axion Electrodynamics*, Phys. Rev. Lett. 58 (1987) 1799.

[375] J. S. M. Ginges, V. V. Flambaum: *Violations of fundamental symmetries in atoms and tests of unification theories of elementary particles*, Phys. Rept. 397 (2004) 63; M. Pospelov, A. Ritz: *Electric dipole moments as probes of new physics*, Annals Phys. 318 (2005) 119; M. Raidal et al.: *Flavour physics of leptons and dipole moments*, Eur. Phys. J. C 57 (2008) 13; M. Pospelov, A. Ritz: *Probing CP violation with electric dipole moments*, in the *"Lepton dipole moments"*, B.L. Roberts and W.J. Marciano (eds.), Advanced series on directions in high energy physics, Vol. 20, World Scientific (2010) p. 439; T. Fukuyama: *Searching for New Physics beyond the Standard Model in Electric Dipole Moment*, Int. J. Mod. Phys. A 27 (2012) 1230015.

[376] J. F. Donoghue: *T Violation in $SU(2) \times U(1)$ Gauge Theories of Leptons*, Phys. Rev. D 18 (1978) 1632.

[377] P.A.M. Dirac: *The quantum theory of the electron*, Proc. R. Soc. Lond. A 117 (1928) 610.

[378] J. Schwinger: *On Quantum electrodynamics and the magnetic moment of the electron*, Phys. Rev. 73 (1948) 416.

[379] D. Hanneke, S. F. Hoogerheide, G. Gabrielse: *Cavity Control of a Single-Electron Quantum Cyclotron: Measuring the Electron Magnetic Moment*, Phys. Rev. A 83 (2011) 052122.

[380] R. H. Parker, C. Yu, W. Zhong, B. Estey, H. Müller: *Measurement of the fine-structure constant as a test of the Standard Model*, Science 360 (2018) 191.

[381] H. Davoudiasl, W. J. Marciano: *Tale of two anomalies*, Phys. Rev. D 98 (2018) 075011.

[382] G.W. Bennett et al.: *Final Report of the Muon E821 Anomalous Magnetic Moment Measurement at BNL*, Phys. Rev. D 73 (2006) 072003.

[383] T. Kinoshita and M. Nio: *Improved α^4 term of the electron anomalous magnetic moment*, Phys. Rev. D 73 (2006) 013003.

[384] T. Kinoshita, M. Nio: *Improved α^4 term of the muon anomalous magnetic moment*, Phys. Rev. D 70 (2004) 113001.

[385] S. Laporta: *High-precision calculation of the 4-loop contribution to the electron g-2 in QED*, Phys. Lett. B 772 (2017) 232.

[386] T. Aoyama, M. Hayakawa, T. Kinoshita, M. Nio: *Complete Tenth-Order QED Contribution to the Muon g-2*, Phys. Rev. Lett. 109 (2012) 111808.

[387] A. Kurz, T. Liu, P. Marquard, A. Smirnov, V. Smirnov, M. Steinhauser: *Electron contribution to the muon anomalous magnetic moment at four loops*, Phys. Rev. D 93 (2016) 053017.

[388] T. Aoyama, M. Hayakawa, T. Kinoshita, M. Nio: *Revised value of the eighth-order electron g-2*, Phys. Rev. Lett. 99 (2007) 110406.

[389] T. Ishikawa, N. Nakazawa, Y. Yasui: *Numerical calculation of the full two-loop electroweak corrections to muon (g-2)*, Phys. Rev. D 99 (2019) 073004.

[390] M. Davier et al.: *A new evaluation of the hadronic vacuum polarisation contributions to the muon anomalous magnetic moment and to* $\boldsymbol{\alpha}(\mathbf{m_Z^2})$, Eur. Phys. J. C 80 (2020) 241; [erratum: Eur. Phys. J. C 80 (2020) 410].

[391] J. Prades et al.: *The Hadronic Light-by-Light Scattering Contribution to the Muon and Electron Anomalous Magnetic Moments*, Adv. Ser. Direct. High Energy Phys. 20 (2009) 303.

[392] B. Krause: *Higher order hadronic contributions to the anomalous magnetic moment of leptons*, Phys. Lett. B 390 (1997) 392.

[393] E. de Rafael: *Hadronic contributions to the muon g-2 and low-energy QCD*, Phys. Lett. B 322 (1994) 239.

[394] W. Gohn: *The MUON g-2 EXPERIMENT at FERMILAB*, arXiv:[hep-ph]/1611.04964.

[395] Fermilab, Muon g-2, The Big Move, Photo Gallery.

[396] J. Grange et al. [Muon g-2]: *Muon (g-2) Technical Design Report*, arXiv:[hep-ph]/1501.06858.

[397] B. Abi et al. [Muon g-2 Coll.]: *Measurement of the Positive Muon Anomalous Magnetic Moment to 0.46 ppm*, arXiv:hep-hex/2104.03281.

[398] H. Iinuma et al.: *Three-dimensional spiral injection scheme for the g-2/EDM experiment at J-PARC*, Nuclear Instruments and Methods in Physics Research Section A: Accelerators, Spectrometers, Detectors and Associated Equipment 832 (2016) 51.

[399] S. Eidelman, M. Passera: *Theory of the tau lepton anomalous magnetic moment*, Mod. Phys. Lett. A 22 (2007) 159.

[400] M. Passera: *Precise mass-dependent QED contributions to leptonic g-2 at order α^2 and α^3*, Phys. Rev. D 75 (2007) 013002.

[401] A. Czarnecki, B. Krause W. J. Marciano: *Electroweak Fermion loop contributions to the muon anomalous magnetic moment*, Phys. Rev. D 52 (1995) 2619; A. Czarnecki,

B. Krause, W. J. Marciano: *Electroweak corrections to the muon anomalous magnetic moment*, Phys. Rev. Lett. 76 (1996) 3267; T. V. Kukhto, E. A. Kuraev, Z. K. Silagadze, A. Schiller: *The Dominant two loop electroweak contributions to the anomalous magnetic moment of the muon*, Nucl. Phys. B 371 (1992) 567.

[402] J. Abdallah et al. [DELPHI]: *Study of tau-pair production in photon-photon collisions at LEP and limits on the anomalous electromagnetic moments of the tau lepton*, Eur. Phys. J. C 35 (2004) 159.

[403] G. A. Gonzalez-Sprinberg, A. Santamaria, J. Vidal: *Model independent bounds on the tau lepton electromagnetic and weak magnetic moments*, Nucl. Phys. B 582 (2000) 3.

[404] A. A. Billur, M. Koksal: *Probe of the electromagnetic moments of the tau lepton in gamma-gamma collisions at the CLIC*, Phys. Rev. D 89 (2014) 3, 037301.

[405] M. Fael, L. Mercolli, M. Passera: *Towards a determination of the tau lepton dipole moments*, Nucl. Phys. B Proc. Suppl. 253-255 (2014) 103.

[406] M. L. Laursen, M. A. Samuel, A. Sen: *Radiation Zeros and a Test for the g Value of the τ Lepton*, Phys. Rev. D 29 (1984) 2652; [Erratum: ibid. D 56 (1997) 3155].

[407] M. A. Samuel, G. W. Li, R. Mendel: *The Anomalous magnetic moment of the tau lepton*, Phys. Rev. Lett. 67 (1991) 668; [Erratum: ibid. 69 (1992) 995].

[408] I. J. Kim: *Magnetic Moment Measurement of Baryons With Heavy Flavored Quarks by Planar Channeling Through Bent Crystal*, Nucl. Phys. B 229 (1983) 251.

[409] D. Chen et al. [E761]: *First observation of magnetic moment precession of channeled particles in bent crystals*, Phys. Rev. Lett. 69 (1992) 3286.

[410] J. Bernabeu, G. A. Gonzalez-Sprinberg, J. Vidal: *Tau spin correlations and the anomalous magnetic moment*, JHEP 0901 (2009) 062.

[411] J. Bernabeu, G. A. Gonzalez-Sprinberg, J. Papavassiliou, J. Vidal: *Tau anomalous magnetic moment form-factor at super B-flavor factories*, Nucl. Phys. B 790 (2008) 160.

[412] M. Pospelov, I. Khriplovich: *Electric dipole moment of the W boson and the electron in the Kobayashi-Maskawa model*, Sov. J. Nucl. Phys. 53 (1991) 638.

[413] D. Ghosh and R. Sato: *Lepton Electric Dipole Moment and Strong CP Violation*, Phys. Lett. B 777 (2018) 335.

[414] L. B. Okun: *On the Electric Dipole Moment of Neutrino*, Sov. J. Nucl. Phys. 44 (1986) 546.

[415] L. B. Okun, M. B. Voloshin, M. I. Vysotsky: *Electromagnetic Properties of Neutrino and Possible Semiannual Variation Cycle of the Solar Neutrino Flux*, Sov. J. Nucl. Phys. 44 (1986), 440.

[416] K. Kirch, P. Schmidt-Wellenburg: *Search for electric dipole moments*, EPJ Web Conf. 234 (2020) 01007.

[417] K. Inami et al. [Belle]: *Search for the electric dipole moment of the tau lepton*, Phys. Lett. B 551 (2003) 16.

[418] S.A. Murthy et al.: *New Limits on the Electron Electric Dipole Moment from Cesium*, Phys. Rev. Lett. 63 (1989) 965.

[419] V. Andreev et al. [ACME]: *Improved limit on the electric dipole moment of the electron*, Nature 562 (2018) 355; J. Baron et al. [ACME]: *Order of Magnitude Smaller Limit on the Electric Dipole Moment of the Electron*, Science 343 (2014) 269.

[420] G. W. Bennett et al. [Muon g-2]: *An Improved Limit on the Muon Electric Dipole Moment*, Phys. Rev. D 80 (2009) 052008.

[421] J. Engel, M. J. Ramsey-Musolf, U. van Kolck: *Electric Dipole Moments of Nucleons, Nuclei, and Atoms: The Standard Model and Beyond*, Prog. Part. Nucl. Phys. 71 (2013) 21; L. Mercolli, C. Smith: *EDM constraints on flavored CP-violating phases*, Nucl. Phys. B 817 (2009) 1; M. Jung, A. Pich: *Electric Dipole Moments in Two-Higgs-Doublet Models*, JHEP 1404 (2014) 076.

[422] J. Bernabeu, G. A. Gonzalez-Sprinberg, J. Vidal: *CP violation and electric-dipole-moment at low energy tau-pair production*, Nucl. Phys. B 701 (2004) 87; J. Bernabeu, G. A. Gonzalez-Sprinberg, J. Vidal: *CP violation and electric-dipole-moment at low energy tau production with polarized electrons*, Nucl. Phys. B 763 (2007) 283.

[423] G. Gonzalez-Sprinberg, J. Bernabeu, J. Vidal: *τ electric dipole moment with polarized beams*, arXiv:0707.1658.

[424] A. Grozin, I. Khriplovich, A. Rudenko: *Electric dipole moments, from e to tau*, Phys. Atom. Nucl. 72 (2009) 1203.

[425] A. Pich: *Precision Tau Physics*, Prog.Part.Nucl.Phys. 75 (2014) 41.

[426] A. Heister et al. [ALEPH]: *Search for anomalous weak dipole moments of the tau lepton*, Eur. Phys. J. C 30 (2003) 291.

[427] M. Acciarri et al. [L3]: *Measurement of the weak dipole moments of the tau lepton*, Phys. Lett. B 426 (1998) 207.

[428] K. Ackerstaff et al. [OPAL]: *Search for CP violation in Z0 — tau+ tau- and an upper limit on the weak dipole moment of the tau lepton*, Z. Phys. C 74 (1997) 403.

[429] J. Bernabeu, G. A. Gonzalez-Sprinberg, M. Tung, J. Vidal: *The tau weak magnetic dipole moment*, Nucl. Phys. B 436 (1995) 474.

[430] W. Bernreuther, U. Low, J. P. Ma, O. Nachtmann: *CP Violation and Z Boson Decays*, Z.Phys. C 43 (1989) 117.

[431] W. Bernreuther, O. Nachtmann, P. Overmann: *The CP violating electric and weak dipole moments of the tau lepton from threshold to 500-GeV*, Phys. Rev. D 48 (1993) 78; W. Bernreuther, A. Brandenburg, P. Overmann: *CP violation beyond the standard model and tau pair production in e^+e^- collisions*, Phys. Lett. B 391 (1997) 413.

[432] J. Bernabeu, G. A. Gonzalez-Sprinberg, J. Vidal: *Normal and transverse single tau polarization at the Z peak*, Phys. Lett. B 326 (1994) 168.

[433] M. Ahmad, et al.: *CEPC-SPPC Preliminary Conceptual Design Report. 1. Physics and Detector*, IHEP-CEPC-DR-2015-01.

[434] P. Abreu et al. [DELPHI]: *A Study of the Lorentz structure in tau decays*, Eur. Phys. J. C 16 (2000) 229.

[435] I.I. Bigi, N.G. Uraltsev: *Induced Multi - Gluon Couplings and the Neutron Electric Dipole Moment*, Nucl. Phys. B 353 (1991) 321.

[436] O. Lebedev et al.: *Probing CP violation with the deuteron electric dipole moment* Phys. Rev. D 70 (2004) 016003; M. Pospelov, A. Ritz: *Electric dipole moments as probes of new physics*, Ann. Phys. 318 (2005) 119.

[437] E. P. Shabalin: *Electric Dipole Moment of Quark in a Gauge Theory with Left-Handed Currents*, Sov. J. Nucl. Phys. 28 (1978) 75 [Yad. Fiz. 28 (1978) 151].

[438] A. Czarnecki, B. Krause: *Neutron electric dipole moment in the standard model: Valence quark contributions*, Phys. Rev. Lett. 78 (1997) 4339.

[439] M. B. Gavela et al.: *CP Violation Induced by Penguin Diagrams and the Neutron Electric Dipole Moment*, Phys. Lett. B 109 (1982) 215.

[440] I. B. Khriplovich and A. R. Zhitnitsky: *What Is the Value of the Neutron Electric Dipole Moment in the Kobayashi-Maskawa Model?*, Phys. Lett. B 109 (1982) 490.

[441] G. Beall and N. G. Deshpande: *Electric Dipole Moment of the Neutron in a Higgs Boson Exchange Model of CP Nonconservation*, Phys. Lett. B 132 (1983) 427.

[442] C. Abel et al. [nEDM]: *Measurement of the permanent electric dipole moment of the neutron*, Phys. Rev. Lett. 124 (2020) 081803.

[443] J. Gunion, D. Wyler: *Inducing a large neutron electric dipole moment via a quark chromoelectric dipole moment*, Phys. Lett. B 248 (1990) 170; D.W. Chang, W.Y. Keung, T.C. Yuan: *Inducing a large neutron electric dipole moment via a quark chromoelectric dipole moment*, Phys. Lett. B 251 (1990) 608.

[444] M.S. Barr, A. Zee: *Electric Dipole Moment of the Electron and of the Neutron*, Phys. Rev. Lett. 65 (1990) 21.

[445] Th. Mannel, N. Uraltsev: *Loop-Less Electric Dipole Moment of the Nucleon in the Standard Model*, Phys. Rev. D 85 (2012) 096002; *Charm CP Violation and the Electric Dipole Moments from the Charm Scale*, JHEP 1303 (2013) 064.

[446] S. M. Barr and A. Zee: *Electric Dipole Moment of the Electron and of the Neutron*, Phys. Rev. Lett. 65 (1990) 21 [erratum: Phys. Rev. Lett. 65 (1990) 2920].

[447] J. F. Gunion, D. Wyler: *Inducing a large neutron electric dipole moment via a quark chromoelectric dipole moment*, Phys. Lett. B 248 (1990) 170.

[448] D. Bowser-Chao, D. Chang, W. Y. Keung: *Electron electric dipole moment from CP violation in the charged Higgs sector*, Phys. Rev. Lett. 79 (1997) 1988-1991.

[449] J. Baron et al. [ACME]: *Order of Magnitude Smaller Limit on the Electric Dipole Moment of the Electron*, Science 343 (2014) 269.

[450] D. Chang, W. Y. Keung, T. C. Yuan: *Chromoelectric dipole moment of light quarks through two loop mechanism*, Phys. Lett. B 251 (1990) 608.

[451] A.A. Anselm, V. Bunakov, V. Gudkov, N.G. Uraltsev: *On the neutron electric dipole moment in the Weinberg CP violation model*, Phys. Lett. B 152 (1985) 116.

[452] C.P. Liu, R.G.E. Timmermans: *P- and T-odd two-nucleon interaction and the deuteron electric dipole moment*, Phys. Rev. C 70 (2004) 055501.

[453] M. Raidal et al.: *Flavour physics of leptons and dipole moments*, Eur. Phys. J. C 57 (2008) 13.

[454] N. Arkani-Hamed, A. G. Cohen, H. Georgi: *(De)constructing dimensions*, Phys. Rev. Lett. 86 (2001) 4757; N. Arkani-Hamed, A. G. Cohen, H. Georgi: *Electroweak symmetry breaking from dimensional deconstruction*, Phys. Lett. B 513 (2001) 232.

[455] H.-C. Cheng, I. Low: *TeV symmetry and the little hierarchy problem*, JHEP 09 (2003) 051; H.-C. Cheng, I. Low: *Little hierarchy, little Higgses, and a little symmetry*, JHEP 08 (2004) 061; I. Low: *T-parity and the Littlest Higgs*, JHEP 10 (2004) 067.

[456] A. Neveu, J.H. Schwarz: *Factorizable dual model of pions*, Nucl. Phys. B 31 (1971) 86; P. Ramond, *Dual Theory for Free Fermions*, Phys. Rev. D 3 (1971) 2415.

[457] Yu.A. Golfand, E.P. Likhtman: *Extension of the Algebra of Poincare Group Generators and Violation of p Invariance*, Sov. Phys.: JETP Lett. 13 (1971) 323.

[458] D.V. Volkov, V.P. Akulov: *Possible universal neutrino interaction*, JETP Lett. 16 (1972) 438.

[459] J. Wess, B. Zumino: *Supergauge Transformations in Four-Dimensions*, Nucl. Phys. B 70 (1974) 39.

[460] P. Fayet, J. Iliopoulos: *Spontaneously Broken Supergauge Symmetries and Goldstone Spinors*, Phys. Lett. B 51 (1974) 461; P. Fayet, S. Ferrara: *Supersymmetry*, Phys. Rept. 32 (1977) 249; P. Fayet: *Mixing Between Gravitational and Weak Interactions Through the Massive Gravitino*, Phys. Lett. B 70 (1977) 461.

[461] R. Barbier et al.: *R-parity violation supersymmetry*, Phys. Rept. 420 (2005) 1.

[462] J.L. Feng: *Naturalness and the Status of Supersymmetry*, Ann. Rev. Nucl. Part. Sci. 63 (2013) 351.

[463] D. Hooper: *Particle Dark Matter*, TASI 2008 Lectures on Dark Matter, arXiv: 0901.4090.

[464] M. Kaku: *Quantum Field Theory – A Modern Introduction*, Oxford: Oxford University Press, 1993, Chapt. 20 "Supersymmetry and Supergravity" and Chapt. 21 "Superstrings".

[465] M. Shifman: *ITEP Lectures on Particle Physics and Field Theory*, World Scientific Lecture Notes in Physics - Vol 62.

[466] B.C. Allanach and H.E. Haber: *110. Supersymmetry, Part I (Theory)*, PDG2019.

[467] O. Buchmuller and P. de Jong: *111. Supersymmetry, Part II (Experiment)*, PDG2019.

[468] S. Coleman, J. Mandula: *All Possible Symmetries of the S Matrix*, Phys. Rev. D 159 (1967) 1251.

[469] S. Weinberg: *The Quantum Theory of Fields, Volume III: Supersymmetry*, Cambridge University Press, Cambridge, UK, 2000/2005; I.J.R. Aitchison: *Supersymmetry in Particle Physics: an elementary introduction*, Cambridge University Press, Cambridge, UK, 2007.

[470] J. Wess, J. Bagger: *Supersymmetry and Supergravity*, Princeton Series in Physics; S.J. Gates (Jr.), M.T. Grisaru, M. Rocek, W. Siegel: *SUPERSPACE or One Thousand and One Lessons in Supersymmetry*, Frontiers in Physics Lecture Note Series 58.

[471] H.P. Nilles: *Supersymmetry, Supergravity and Particle Physics*, Phys. Rep. 110 (1984) 1.

[472] St. P. Martin: *A Supersymmetry Primer*, Adv. Ser. Direct. High Energy Phys. 21 (2010) 1.

[473] Y. Shirman: *TASI 2008 Lectures: Introduction to Supersymmetry and Supersymmetry Breaking*, arXiv:0907.0039.

[474] L. Alvarez-Gaume, M. Claudson, M.B. Wise: *Low-Energy Supersymmetry*, Nucl. Phys. B 207 (1982) 96; M. Dine, W. Fischler: *A Phenomenological Model of Particle Physics Based on Supersymmetry*, Phys. Lett. B 110 (1982) 227.

[475] L. Randall, R. Sundrum: *Out of this world supersymmetry breaking*, Nucl. Phys. B 557 (1999) 79; G.F. Giudice et al.: *Gaugino mass without singlets*, JHEP 9812 (1998) 027.

[476] D.E. Kaplan, G.D. Kribs, M. Schmaltz: *Supersymmetry breaking through transparent extra dimensions*, Phys. Rev. D 62 (2000) 035010; Z. Chacko et al.: *Gaugino mediated supersymmetry breaking*, JHEP 01 (2000) 003.

[477] G.G. Ross, K. Schmidt-Hoberg, F. Staub: *Revisiting fine-tuning in the MSSM*, JHEP 03 (2017) 021.

[478] S. Bertolini et al.: *Effects of supergravity induced electroweak breaking on rare B decays and mixings*, Nucl. Phys. B 353 (1991) 591.

[479] A. Masiero, O. Vives, in: *Proceedings of the International School of Physics "Enrico Fermi"*, M. Giorgi et al. (eds.), Course CLXIII, IOS Press, Amsterdam, SIF, Bologna, 2006, p. 133; J.L. Hewett et al. [arXiv:[hep-ph]/0503261], p. 391ff.

[480] S. Herrlich, U. Nierste: *Enhancement of the $K_L - K_S$ mass difference by short distance QCD corrections beyond leading logarithms*, Nucl. Phys. B 419 (1994) 292.

[481] J. Brod and M. Gorbahn: *Next-to-Next-to-Leading-Order Charm-Quark Contribution to the CP Violation Parameter ϵ_K and ΔM_K*, Phys. Rev. Lett. 108 (2012) 121801.

[482] N. H. Christ et al. [RBC and UKQCD]: *Long distance contribution to the $K_L - K_S$ mass difference*, Phys. Rev. D 88 (2013) 014508.

[483] Z. Bai, N. H. Christ, C. T. Sachrajda: *The $K_L - K_S$ Mass Difference*, EPJ Web Conf. 175 (2018) 13017.

[484] N. H. Christ, X. Feng, G. Martinelli, C. T. Sachrajda: *Effects of finite volume on the K_L-K_S mass difference*, Phys. Rev. D 91 (2015) 114510.

[485] Z. Bai, N. H. Christ, T. Izubuchi, C. T. Sachrajda, A. Soni, J. Yu: *$K_L - K_S$ Mass Difference from Lattice QCD*, Phys. Rev. Lett. 113 (2014) 112003.

[486] F. Gabbiani, E. Gabrielli, A. Masiero, L. Silvestrini: *A Complete analysis of FCNC and CP constraints in general SUSY extensions of the standard model*, Nucl. Phys. B 477 (1996) 321.

[487] V. Bertone et al. [ETM]: *Kaon Mixing Beyond the SM from N_f=2 tmQCD and model independent constraints from the UTA*, JHEP 03 (2013) 089.

[488] P. A. Boyle et al. [RBC and UKQCD]: *Neutral kaon mixing beyond the standard model with $n_f = 2 + 1$ chiral fermions*, Phys. Rev. D 86 (2012) 054028.

[489] P. Dimopoulos et al. [ALPHA]: *Non-Perturbative Renormalisation and Running of BSM Four-Quark Operators in $N_f = 2$ QCD*, Eur. Phys. J. C 78 (2018) 579.

[490] G. Isidori, F. Mescia, P. Paradisi, C. Smith, S. Trine: *Exploring the flavour structure of the MSSM with rare K decays*, JHEP 08 (2006) 064.

[491] W. Altmannshofer, P. S. Bhupal Dev, A. Soni: *$R_{D^{(*)}}$ anomaly: A possible hint for natural supersymmetry with R-parity violation*, Phys. Rev. D 96 (2017) 095010.

[492] Q. Y. Hu, X. Q. Li, Y. Muramatsu, Y. D. Yang: *R-parity violating solutions to the $R_{D^{(*)}}$ anomaly and their GUT-scale unifications*, Phys. Rev. D 99 (2019) 015008.

[493] J. Albrecht, S. Reichert, D. van Dyk: *Status of rare exclusive B meson decays in 2018*, Int. J. Mod. Phys. A 33 (2018) 1830016.

[494] I. I. Y. Bigi, F. Gabbiani: *Impact of different classes of supersymmetric models on rare B decays, B^0-$\bar{B}^0$ mixing and CP violation*, Nucl. Phys. B 352 (1991) 309.

[495] A. Dery, Y. Nir: *Implications of the LHCb discovery of CP violation in charm decays*, JHEP 12 (2019) 104.

[496] M. Chakraborti, S. Heinemeyer, I. Saha: *Improved $(g-2)_\mu$ Measurements and Supersymmetry*, Eur. Phys. J. C 80 (2020) 984.

[497] A. Bartl, T. Gajdosik, E. Lunghi, A. Masiero, W. Porod, H. Stremnitzer, O. Vives: *General flavor blind MSSM and CP violation*, Phys. Rev. D 64 (2001) 076009.

[498] F. del Aguila, M. B. Gavela, J. A. Grifols, A. Mendez: *Specifically Supersymmetric Contribution to Electric Dipole Moments*, Phys. Lett. B 126 (1983) 71 [Erratum: Phys. Lett. B 129 (1983) 473].

[499] D. Chang, W. Y. Keung, A. Pilaftsis: *New two loop contribution to electric dipole moment in supersymmetric theories*, Phys. Rev. Lett. 82 (1999) 900. [Erratum: Phys. Rev. Lett. 83 (1999) 3972].

[500] Y. Li, S. Profumo, M. Ramsey-Musolf: *Higgs-Higgsino-Gaugino Induced Two Loop Electric Dipole Moments*, Phys. Rev. D 78 (2008) 075009.

[501] N. Arkani-Hamed, S. Dimopoulos, G. F. Giudice, A. Romanino: *Aspects of split supersymmetry*, Nucl. Phys. B 709 (2005) 3.

[502] K. Blum, C. Delaunay, M. Losada, Y. Nir, S. Tulin: *CP violation Beyond the MSSM: Baryogenesis and Electric Dipole Moments*, JHEP 05 (2010) 101.

[503] B. Famaey and S. McGaugh, *Challenges for Lambda-CDM and MOND*, J. Phys. Conf. Ser. 437 (2013) 012001; *Modified Newtonian Dynamics (MOND): Observational Phenomenology and Relativistic Extensions*, Living Rev. Rel. 15 (2012) 10.

[504] S. W. Randall et al.: *Constraints on the Self-Interaction Cross-Section of Dark Matter from Numerical Simulations of the Merging Galaxy Cluster 1E 0657-56*, Astrophys. J. 679, (2008) 1173.

[505] R. H. Sanders, *Does GW170817 falsify MOND?*, Int. J. Mod. Phys. D 27, (2018) 14.

[506] G. Bertone, D. Hooper, J. Silk: *Particle dark matter: Evidence, candidates and constraints* Phys. Rept. 405 (2005) 279.

[507] G. Bertone and D. Hooper: *A history of Dark Matter*, Rev. Mod. Phys. 90 (2018) 4.

[508] *Testing WIMPs to the limit*, CERN Courier, March 2017, pp. 35 - 38.

[509] S. D. McDermott, H.-B. Yu and K. M. Zurek: *Turning off the Lights: How Dark is Dark Matter?*, Phys. Rev. D 83, (2011) 063509.

[510] J. Buch, S. C. J. Leung and J. Fan: *Using Gaia DR2 to Constrain Local Dark Matter Density and Thin Dark Disk*, JCAP 04 (2019) 026.

[511] S. Tremaine and J. E. Gunn: *Dynamical Role of Light Neutral Leptons in Cosmology*, Phys. Rev. Lett. 42 (1979) 407.

[512] L. Randall, J. Scholtz and J. Unwin: *Cores in Dwarf Galaxies from Fermi Repulsion*, Mon. Not. Roy. Astron. Soc. 467 (2017) 1515.

[513] R. Hlozek et al., Phys. Rev. D 91 (2015) 103512 [arXiv:[hep-ph]/1410.2896]; E. Armengaud et al., Mon. Not. Roy. Astron. Soc. 471 (2017) 4606 [arXiv:[hep-ph]/1703.09126]; M. Nori et al., Mon. Not. Roy. Astron. Soc. 482 (2019) 3227 [arXiv:[hep-ph]/1809.09619].

[514] Fr. Zwicky, *Die Rotverschiebung von extragalaktischen Nebeln*, Helvetica Physica Acta 6 (1933) 110.

[515] J. de Swart: *Zwicky: new lens on an elusive astrophysicist*, Nature 573 (2019) 33.

[516] N. Trevisani [ATLAS,CMS]: *Collider Searches for Dark Matter (ATLAS + CMS)*, Universe 4 (2018) 131.

[517] L. Baudis and S. Profumo: *Dark matter*, PDG2020.

[518] E. Aprile et al. [XENON]: *Dark Matter Search Results from a One Ton-Year Exposure of XENON1T*, Phys. Rev. Lett. 121 (2018) 111302.

[519] P. Agnes et al. [DARKSIDE]: *Low-Mass Dark Matter Search with the DarkSide-50 Experiment*, Phys. Rev. Lett. 121 (2018) 081307.

[520] A. H. Abdelhameed et al. [CRESST]: *First results from the CRESST-III low-mass dark matter program*, Phys. Rev. D 100 (2019) 102002.

[521] C. Amole et al. (PICO), Phys. Rev. D 100 (2019) 022001 [arXiv:[hep-ph]/1902.04031].

[522] E. Aprile et al. [XENON]: *Constraining the spin-dependent WIMP-nucleon cross sections with XENON1T*, Phys. Rev. Lett. 122 (2019) 141301.

[523] R. Bernabei et al. [DAMA]: *Search for WIMP annual modulation signature: Results from DAMA/NaI-3 and DAMA/NaI-4 and the global combined analysis*, Phys. Lett. B 480 (2000) 23.

[524] K. Freese et al.: *Signal Modulation in Cold Dark Matter Detection*, Phys. Rev. D 37 (1988) 3388; R. Bernabei et. al. [DAMA]: *New limits on WIMP search with large-mass low-radioactivity NaI(Tl) set-up at Gran Sasso*, Phys. Lett. B 389 (1996) 757; ibid B 408 (1997) 439; ibid B 424 (1998) 195; ibid B 450 (1999) 448.

[525] R. Bernabei et al., *First model independent results from DAMA/LIBRA-phase2*, Nucl. Phys. Atom. Energy 19 (2018) 307.

[526] M. Ackermann et al. [Fermi-LAT]: *Search for Gamma-Ray Emission from DES Dwarf Spheroidal Galaxy Candidates with Fermi-LAT Data*, Phys. Rev. Lett. 115 (2015) 231301.

[527] A. W. Strong, I. V. Moskalenko and V. S. Ptuskin, Ann. Rev. Nucl. Part. Sci. 57 (2007) 285 [arXiv:astro-ph/0701517].

[528] A. Coogan and S. Profumo, Phys. Rev. D96, 8, 083020 (2017) [arXiv:[hep-ph]/1705.09664].

[529] O. Adriani et al. [PAMELA]: *Cosmic-Ray Positron Energy Spectrum Measured by PAMELA*, Phys. Rev. Lett. 111 (2013) 081102.

[530] M. Aguilar et al. [AMS]: *First Result from the Alpha Magnetic Spectrometer on the International Space Station: Precision Measurement of the Positron Fraction in Primary Cosmic Rays of 0.5–350 GeV*, Phys. Rev. Lett. 110 (2013) 141102.

[531] S. Profumo, F. Queiroz and C. Siqueira: *Has AMS-02 Observed Two-Component Dark Matter?* (2019) [arXiv:[hep-ph]/1903.07638]; D. Hooper, P. Blasi and P. D. Serpico: *Pulsars as the Sources of High Energy Cosmic Ray Positrons*, JCAP 0901 (2009) 025 [arXiv:0810.1527]; S. Profumo: *Dissecting cosmic-ray electron-positron data with Occam's Razor: the role of known pulsars*, Central Eur. J. Phys. 10 (2011) 1 [arXiv:0812.4457].

[532] O. Adriani et al. [PAMELA]: *PAMELA results on the cosmic-ray antiproton flux from 60 MeV to 180 GeV in kinetic energy*, Phys. Rev. Lett. 105 (2010) 121101.

[533] M. Aguilar et al. [AMS-02]: *Antiproton Flux, Antiproton-to-Proton Flux Ratio, and Properties of Elementary Particle Fluxes in Primary Cosmic Rays Measured with the Alpha Magnetic Spectrometer on the International Space Station*, Phys. Rev. Lett. 117, 9, 091103 (2016).

[534] M.-Y. Cui et al.: *Possible Dark Matter Annihilation Signal in the AMS-02 Antiproton Data*, Phys. Rev. Lett. 118 (2017) 191101 [arXiv:[hep-ph]/1610.03840]; A. Reinert and M. W. Winkler: *A precision search for WIMPs with charged cosmic rays*, JCAP 1801 (2018) 055 [arXiv:[hep-ph]/1712.00002]; M. W. Winkler: *Cosmic ray antiprotons at high energies*, JCAP 1702 (2017) 048 [arXiv:[hep-ph]/1701.04866].

[535] R. Bird et al., in "36th International Cosmic Ray Conference (ICRC 2019) Madison, Wisconsin, USA, July 24-August 1, 2019," (2019) [arXiv:[hep-ph]/1908.03154].

[536] Sz. Borsanyi et al.: *Calculation of the axion mass based on high-temperature lattice quantum chromodynamics*, Nature 539 (2016) 69.

[537] A. Ringwalk: *Alternative dark matter candidate: Axions*, PoS NOW2016 (2016) 081.

[538] D. Budker et al.: *Proposal for a Cosmic Axion Spin Precession Experiment (CASPEr)*, Phys. Rev. X 4 (2014) 021030.

[539] A. Caldwell et al. [MADMAX]: *Dielectric Haloscopes: A New Way to Detect Axion Dark Matter*, Phys. Rev. Lett. 118 (2017) 091801.

[540] Y. Kahn, B.R. Safdi, J. Thaler, *Broadband and Resonant Approaches to Axion Dark Matter Detection*, Phys. Rev. Lett. 117 (2016) 141801.

[541] I. Stern [ADMX]: *ADMX Status*, PoS ICHEP2016 (2016) 198.

[542] G. Rybka et al.: *Search for dark matter axions with the Orpheus experiment*, Phys. Rev. D 91 (2015) 011701.

[543] B.M. Brubaker et al.: *First results from a microwave cavity axion search at 24 μeV*, Phys. Rev. Lett. 118 (2017) 6, 061302.

[544] V. Anastassopoulos et al. [CAST]: *New CAST Limit on the Axion-Photon Interaction*, Nature Phys. 13 (2017) 584.

[545] R. Barbieri et al.: *Searching for galactic axions through magnetized media: the QUAX proposal*, Phys. Dark Univ. 15 (2017) 135.

[546] L.M. Krauss et al.: *Spin coupled axion detections*, HUTP-85/A006 (1985); R. Barbieri et al.: *Axion to magnon conversion. A scheme for the detection of galactic axions*, Phys. Lett. B 226 (1989) 357; F. Caspers, Y. Semertzidis: *Ferri-magnetic resonance, magnetostatic waves and open resonators for axion detection*, Proc. of the Workshop on Cosmic Axions, World Scientific (1990) p. 173; A.I. Kakhizde, I.V. Kolokolov: *Antiferromagnetic axion detector*, Sov. Phys. JETP 72 (1991) 598; P.V. Vorob'ev et al.: Phys. Atom. Nuclei 58 (1995) 959.

[547] N. Crescini et al. [QUAX]: *Axion search with a quantum-limited ferromagnetic haloscope'*, Phys. Rev. Lett 124 (2020) 171801.

[548] G. Ballesteros, J. Redondo, A. Ringwald, C. Tamarit: *Unifying inflation with the axion, dark matter, baryogenesis and the seesaw mechanism*, Phys. Rev. Lett. 118 (2017) 071802 [arXiv:[hep-ph]/1608.05414 [hep-ph]].

[549] M. Y. Khlopov, A. D. Linde: *Is It Easy to Save the Gravitino?*, Phys. Lett. B 138 (1984) 265.

[550] G. Jungman, M. Kamionkowski, K. Griest: *Supersymmetric Dark Matter*, Physics Reports 267 (1996) 195.

[551] T. Falk, K. A. Olive, M. Srednicki: *Heavy sneutrinos as dark matter*, Phys. Lett. B 339 (1994) 248.

[552] C. Patrignani et al. [PDG] *The Review of Particle Physics*, Chin. Phys. 40 (2016) 100001, in the Sect.26. Dark Matter.

[553] K. Petraki, R.R. Volkas: *Review of asymmetric dark matter*, Int. J. Mod. Phys. A 28 (2013) 1330028 [arXiv:[hep-ph]/1305.4939].

[554] A. Vincent, P. Scott, A. Serenelli: *Possible Indication of Momentum-Depending Asymmetric Dark Matter in the Sun*, Phys. Rev. Lett. 114 (2015) 081302.

[555] R.T. D'Agnolo, J.T. Ruderman: *Light Dark Matter from Forbidden Dark Matter*, Phys. Rev. Lett. 115 (2015) 061301.

[556] E.W. Kolb, M.S. Turner: *The Early Universe*, Front. Phys. 69 (1990) 1.

[557] M. Pospelov: *Dark sectors and their signatures in Kaon physics*, J. Phys. Conf. Series 800 (2017) 012015.

[558] M. Fabbrichesi, E. Gabrielli, B. Mele: *Hunting Down Massless Dark Photons in Kaon Physics*, Phys. Rev. Lett. 119 (2017) 031801.

[559] C. Csaki: *TASI lectures on extra dimensions and branes*, arXiv:[hep-ph]/0404096; R. Sundrum: *BFKL resummation effects in* $\gamma^*\gamma^* \to \rho\rho$, arXiv:[hep-ph]/0508134; G. Kribs: *TASI 2004 lectures on the phenomenology of extra dimensions*, arXiv: [hep-ph]/0605325.

[560] L. Randall: *Warped Passages unraveling the Mysteries of the Universe's Hidden Dimensions*, Harper Perennial (2006).

[561] N. Arkani-Hamed, S. Dimopoulos, G. R. Dvali: *The Hierarchy problem and new dimensions at a millimeter*, Phys. Lett. B 429 (1998) 263; *Phenomenology, astrophysics and cosmology of theories with submillimeter dimensions and TeV scale quantum gravity*, Phys. Rev. D 59 (1999) 086004.

[562] L. Randall, R. Sundrum: *A Large mass hierarchy from a small extra dimension*, Phys. Rev. Lett. 83 (1999) 3370.

[563] T. Appelquist, H.C. Cheng, B.A. Dobrescu: *Bounds on universal extra dimensions*, Phys. Rev. D 64 (2001) 035002.

[564] A. Buras et al.: *The Impact of universal extra dimensions on* $B \to X(s)\gamma$, $B \to X(s)g$, $B \to X(s)\mu^+\mu^-$, $K_L \to \pi^0 e^+ e^-$ *and* ϵ'/ϵ, Nucl. Phys. B 678 (2004) 455.

[565] N. Arkani-Hamed, M. Schmaltz: *Hierarchies without symmetries from extra dimensions*, Phys. Rev. D 61 (2000) 033005.

[566] B. Lillie in: *The Discovery Potential of a Super B Factory*, J. Hewett and D. Hitlin (eds.) [arXiv:[hep-ph]/0503261].

[567] B. P. Abbott et al. [LIGO Scientific and Virgo]: *Observation of Gravitational Waves from a Binary Black Hole Merger*, Phys. Rev. Lett. 116 (2016) 061102.

[568] K. Pardo, M. Fishbach, D. E. Holz, D. N. Spergel: *Limits on the number of spacetime dimensions from GW170817*, JCAP 07 (2018) 048.

[569] A.D. Dolgov, Ya.B. Zel′dovich: *Cosmology and Elementary Particles*, Rev. Mod. Phys. 53 (1981) 1.

[570] A.D. Sakharov: *Quark - Muonic currents and violation of CP invariance*, JETP Lett. 5 (1967) 27.

[571] A. D. Sakharov: *Cosmological Transitions With A Change In Metric Signature, Sov. Phys. JETP* 60 (1984) 214; another 'personal' comment: Sakharov in the end of this paper thanked his wife E.G. Bonner for her help.

[572] V.A. Kuzmin: *CP violation and baryon asymmetry of the universe*, Pis′ma Z. Eksp. Teor. Fiz. 12 (1970) 335.

[573] J.C. Pati, A. Salam: *Lepton Number as the Fourth Color*, Phys. Rev. D 10 (1974) 275; H. Georgi, S.L. Glashow: *Unity of All Elementary Particle Forces*, Phys. Rev. Lett. 32 (1974) 438.

[574] M. Yoshimura: *Unified Gauge Theories and the Baryon Number of the Universe*, Phys. Rev. Lett. 41 (1978) 281.

[575] D. Touissaint, S. B. Treiman, F. Wilczek, A. Zee: *Matter - Antimatter Accounting, Thermodynamics, and Black Hole Radiation*, Phys. Rev. D 19 (1979) 1036; S. Weinberg: *Cosmological Production of Baryons*, Phys. Rev. Lett. 42 (1979) 850; M. Yoshimura: *Origin of Cosmological Baryon Asymmetry*, Phys. Lett. B 88 (1979) 294.

[576] For a review: A.D. Dolgov, in: *Proceedings of the XXVth ITEP Winterschool of Physics*, Feb. 18 - 27 (1997) Moscow, Russia [arXiv:[hep-ph]/9707419].

[577] W.-Ch. Huang, H. Päs, S. Zeissner: *Neutrino assisted GUT baryogenesis - revisited*, Phys. Rev. D 97 (2018) 055040.

[578] V. A. Kuzmin, V. A. Rubakov, M. E. Shaposhnikov: *On the Anomalous Electroweak Baryon Number Nonconservation in the Early Universe*, Phys. Lett. B 155 (1985) 36.

[579] Sz. Borsanyi et al.: *Axion cosmology, lattice QCD and the dilute instanton gas*, Phys. Lett. B 752 (2016) 175.

[580] C. Bonati et al.: *Recent progress on QCD inputs for axion phenomenology*, EPJ Web Conf. 137 (2017) 08004.

[581] Fr. Klinkhamer: *Sphalerons and anomalies (an introduction)*, Physics Colloquium, 2015. [These slides from the internet are indeed a good introduction with a short list of references.]

[582] V. A. Rubakov, M. E. Shaposhnikov: *Electroweak baryon number nonconservation in the early universe and in high-energy collisions*, Usp. Fiz. Nauk 166 (1996) 493.

[583] G.D. Moore: *Sphaleron rate in the symmetric electroweak phase*, Phys. Rev. D 62 (2000) 085011.

[584] K. Kajantie et al.: *The Electroweak phase transition: A Nonperturbative analysis*, Nucl. Phys. B 466 (1996) 189.

[585] M. D'Onofrio, K. Rummukainen, A. Tranberg: *Sphaleron Rate in the Minimal Standard Model*, Phys. Rev. Lett. 113 (2014) 141602.

[586] A.H. Guth: *The Inflationary Universe: A Possible Solution to the Horizon and Flatness Problems*, Phys. Rev. D 23 (1981) 347.

[587] A. Vilenkin: *Creation of Universes from Nothing*, Phys. Lett. B 117 (1982) 25.

[588] J.B. Hartle, S.W. Hawking: *Wave Function of the Universe*, Phys. Rev. D 28 (1983) 2960.

[589] A. Linde: *The new inflationary universe scenario*, https://web.stanford.edu/~alinde/LindeNuffield.pdf.

[590] P.J. Steinhardt, in: Cambridge 1982, Proceedings, *The Very Early Universe*, 251-266.

[591] W. Fischler, Bh. Ratra, L. Susskind: *Quantum Mechanics of Inflation*, Nucl. Phys. B 259 (1985) 730; Erratum: Nucl. Phys. B 268 (1986) 747.

[592] A. G. Riess et al.[Supernova Search Team Collaboration]: *Observational evidence from supernovae for an accelerating universe and a cosmological constant*, Astron. J. 116 (1998) 1009.

[593] S. Perlmutter et al.[Supernova Cosmology Project Collaboration]: *Measurements of Omega and Lambda from 42 high redshift supernovae*, Astrophys. J. 517 (1999) 565.

[594] P. A. R. Ade et al. [Planck]: *Planck 2013 results. XVI. Cosmological parameters*, Astron. Astrophys. 571 (2014) A16.

[595] P. A. R. Ade et al. [Planck]: *Planck 2013 results. XXII. Constraints on inflation*, Astron. Astrophys. 571 (2014) A22.

[596] *Centre for Theoretical Cosmology*, www.ctc.cam.ac.uk.

[597] R. Bousso, L. Susskind: *The Multiverse Interpretation of Quantum Mechanics*, Phys. Rev. D 85 (2012) 045007.

[598] A. Linde: *A brief history of the multiverse*, Rep. Prog. Phys. 80 (2017) 022001.

[599] P.C.W. Davies, S.D. Unwin: *Why is the Cosmological Constant so Small?*, Proc. Roy. Soc. 377 (1981) 147.

[600] S. Weinberg: *Anthropic Bound on the Cosmological Constant*, Phys. Rev. Lett. 59 (1987) 2607.

[601] I. Affleck, M. Dine: *A New Mechanism for Baryogenesis*, Nucl. Phys. B 249 (1985) 361.

[602] A. Ijjas, P.J. Steinhardt, A. Loeb: *Cosmic inflation theory faces challenges*, Scientific American (2017).

[603] A. Ijjas, P.J. Steinhardt, A. Loeb: *Inflationary paradigm in trouble after Planck2013*, Phys. Lett. B 723 (2013) 261.

[604] A. Ijjas, P.J. Steinhardt: *Fully stable cosmological solutions with a non-singular classical bounce*, Phys. Lett. B 764 (2017) 289.

[605] B. Aubert et al. [BaBar]: *Measurement of the CP-violating asymmetry amplitude* $\sin 2\beta$, Phys. Rev. Lett. 89 (2002) 201802.

[606] K. Abe et al. [Belle]: *An Improved measurement of mixing induced CP violation in the neutral B meson system*, Phys. Rev. D 66 (2002) 071102.

[607] A. Alavi-Harati et al. [KTeV]: *Measurements of direct CP violation, CPT symmetry, and other parameters in the neutral kaon system*, Phys. Rev. D 67 (2003) 012005.

[608] E. Cortina Gil et al. [NA62]: *Search for a feebly interacting particle X in the decay* $K^+ \to \pi^+ X$, arXiv:2011.11329.

[609] R. Aaij et al. [LHCb]: *Observation of Several Source of* CP *Violation in* $B^+ \to \pi^+\pi^+\pi^-$ *Decays*, Phys. Rev. Lett. 124 (2020) 031801.

[610] H. Albrecht et al. [ARGUS]: *Observation of* $B^0 - \bar{B}^0$ *Mixing*, Phys. Lett. B 192 (1987) 245.

[611] M. Artuso et al.: B^0-$\bar{B}^0$ *Mixing at the Upsilon (4S)*, Phys. Rev. Lett. 62 (1989) 2233.

[612] The Event Horizon Telescope Collaboration: *First M87 Event Horizon Telescope Results. I. The Shadow of the Supermassive Black Hole*, The Astrophysical Journal Letters (2019) 875.

[613] R. Aaij et al. [LHCb]: *Measurement of matter-antimatter differences in beauty baryon decays*, Nature Phys. 13 (2017) 391.

[614] P. Adlarson [BESIII]: *Hyperon Structure and CP tests at BESIII*, J. Phys. Conf. Ser. 1435 (2020) 012031.

[615] R. Aaij et al. [LHCb]: *A measurement of the CP asymmetry difference in* $\Lambda_c^+ \to pK^-K^+$ *and* $p\pi^-\pi^+$ *decays*, JHEP 03 (2018) 182.

[616] I. Brivio et al.: *O new physics, where art thou? A global search in the top sector*, JHEP 02 (2020) 131.

[617] N. Castro et al.: *Novel flavour-changing neutral currents in the top quark sector*, JHEP 10 (2020) 038.

[618] M. Aaboud et al. [ATLAS]: *Observation of Higgs boson production in association with a top quark pair at the LHC with the ATLAS detector*, Phys. Lett. B 784 (2018) 173.

[619] M. Aaboud et al. [ATLAS]: *Search for charged Higgs bosons decaying into top and bottom quarks at* $\sqrt{s} = 13$ *TeV with the ATLAS detector*, JHEP 11 (2018) 085.

[620] A. M. Sirunyan et al. [CMS]: *Measurements of* $\mathrm{t\bar{t}}H$ *Production and the CP Structure of the Yukawa Interaction between the Higgs Boson and Top Quark in the Diphoton Decay Channel*, Phys. Rev. Lett. 125 (2020) 061801.

[621] M.C. Gonzalez-Garcia and M. Yokoyama: *14. Neutrino Masses, Mixing, and Oscillations*, PDG2020.

[622] B. Abi et al. [DUNE]: *Volume I. Introduction to DUNE*, JINST 15 (2020) 08, T08008.

[623] M. Yokoyama et al.: *The Hyper-Kamiokande Experiment*, arXiv:[hep-ph]/1705.00306 [hep-ex]; Hyper-Kamiokande Proto-Collab.: *Physics Potentials with the Second Hyper-Kamiokande Detector in Korea*, arXiv:[hep-ph]/1611.00118.

[624] A. Gando et al.[KamLAND-Zen]: *Search for Majorana Neutrinos near the Inverted Mass Hierarchy Region with KamLAND-Zen*, Phys. Rev. Lett. 117 (2016) 8, 082503; [Addendum: Phys. Rev. Lett. 117 (2016) 109903].

[625] M. Agostini et al.[GERDA]: *Final Results of GERDA on the Search for Neutrinoless Double-β Decay*, Phys. Rev. Lett. 125 (2020) 252502.

[626] D.Q. Adams et al. [CUORE]: *Improved Limit on Neutrinoless Double-Beta Decay in ^{130}Te with CUORE*, Phys. Rev. Lett. 124 (2020) 12501.

[627] S. Tapia and J. Zamora-Saá: *Exploring CP-Violating heavy neutrino oscillations in rare tau decays at Belle II*, Nucl. Phys. B 952 (2020) 114936.

[628] A. M. Sirunyan et al. [CMS]: *Observation of* $\mathrm{t\bar{t}}H$ *production*, Phys. Rev. Lett. 120 (2018) 231801.

[629] M. Duff, L. Okun, G. Veneziano: *Trialogue on the number of fundamental constants*, JHEP 0203 (2002) 023.

[630] S. Weinberg, H.B. Nielsen, J.G. Taylor: *Overview of Theoretical Prospects for Understanding the Values of Fundamental Constants [and Discussion]*, Phil. Trans. R. Soc. London, Series A, Math. Phys. Sciences 310 (1983) 249.

[631] J.K. Webb et al.: *Further evidence for cosmological evolution of the fine structure constant*, Phys. Rev. Lett. 87 (2001) 091301.

Index